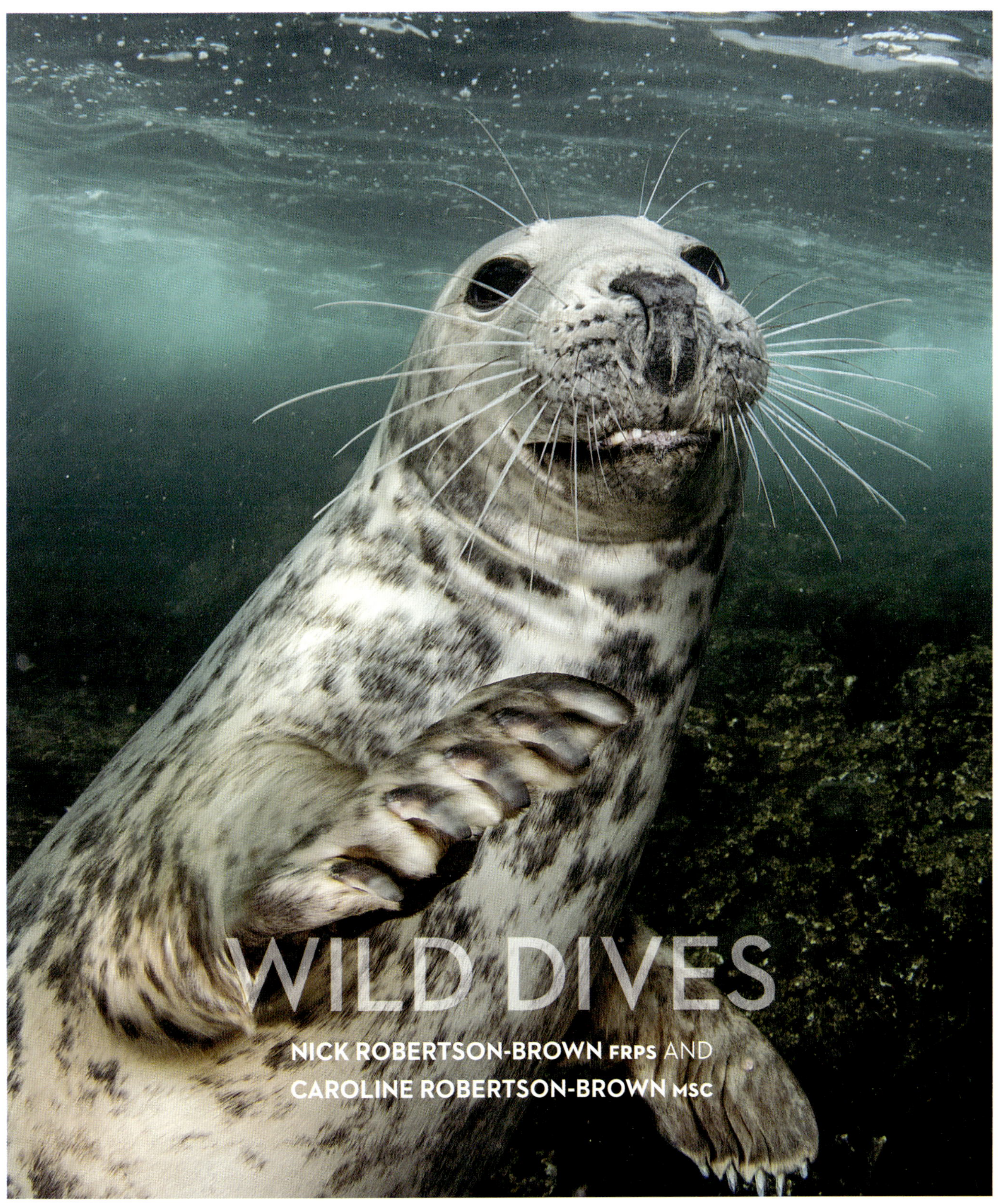

WILD DIVES

NICK ROBERTSON-BROWN FRPS AND
CAROLINE ROBERTSON-BROWN MSC

First published in 2019 by Reed New Holland Publishers
London • Sydney • Auckland

Bentinck House, 3–8 Bolsover Street, London W1W 5BB, UK
1/66 Gibbes Street, Chatswood, NSW 2067, Australia
5/39 Woodside Avenue, Northcote, Auckland 0627, New Zealand

newhollandpublishers.com

A record of this book is held at the British Library and the National Library of Australia.

ISBN 978 1 92554 642 2

Group Managing Director: Fiona Schultz
Publisher and Project Editor: Simon Papps
Designer: Sean Robertson
Production Director: Arlene Gippert
Printer: Toppan Leefung Printing Limited

Image credits: All images by Nick Robertson-Brown and Caroline Robertson-Brown except for those credited on page 223

10 9 8 7 6 5 4 3 2 1

Keep up with Reed New Holland and New Holland Publishers on Facebook
facebook.com/ReedNewHolland
facebook.com/NewHollandPublishers

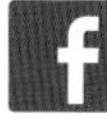

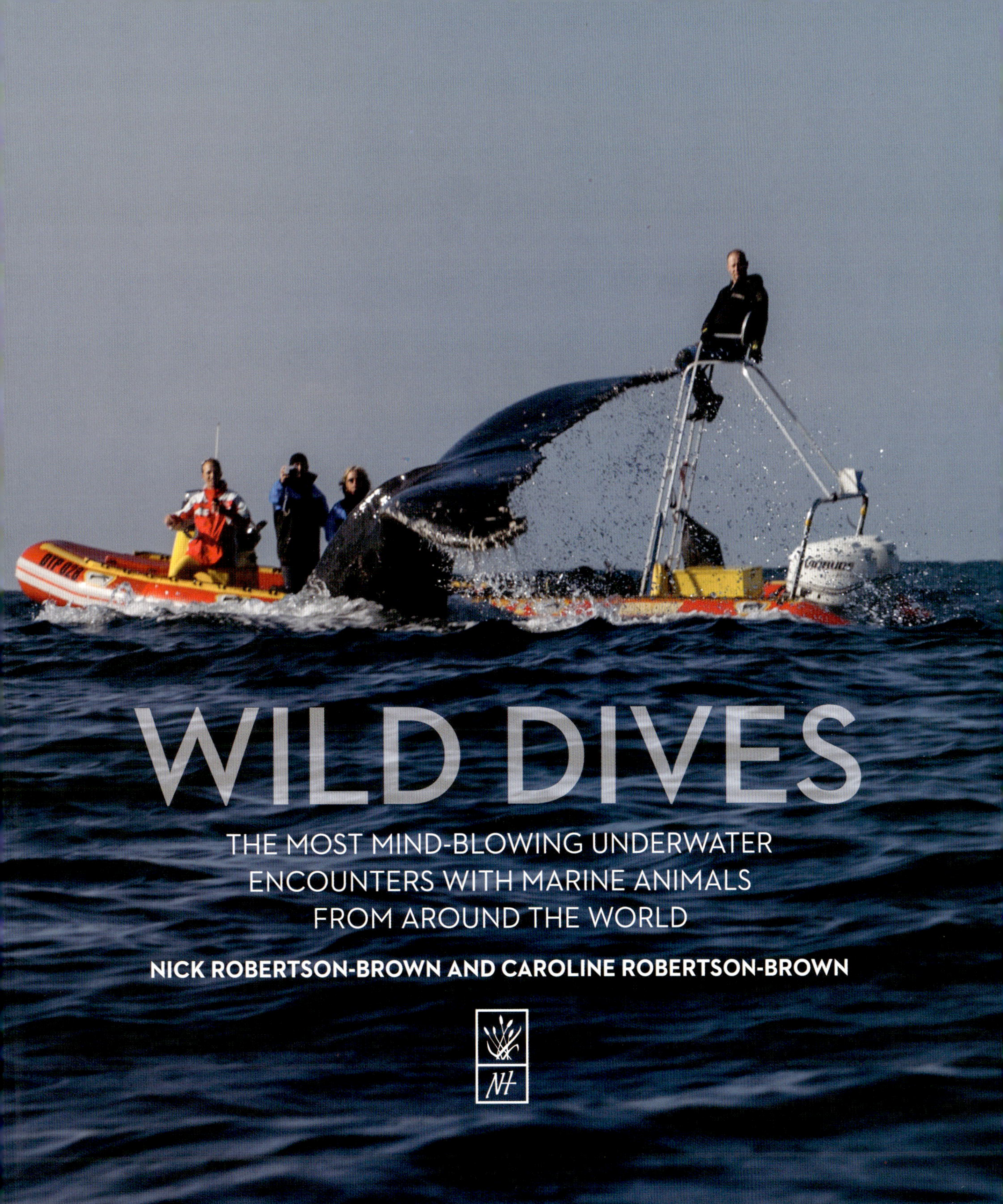

WILD DIVES

THE MOST MIND-BLOWING UNDERWATER ENCOUNTERS WITH MARINE ANIMALS FROM AROUND THE WORLD

NICK ROBERTSON-BROWN AND CAROLINE ROBERTSON-BROWN

NH

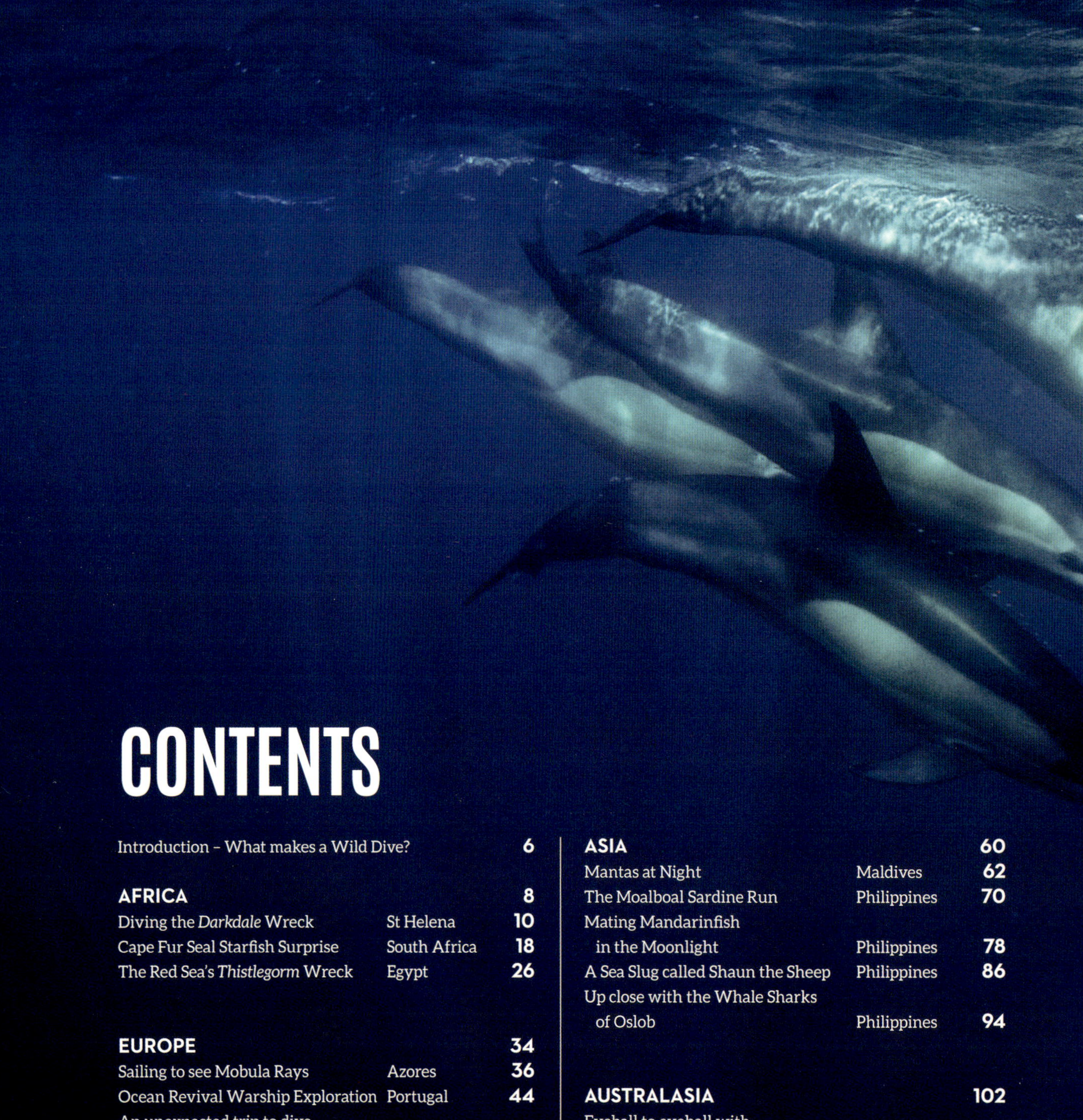

CONTENTS

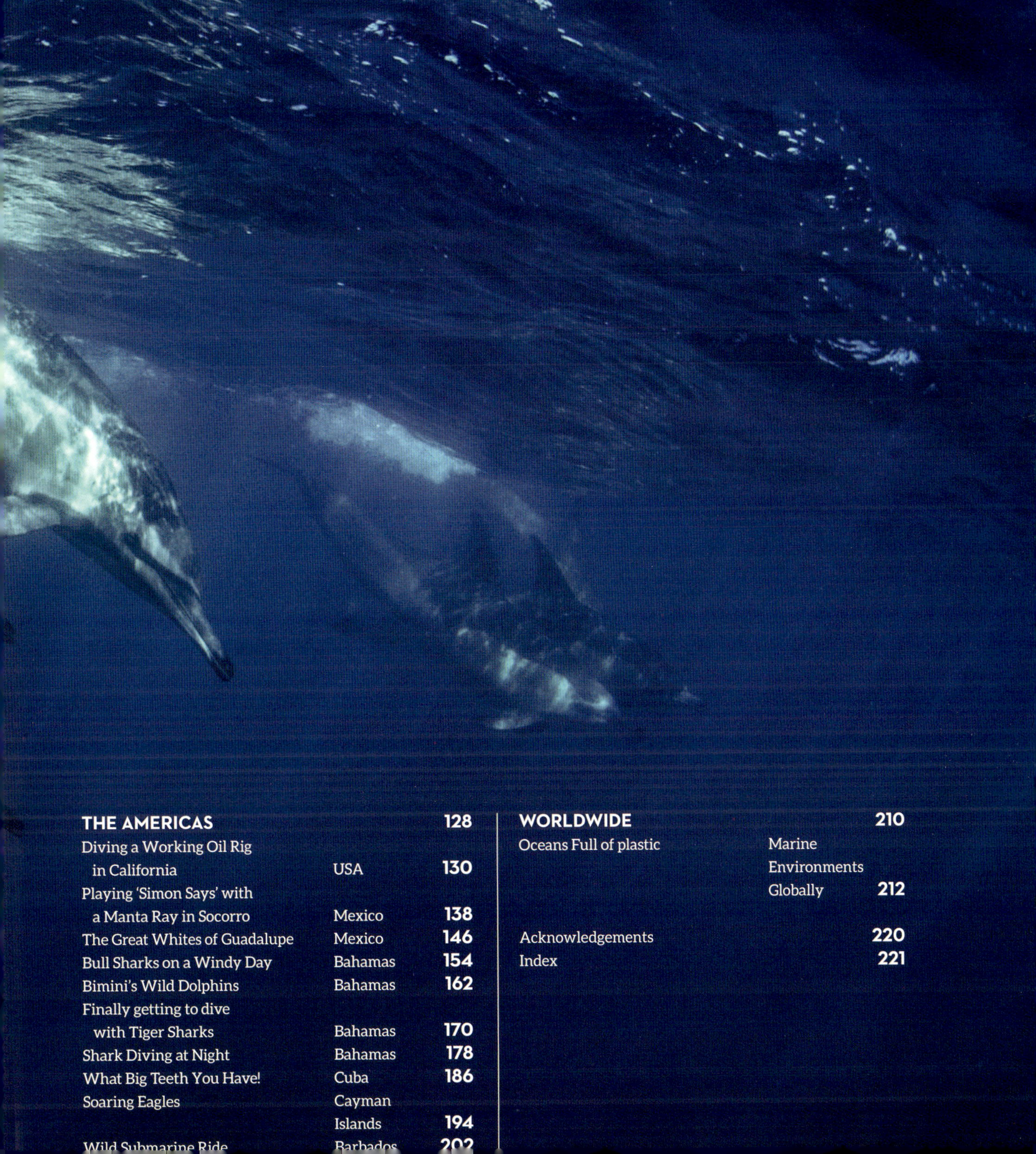

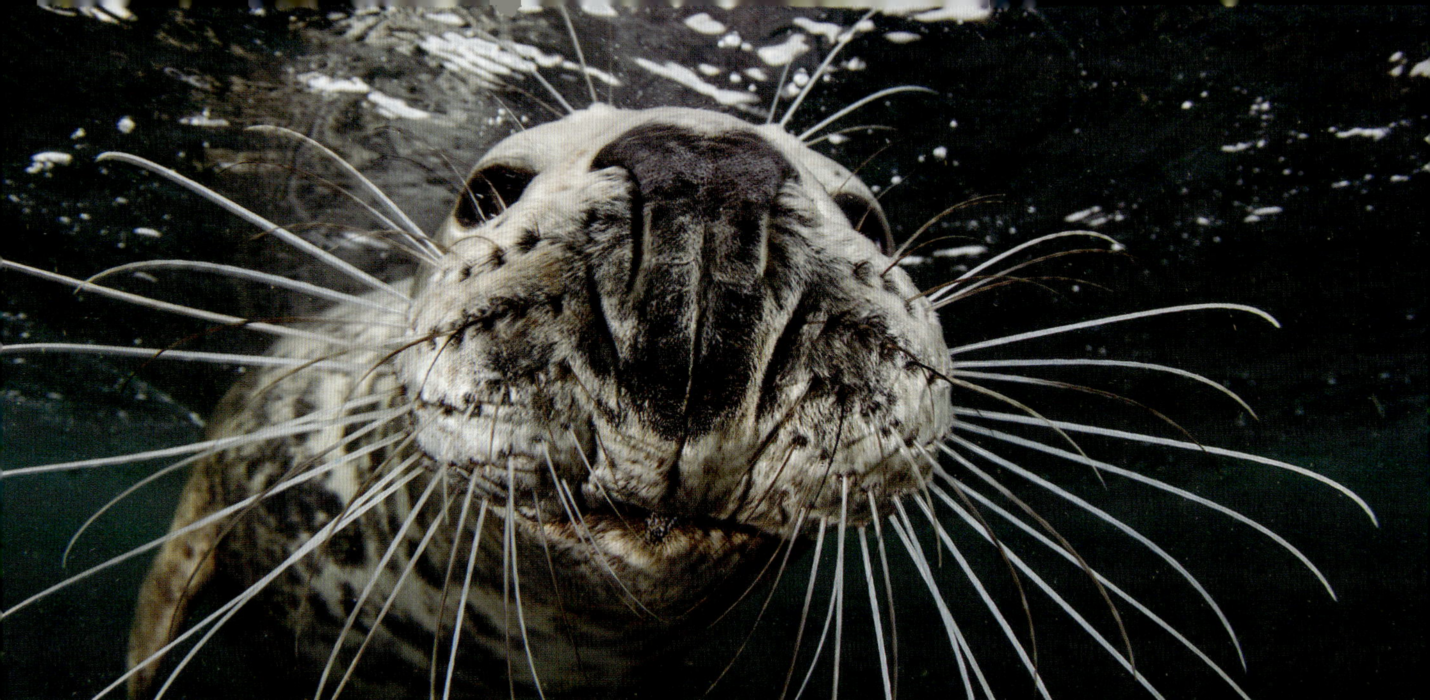

INTRODUCTION
WHAT MAKES A WILD DIVE?

While we were developing the concept of a book entitled *Wild Dives* we weren't completely sure what this was actually going to involve. Of course there are all the synonyms for wild, such as desolate, barren, rough, remote and uninhabited to mention but a few, but in a modern context wild can be used to describe something really unusual, cool or thought-provoking, or even something with a design so far outside the bell curve of normal that wild is an apt description.

So, looking at these alternative uses for the word wild, we started to look at the dives we have done and the dives we wanted to do that fitted these wider definitions. It took us on a journey to diving destinations we had longed to visit, and in the process up close with some incredible wildlife.

We have included dives that had us looking for specific species, both big and small; dives that took us to some of the most remote places in the world; dives (and sometime snorkels) that saw us experience amazing encounters with wild marine animals; dives on wrecks with a particular wow-factor; dives in unusual locations; and even a submarine ride.

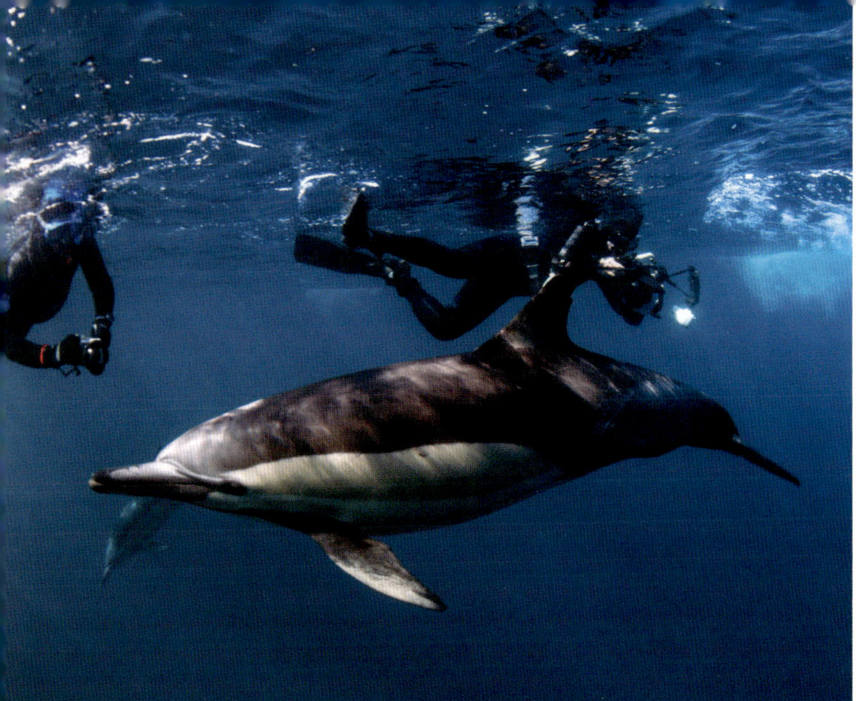

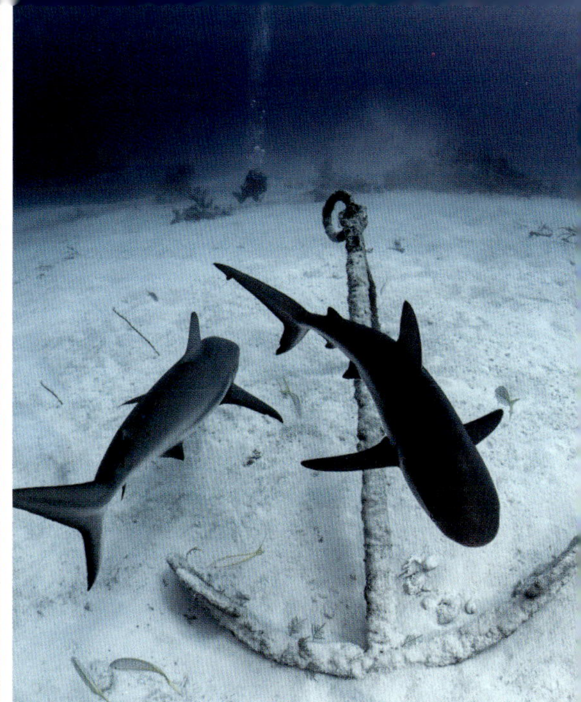

The images we captured for this book have been taken all over the world, including the Caribbean and other parts of the Americas, Africa, Europe, Oceania, South Asia, the Middle East and the South Atlantic. Equally though, a wild dive can take place close to home. A remarkable encounter with a Grey Seal in the Farne Islands, not so far from where we live in the UK, springs to mind. Caroline takes up the story:

"On this particular dive a seal was bold enough to be lying next to me and I finally got a chance to try a selfie, or a 'sealfie' as I now like to call it. It pressed its nose up against my hood and then used its long whiskers to explore my camera housing. After quite some time I decided to head back to the boat, but I could feel the seal gripping the handles on the camera housing over my own gloved hands and it was pulling the camera from me. It wanted to steal my most prized possession and I was not going to let it win. We had a stand-off for a few moments, each pulling at the camera to try to break it free from the other's grasp, but then suddenly to my huge relief the seal let go and headed back into deeper water, turning occasionally to see if I would follow. That's what I call a wild dive."

Wild Dives aims to bring diving adventures to a wider audience and to inspire readers to get the travel bug and go and try out some wild dives of their own. We believe that thinking about the connotations of the broader definitions of the word wild makes for an interesting and enjoyable read, and we hope that readers will agree.

Nick Robertson-Brown and Caroline Robertson-Brown

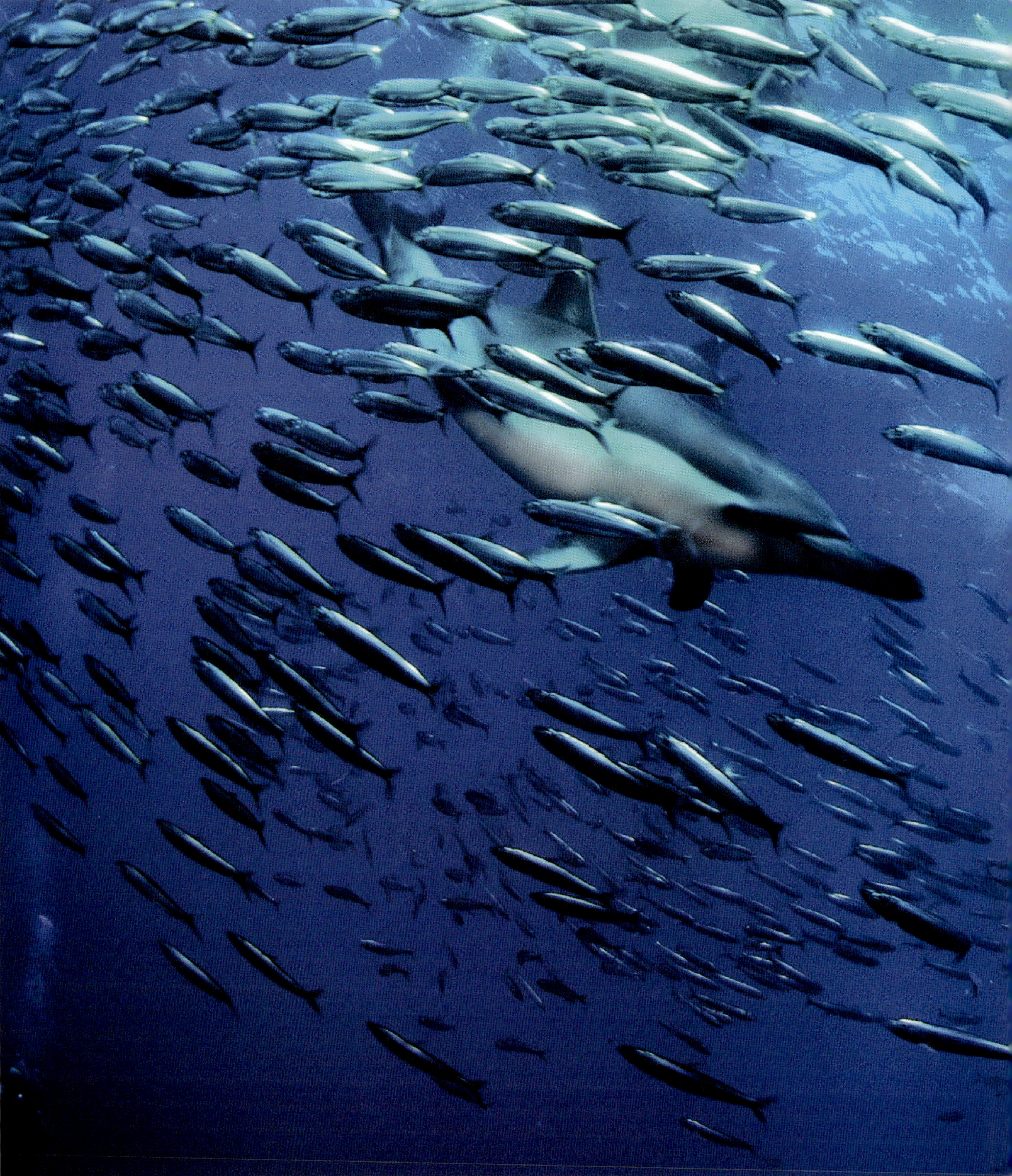

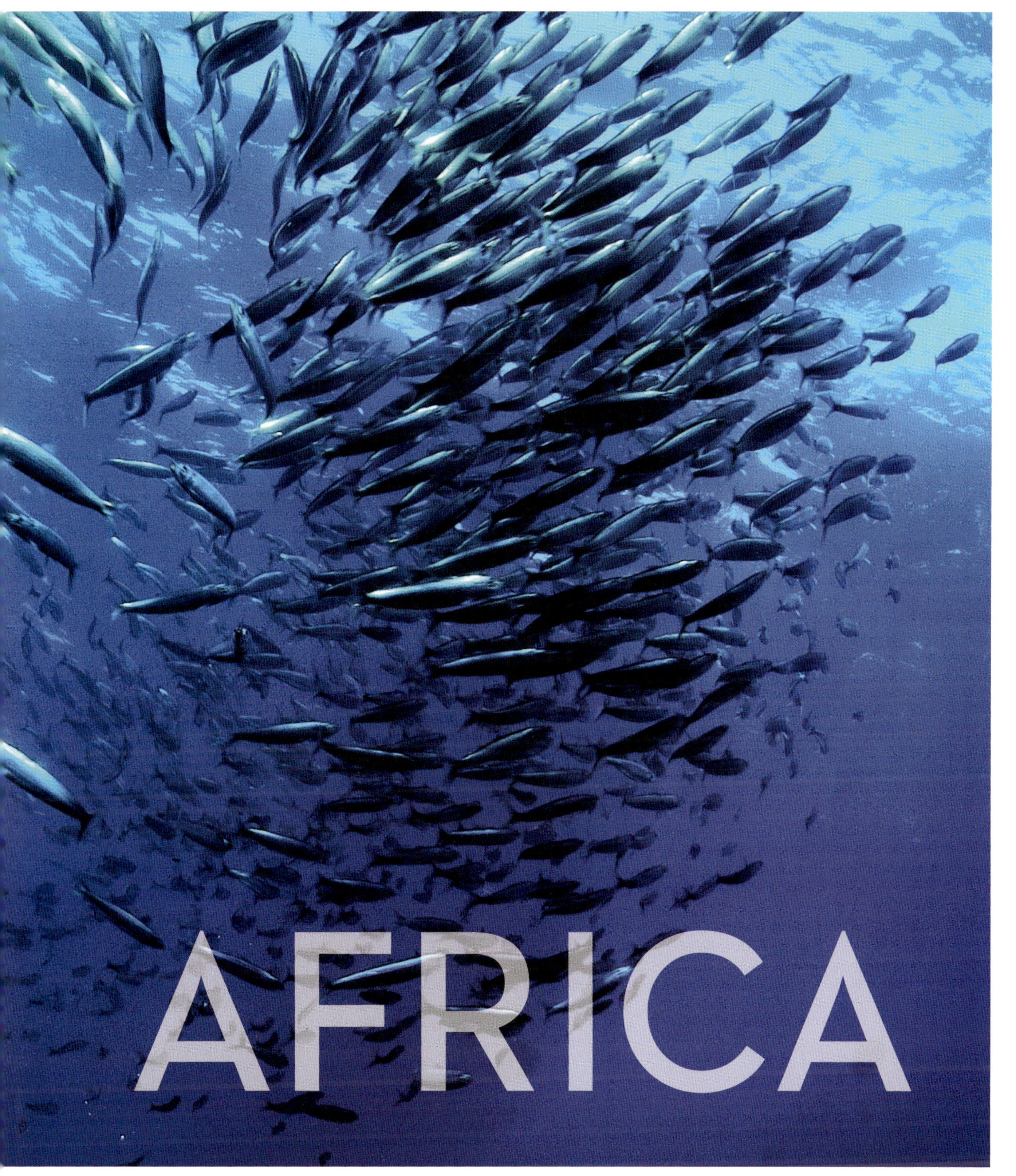

AFRICA

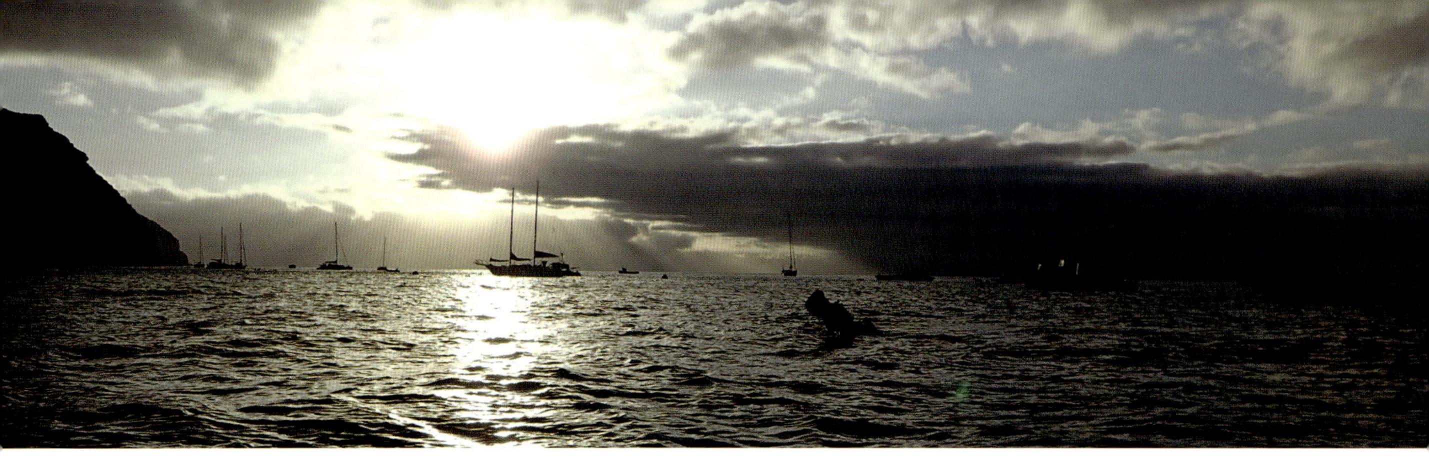

01 DIVING THE *DARKDALE* WRECK

ST HELENA

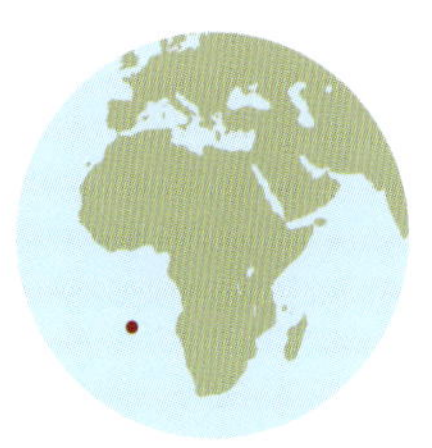

OPPOSITE Many of the wrecks of St Helena are now encrusted in marine life.

ST HELENA IS KNOWN as the second most remote island in the world – the most remote being its distant neighbour Tristan da Cunha – and it sits in the South Atlantic Ocean, some 4,000km (2,500 miles) east of Rio de Janeiro and 1,900km (1,200 miles) from the coast of Namibia. The island has an incredible history, being the place of exile for Napoléon Bonaparte as well as Dinuzulu kaCetshwayo and thousands of Boer prisoners of war. It used to be a very busy shipping port until the Suez Canal opened new and quicker routes for transporting goods by sea. Today about 5,000 people live here, and they have a rich culture and history to share with visitors.

The island is volcanic, with barren moonscapes, lush forests and cliffs that plummet into the Southern Atlantic Ocean below. The landscape seems to change with every turn you take on its winding and steep roads. It is home to a host of endemic wildlife and marine life and is a paradise for those who love to hike.

Visiting St Helena is a little bit like traveling back in time. This quiet British Overseas Territory has virtually no crime and the local people are friendly and happy to give strangers a welcoming smile and a wave, while the internet has only recently arrived and is still slow enough and expensive enough to encourage you to put down your phone and look at the beauty around you instead. A good night out here will see people from the whole community – young and old – gather in

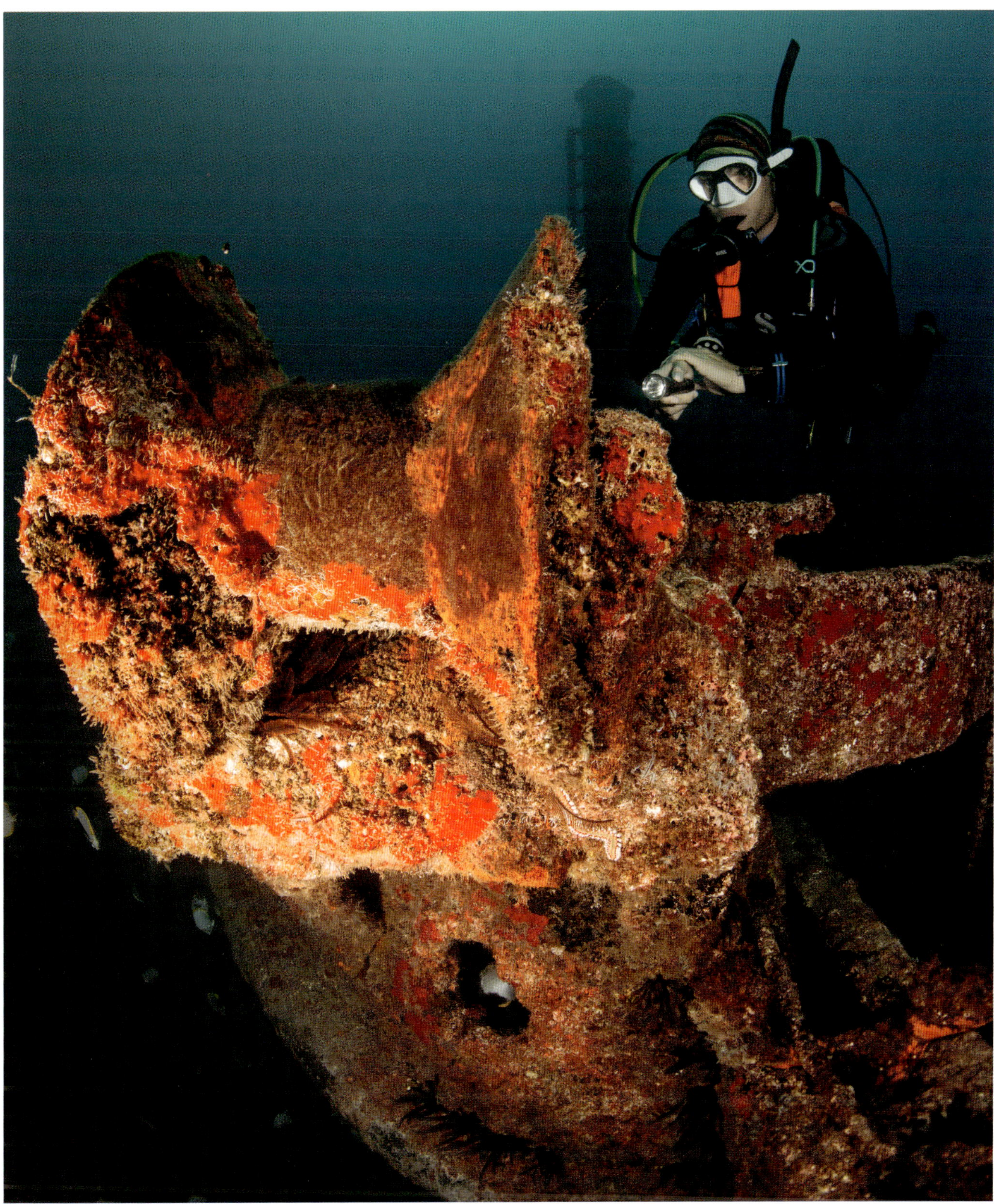

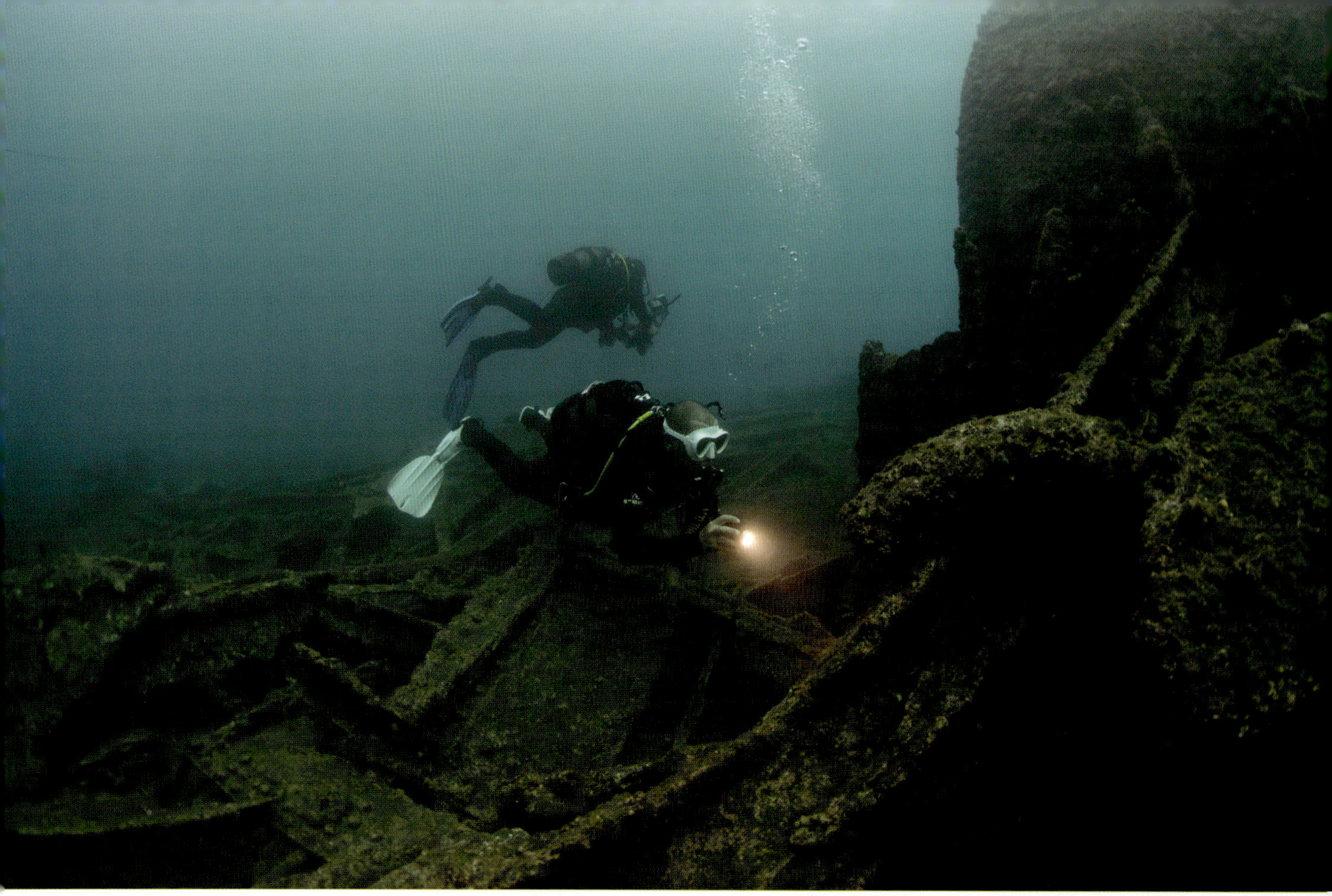

ABOVE Divers explore the *Papanui* wreck in St Helena.

RIGHT Fish life is prolific on all the wrecks.

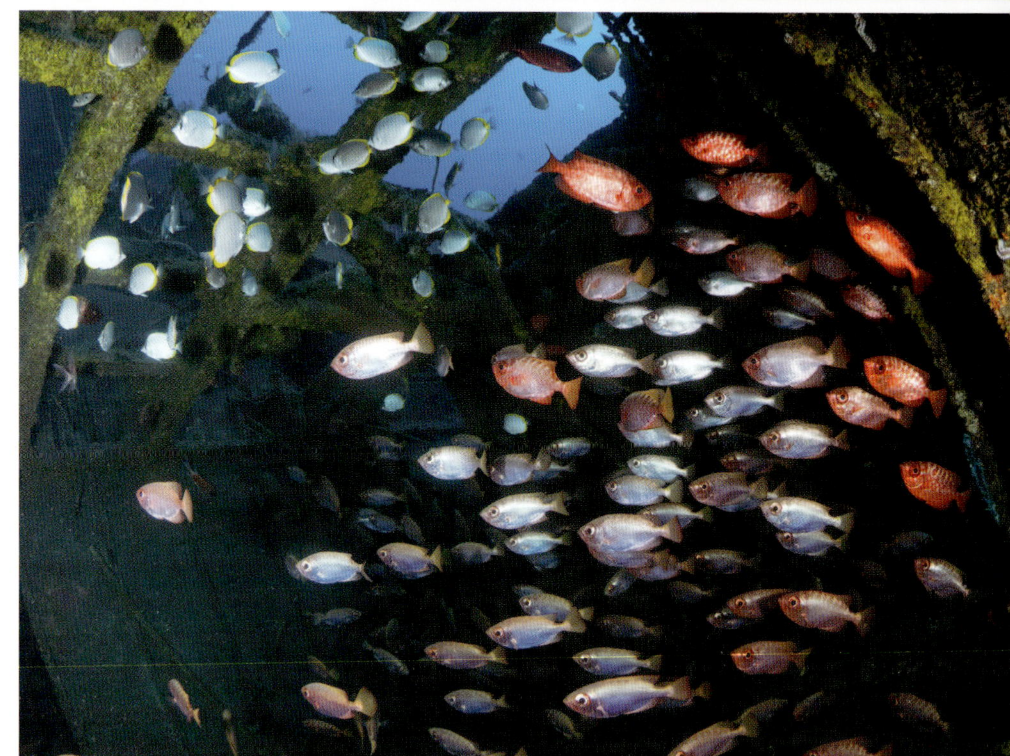

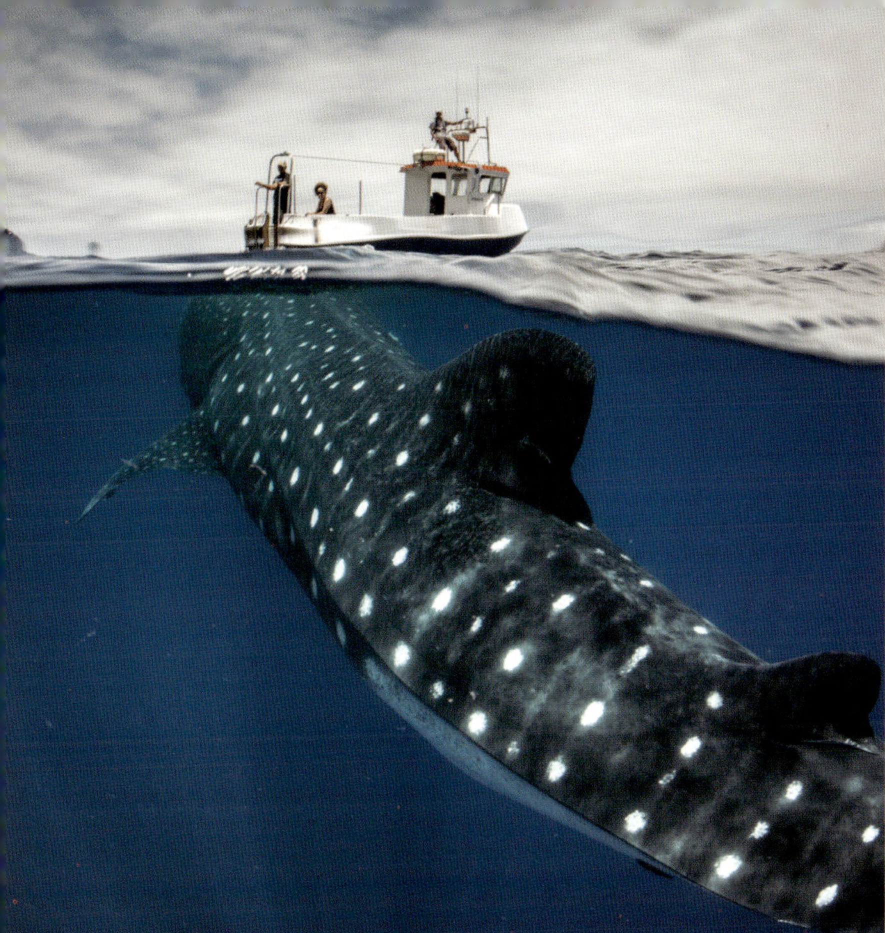

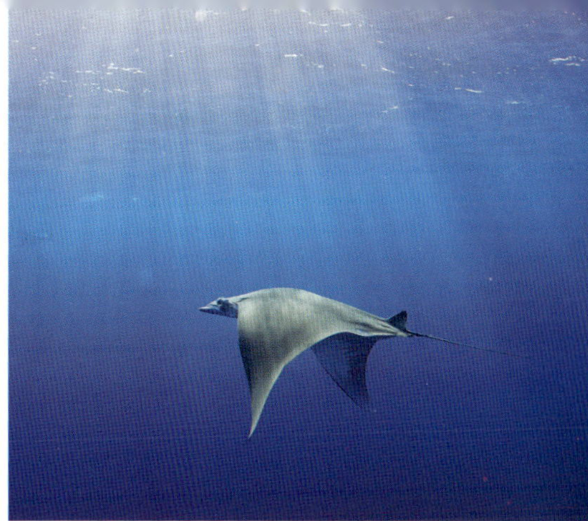

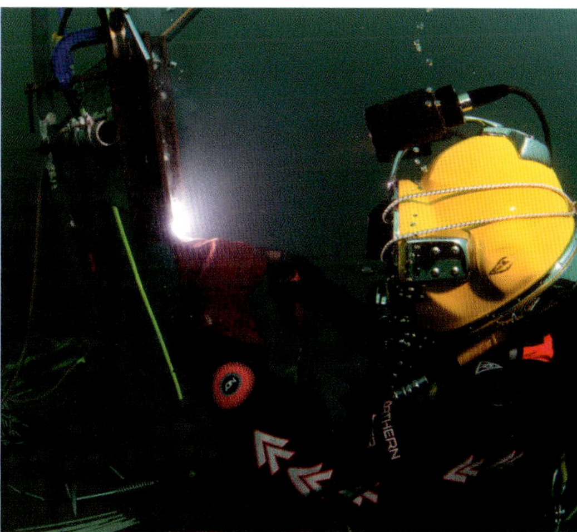

one place to party into the small hours. Of all the places we have visited around the world, St Helena really stands out as being delightfully unique.

GENERAL DIVING

As the land gives way to the sea, the rugged rock formations continue below the waves. Huge steps seem to have been carved into the rocks and thousands of fish hug the underwater seascapes, making the very reef seem alive. Mobula rays visit the shallower sites, sometimes allowing their curious nature to get the better of them, approaching divers and swooping between them. In the summer months, from January to March, huge aggregations of Whale Sharks frequent these waters, unusually with both male and female sharks swimming together in these groups. It is believed that research in these waters might finally lead to more information about the lives of these elusive giants. Here, the biggest fish in the sea seems content to approach snorkelers and boats, making this a top destination to see them.

ABOVE LEFT A huge Whale Shark approaches a boat in St Helena.

TOP RIGHT: Mobula rays swim through the clear blue water.

ABOVE A salvage diver training in the UK before the *Darkdale* mission.

Over the years a number of ships have sunk in this area and their structures remain relatively untouched, making this a top wreck diving destination too. The *Papanui*, sunk in 1911, lies in the shallows of the harbour and part of the wreck juts out of the sea, making this a site that can be snorkelled as well as dived. And yet, it still has plates and bottles that sunk with it visible to the visiting diver. The structure of the wreck is broken up, but this has produced an ideal haven for marine life and it is a wonderful dive to do as the sun is setting and the orange light of sunset is glinting on the surface.

As if that was not enough, divers can also explore caverns and swim-throughs, which have been cut into the rocks over millions of years of relentless wave action from the Atlantic Ocean. The insides of these caverns and overhangs are filled with countless crustaceans and shoaling fish. It feels as if Poseidon himself has created a playground for marine life and divers alike. Bright red soldierfish and snapper form tight groups, sheltering from the surge in the darkness.

THE WILD DIVE

Our visit to St Helena was not a random one. Some years ago we were asked to assist a UK Ministry of Defence (MoD) team of salvage divers with their underwater photography. At first the particular mission they wanted this tuition for was a secret, but we later found out that it was to document their work on the wreck of the RFA *Darkdale* which was sunk by a German U-boat in 1941 in James Bay, just off Jamestown in St Helena. They removed fuel oil and munitions from the wreck, making it safe for the islanders and visiting divers alike.

We told this story to a representative of St Helena Tourism, who immediately said that we should join a trip out to this remote island to make the story come full circle. Who were we to refuse! This is how we found ourselves kitted up on the side of a RIB about to roll into the South Atlantic Ocean to dive a war grave that sits between 35–45m (115–150ft) below the surface of the water. The wreck is just a few minutes by boat from the harbour and is marked by mooring buoy. I rolled into the water and made my way to the descent line, and even though there was no current, followed the line down, so as not to miss the wreck. As we were performing a no-decompression dive, we were not going to get much time on the wreck structure

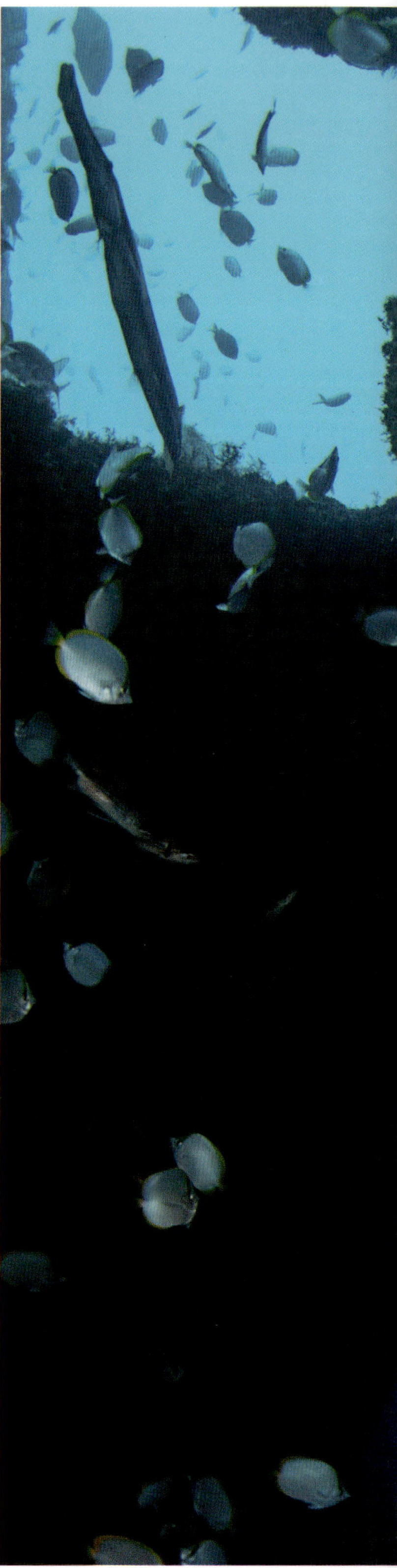

RIGHT A diver inspects the prop of one of the many wrecks on the island.

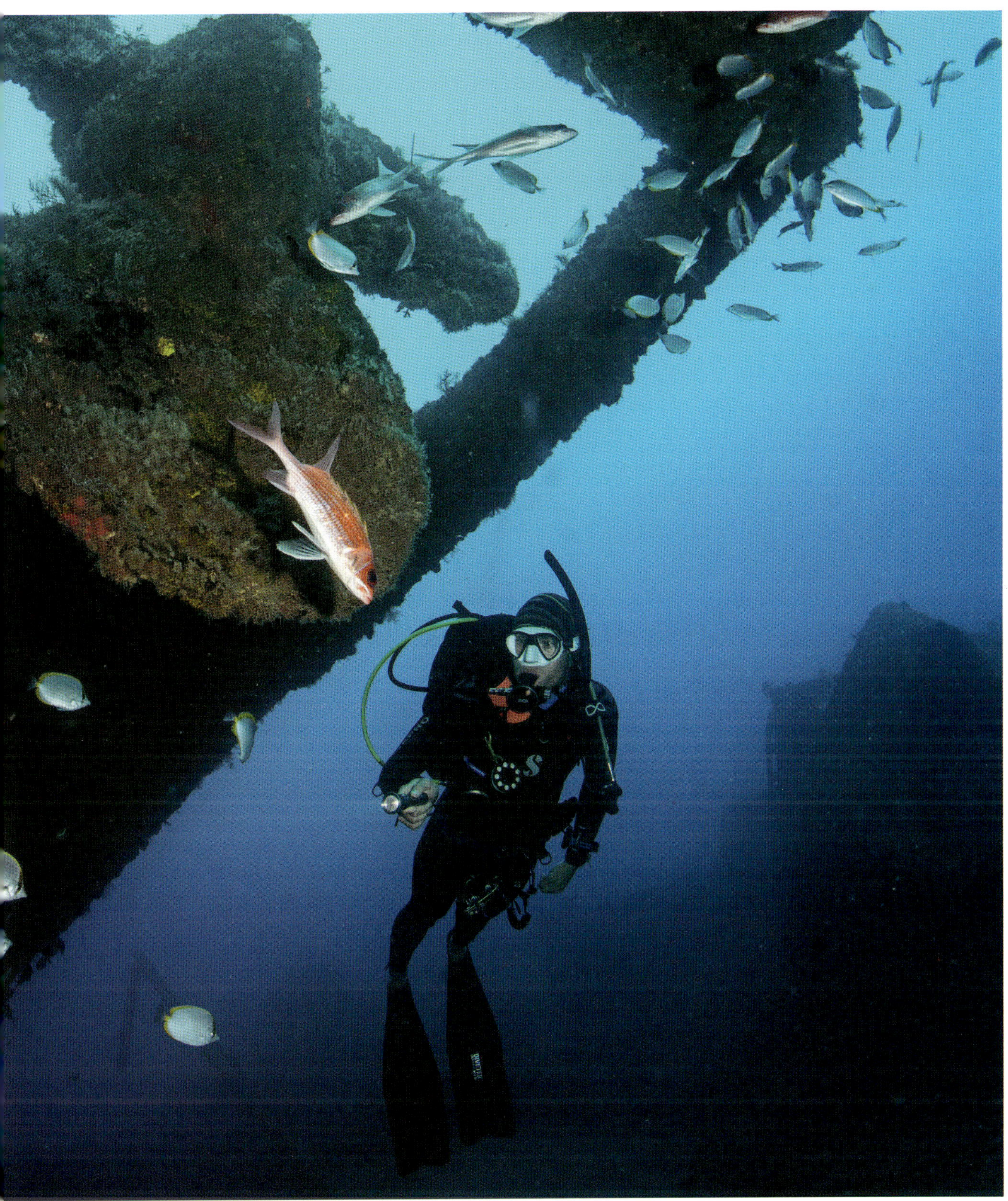

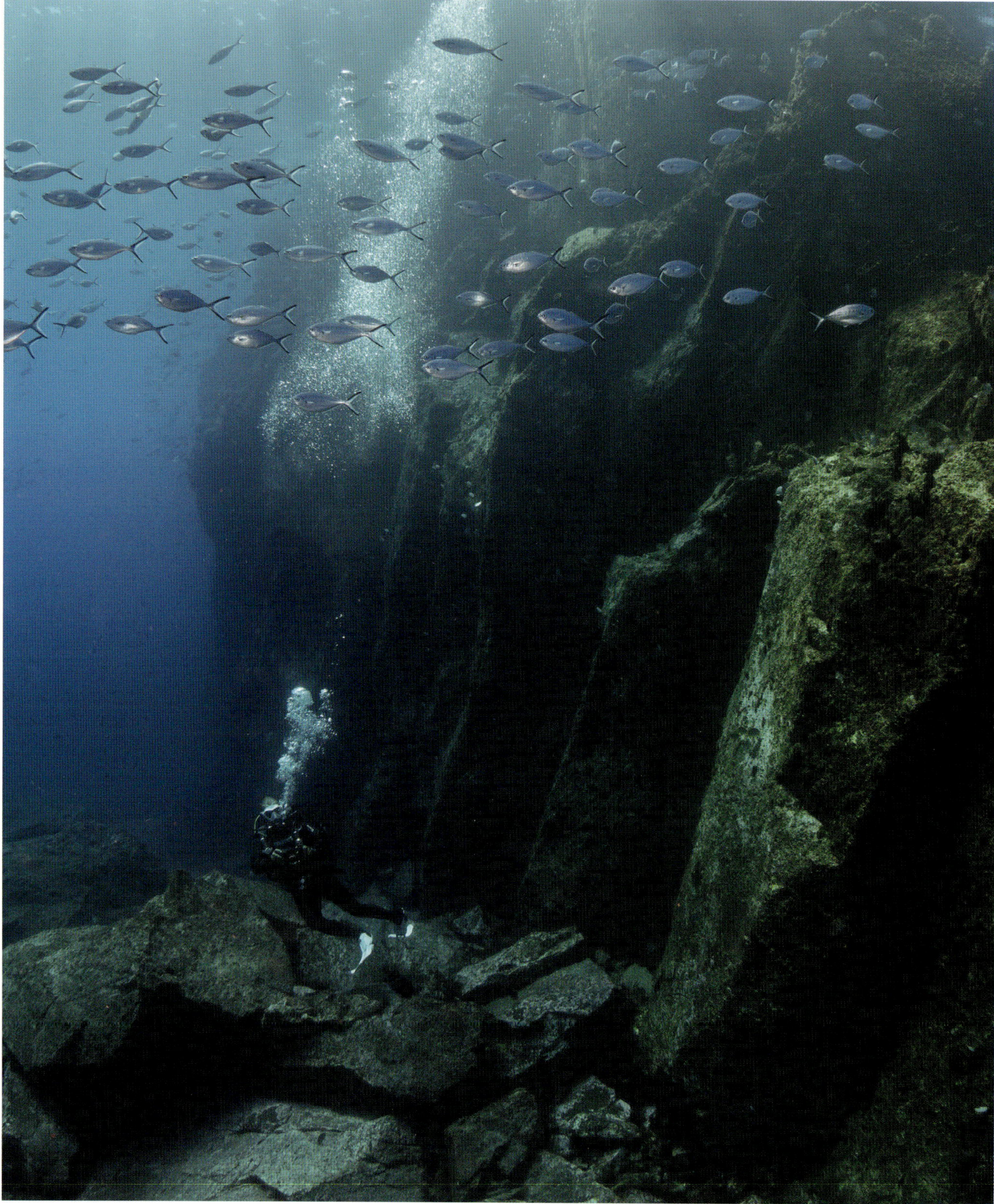

itself, just a few minutes at the shallowest part of the wreck. It was a poignant dive. It was the first British ship sunk south of the equator during the war and 41 of the crew were lost. As she loomed out of the dark blue water, her size became instantly apparent. As with all the dives we had done in St Helena, thousands of the endemic white cunningfish coated her structure. A few bigger jacks joined us on the dive too. I hovered in silence, taking in the wreck before me, and then started to make my way back up to the surface. As we got to shore, I made my way along the harbour to the memorial that stands there to remember those lost on the *Darkdale*. A fitting way to close the circle of this unusual diving story.

OPPOSITE Huge rock formations make impressive backdrops to the dives.

BELOW Whale Sharks come here between January and March.

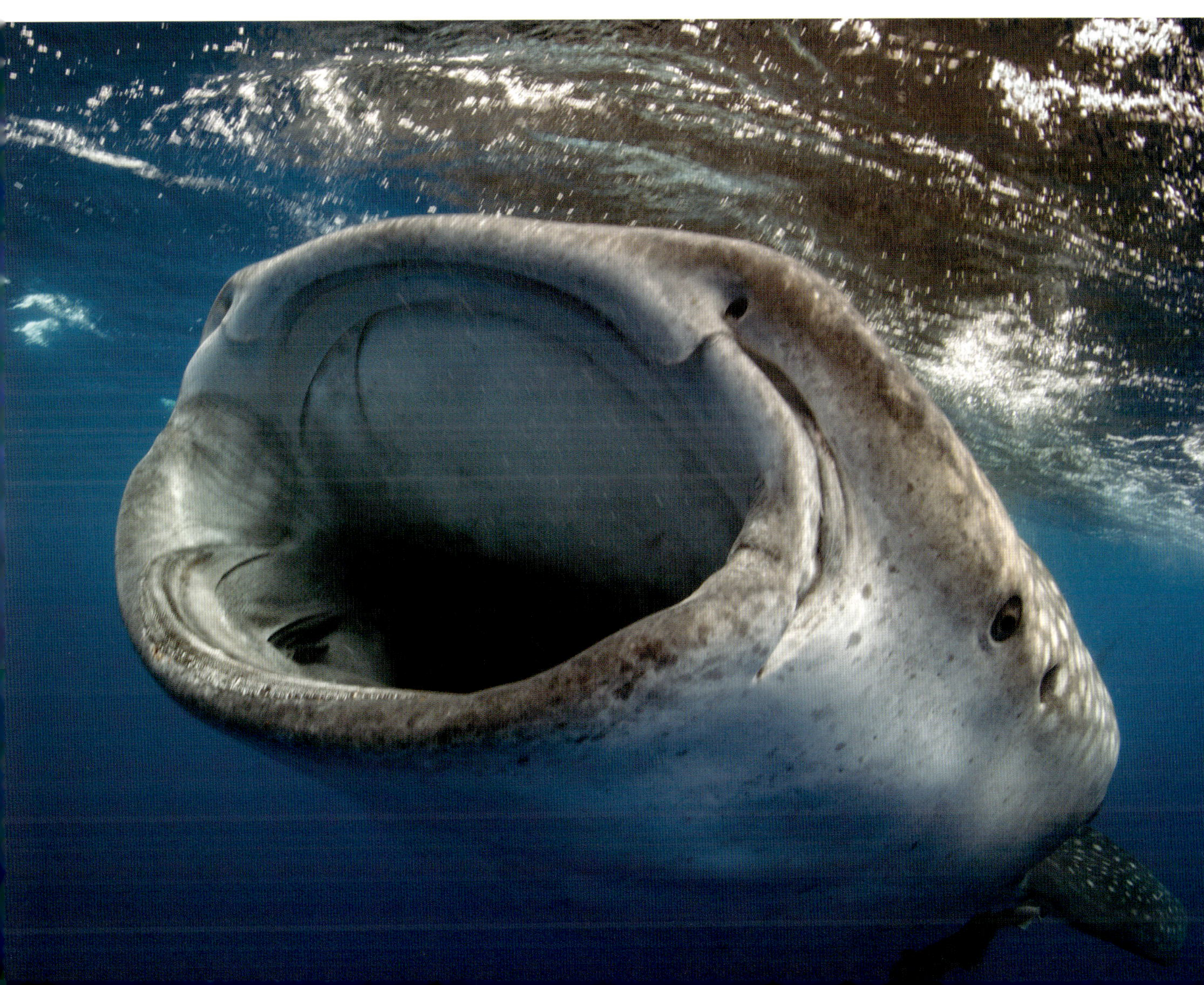

02 CAPE FUR SEAL STARFISH SURPRISE

SOUTH AFRICA

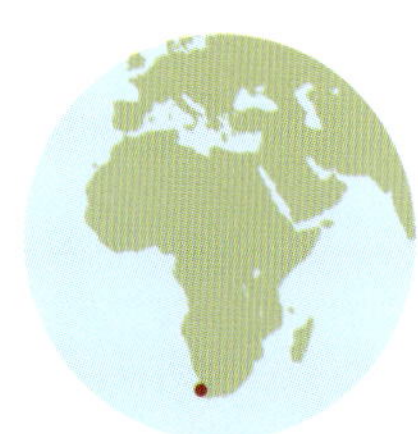

OPPOSITE It was like playing with my Golden Retriever puppy.

BELOW Three Cape Fur Seals play on the reef

SIMON'S TOWN IS SITUATED not far from Cape Town in South Africa. It sits on the banks of False Bay on the eastern side of the Cape Peninsula and has a rich naval history. It is a picturesque town that is popular with tourist and locals alike. Being based in Cape Town allows you to experience much more than just the diving this coastline has to offer. You can take a cable car up to the top of Table Mountain, which is sometimes enveloped in clouds but most of the time offers stunning views of the city, shore and ocean below. This is also a wine region, with some of the most famous South African vineyards just a short excursion from the capital, making a lovely way to spend time out on a non-diving day. You might also want to consider visiting Robben Island, the notorious prison where Nelson Mandela was held for 27 years. The city is packed with great places to eat and drink after a day out on the water or exploring the history, culture or surrounding landscape.

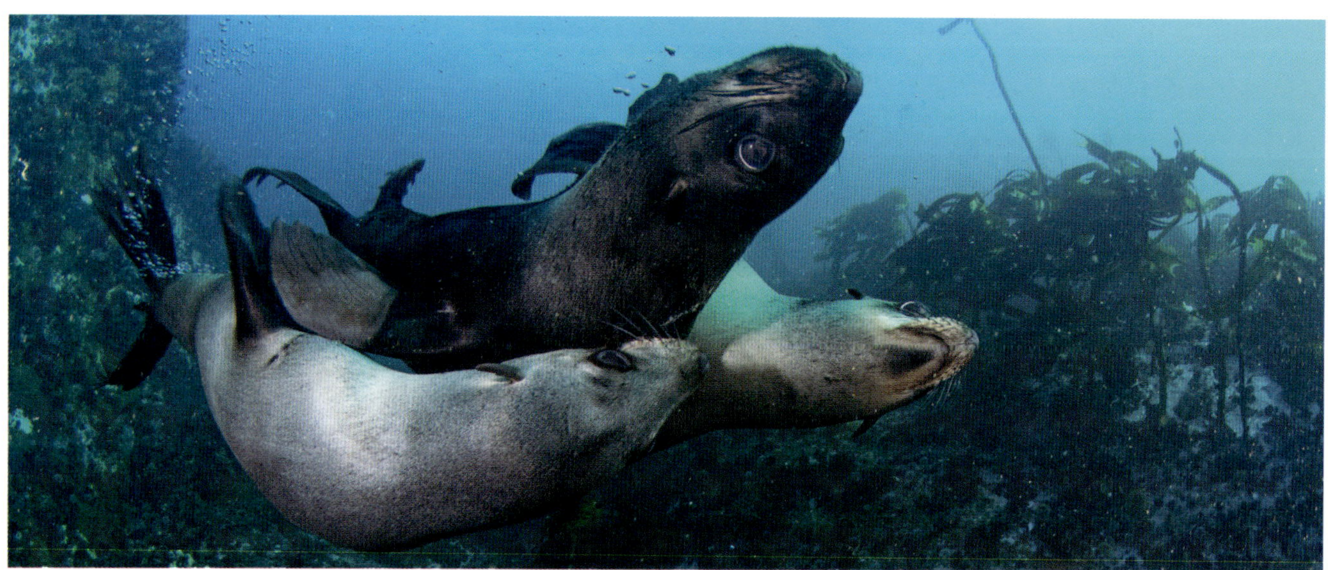

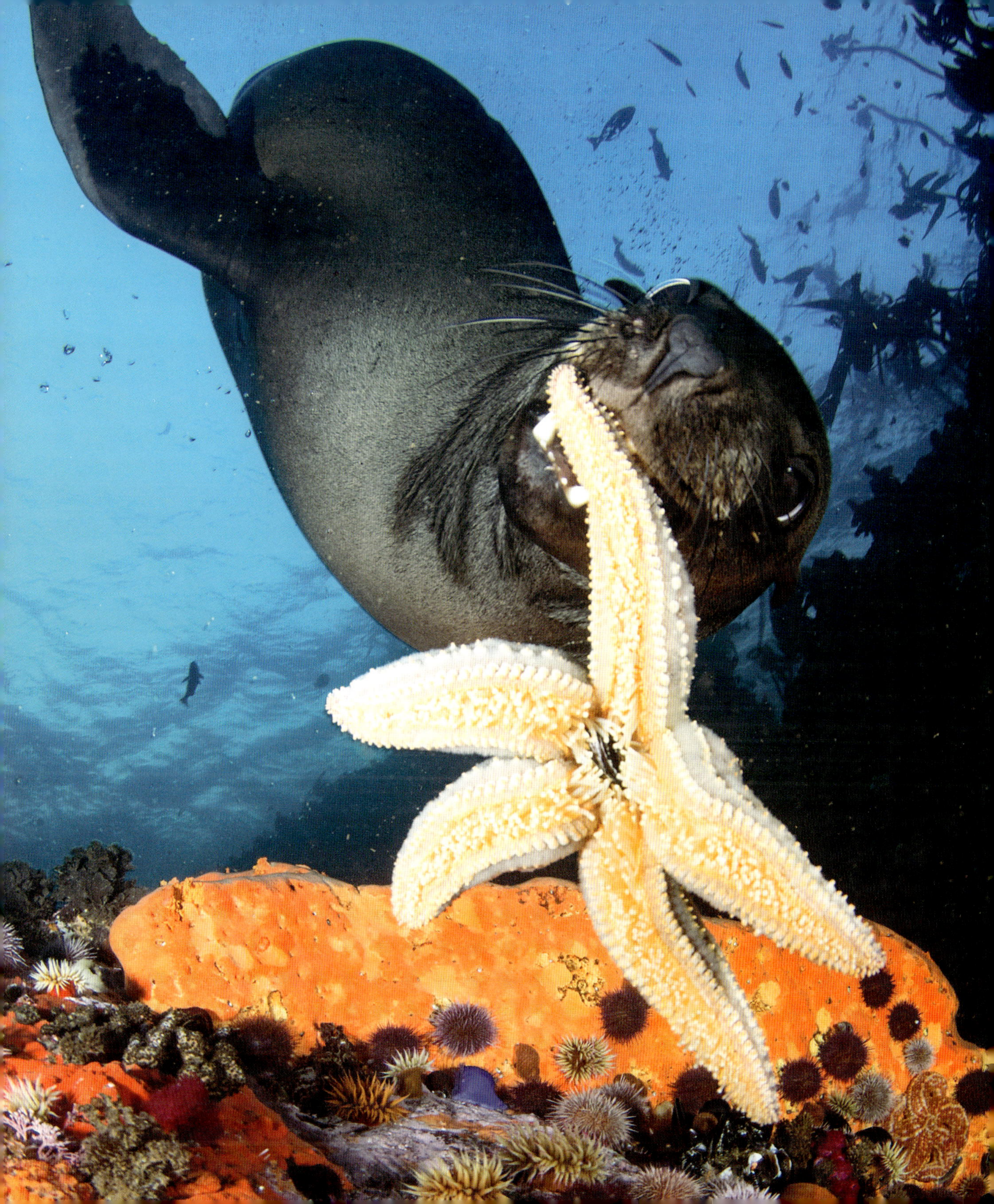

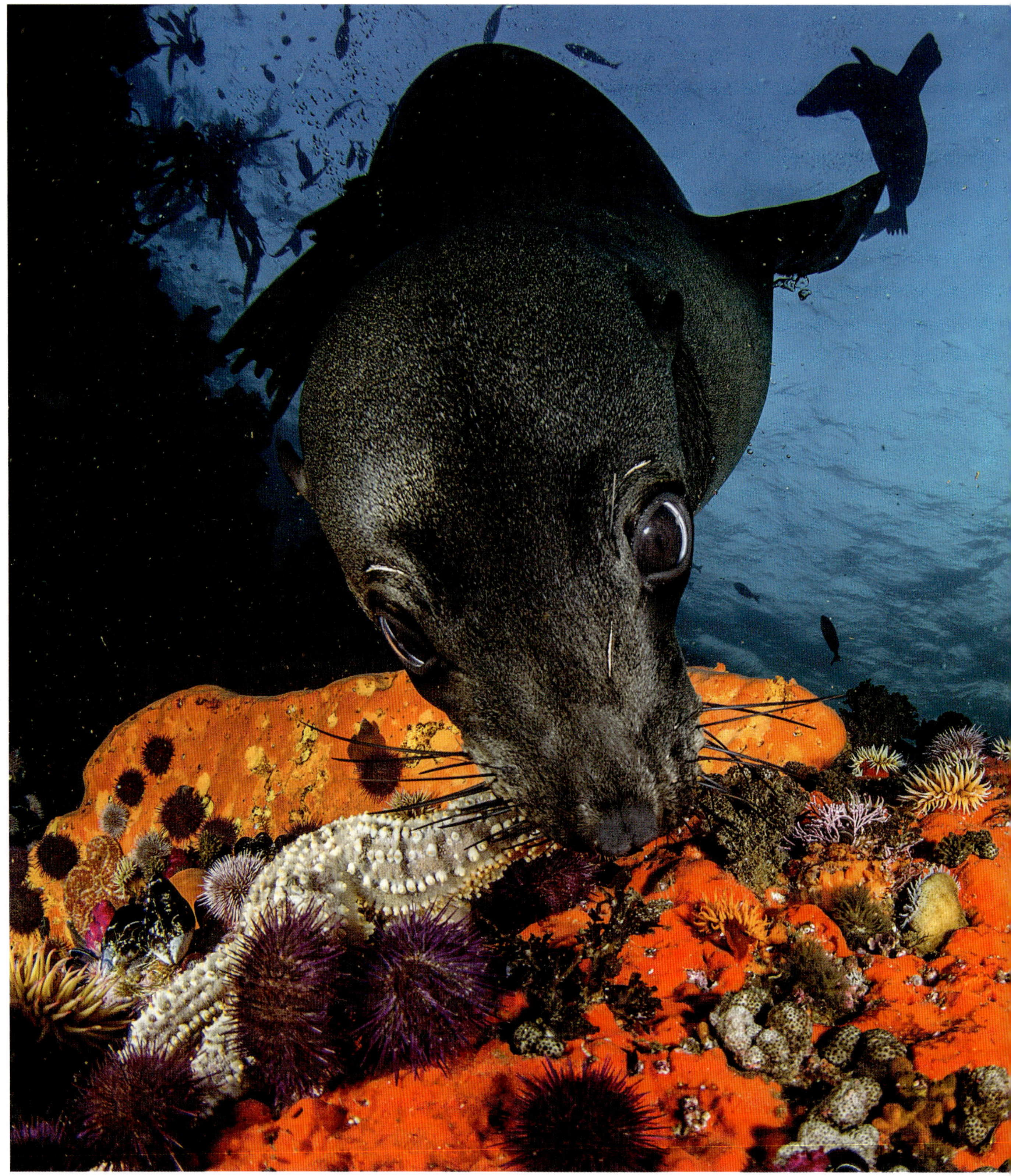

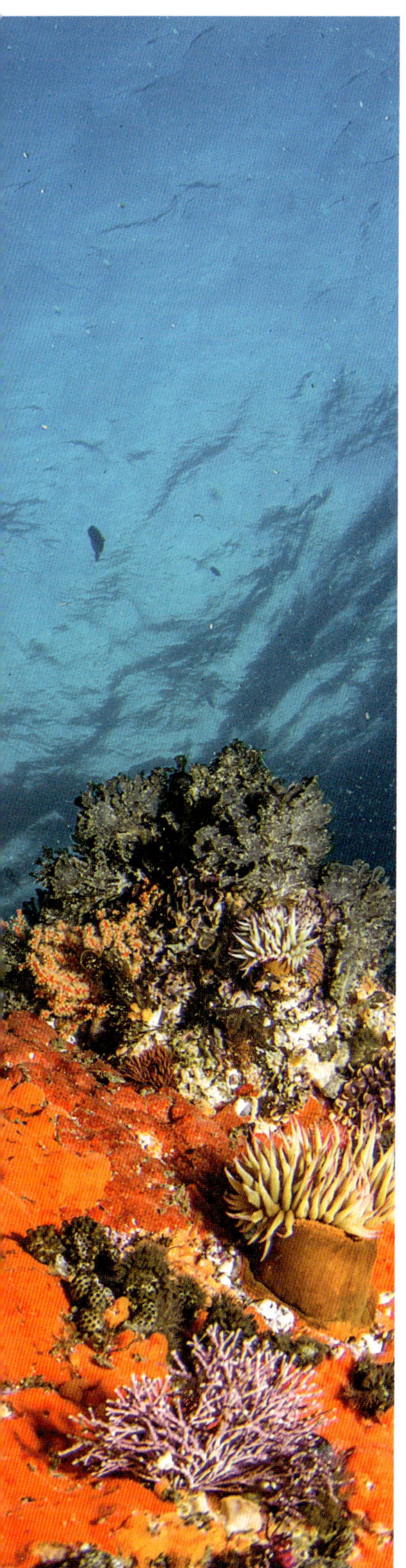

GENERAL DIVING

The South African coastline offers some of the best diving in the world. The productive Indian Ocean waters bring in a huge array of pelagic animals that give adrenalin-packed dives and Simon's Town is no different. Cage diving with Great White Sharks is a popular experience both here and at Gansbaai further up the coast. Featured on many bucket lists, you do not have to be a diver to come face to face with a Great White, as the cages are at the surface and you just have to wear a mask and pop your head underwater to see them swim past the cage. It is also possible to see several other shark species in these waters. Other options here include open water snorkelling with Blue and Mako Sharks and exploring the kelp forest for the elusive Broadnose Sevengill Shark as well as much smaller species with delightful names such as the Pyjama Shark and Shy Shark. Penguins can be seen heading out to sea in the mornings and returning at the end of the day to the famous Boulders Beach colony. The water here is cool and rich in nutrients but can also be tumultuous, so be prepared to add extra diving days just in case the weather is too lively to get out on the water.

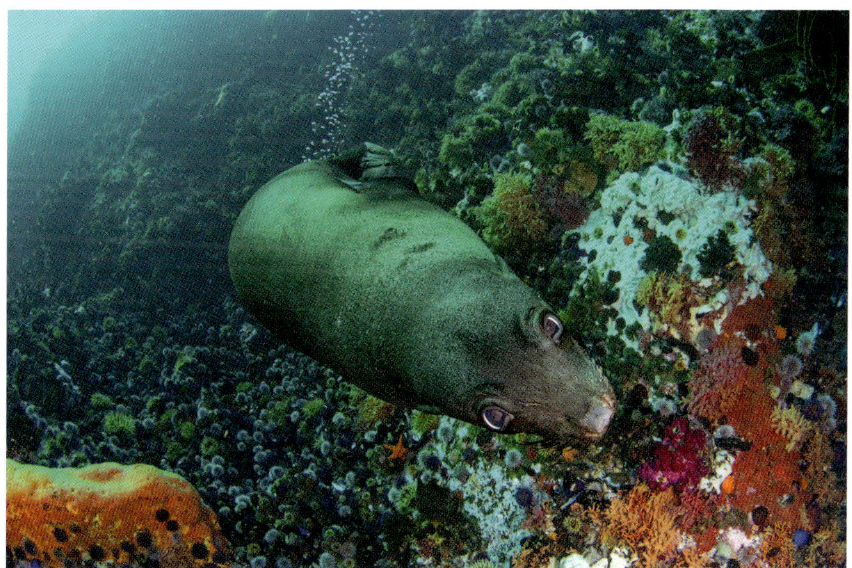

ABOVE 'It wasn't me' – a guilty looking seal.

LEFT The fur seal takes a starfish from the seabed.

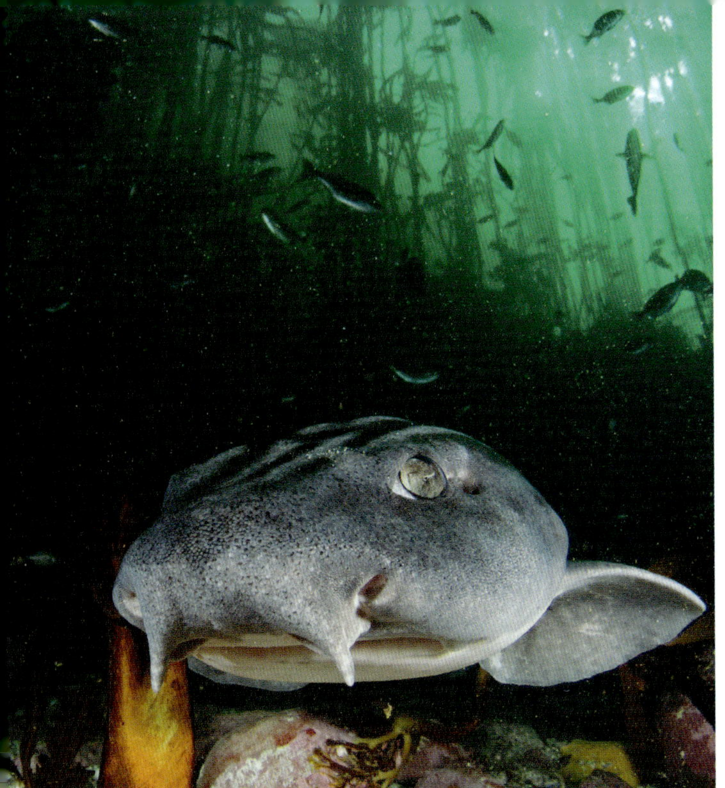

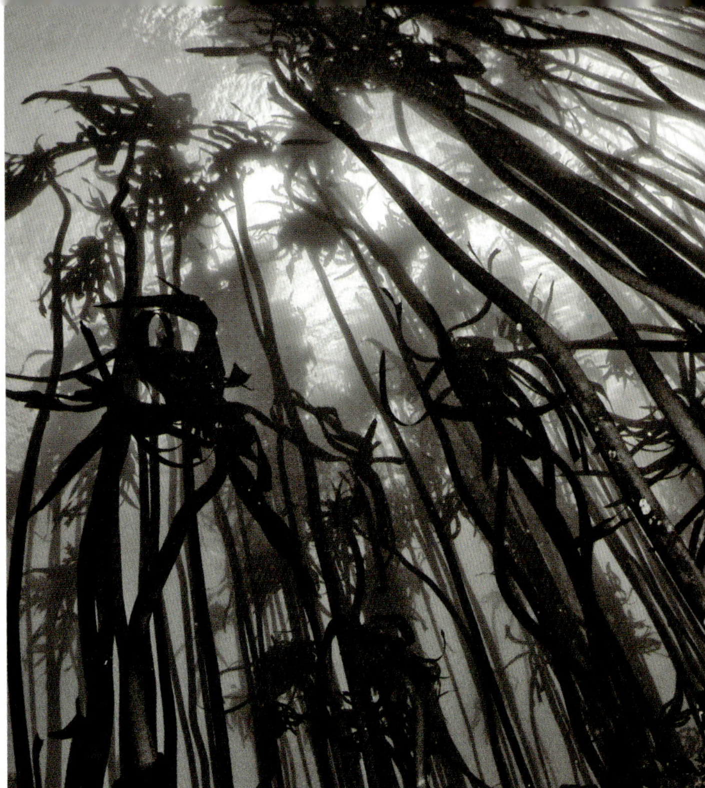

THE WILD DIVE

After a morning of diving with sharks in the kelp forest, we finished off our diving day moored up next to a rocky outcrop populated with Cape Fur Seals. The sun was already low in the sky and the light was sparkling in the blue water around us. As we got ready, the seals seemed to know it was time to play and started wobbling their ungainly way from the rocks into the water. As I dropped under the waves I saw a gully between the rocks and headed towards it. The seabed was covered in bright urchins, while orange sponges and kelp hung onto the rocks and stretched up towards the surface.

I felt a tug at my fins and turned, expecting to find one of my dive buddies messing around, but instead came face to face with a Cape Fur Seal. It turned, showing incredible agility and zoomed off into the kelp. I hung out near the same patch of kelp and soon had three of these delightful pinnipeds darting around me, grabbing kelp fronds and stopping to look at themselves in the refection of the camera lens. I turned to see that each of my dive group were having a similar experience, these playful creatures pulling at snorkels and fins and mouthing neoprene hoods.

The water was cool and as the sun dropped further it seemed to get colder, forcing some of the group back to the surface. I decided to head to the bottom of the gully, where the vibrant seabed would make for a nice close-focus, wide-angle

ABOVE LEFT The delightfully named Pyjama Shark.

ABOVE RIGHT Kelp forests are an important part of the marine ecosystem.

OPPOSITE Three or four seals would come and investigate the divers.

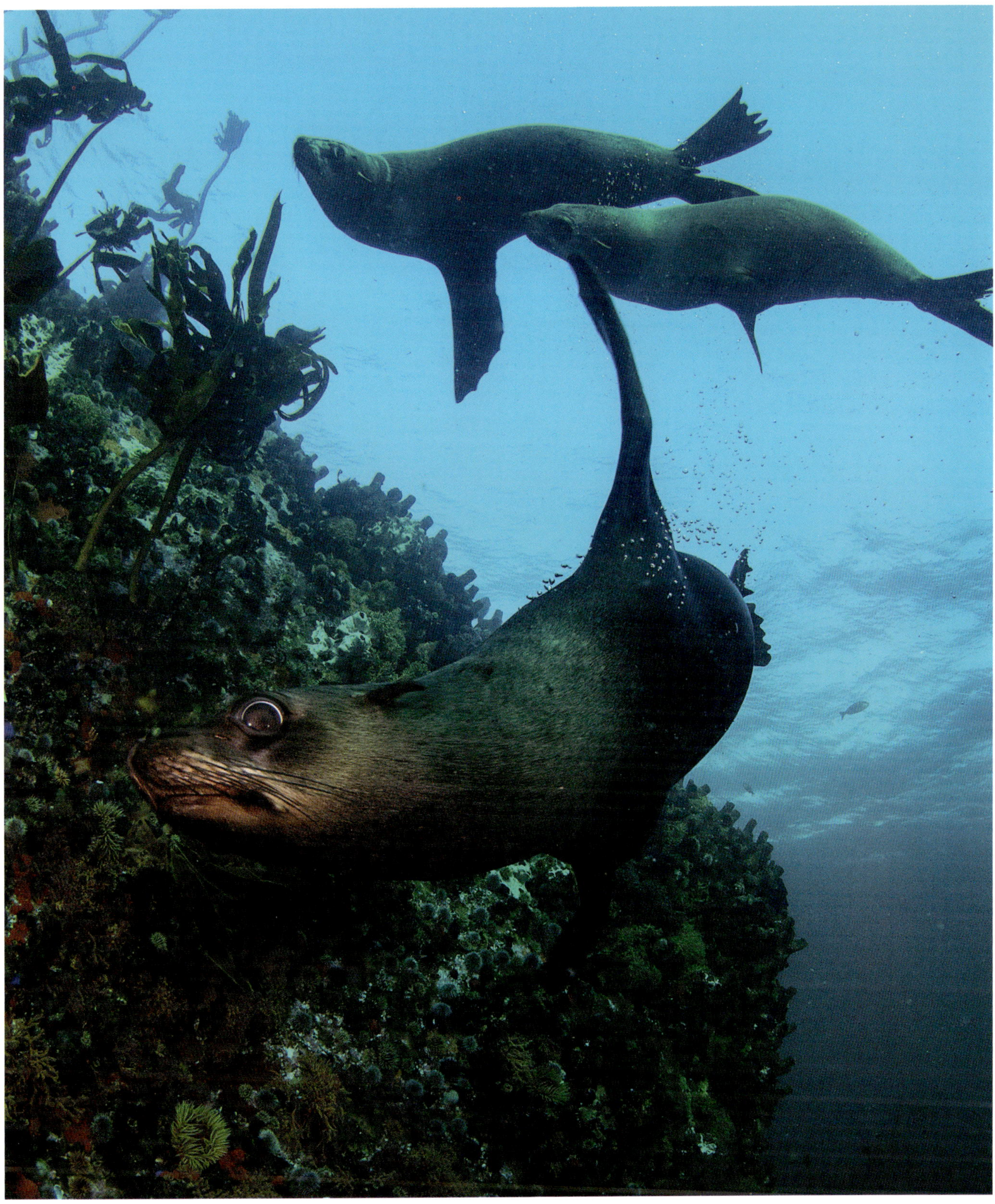

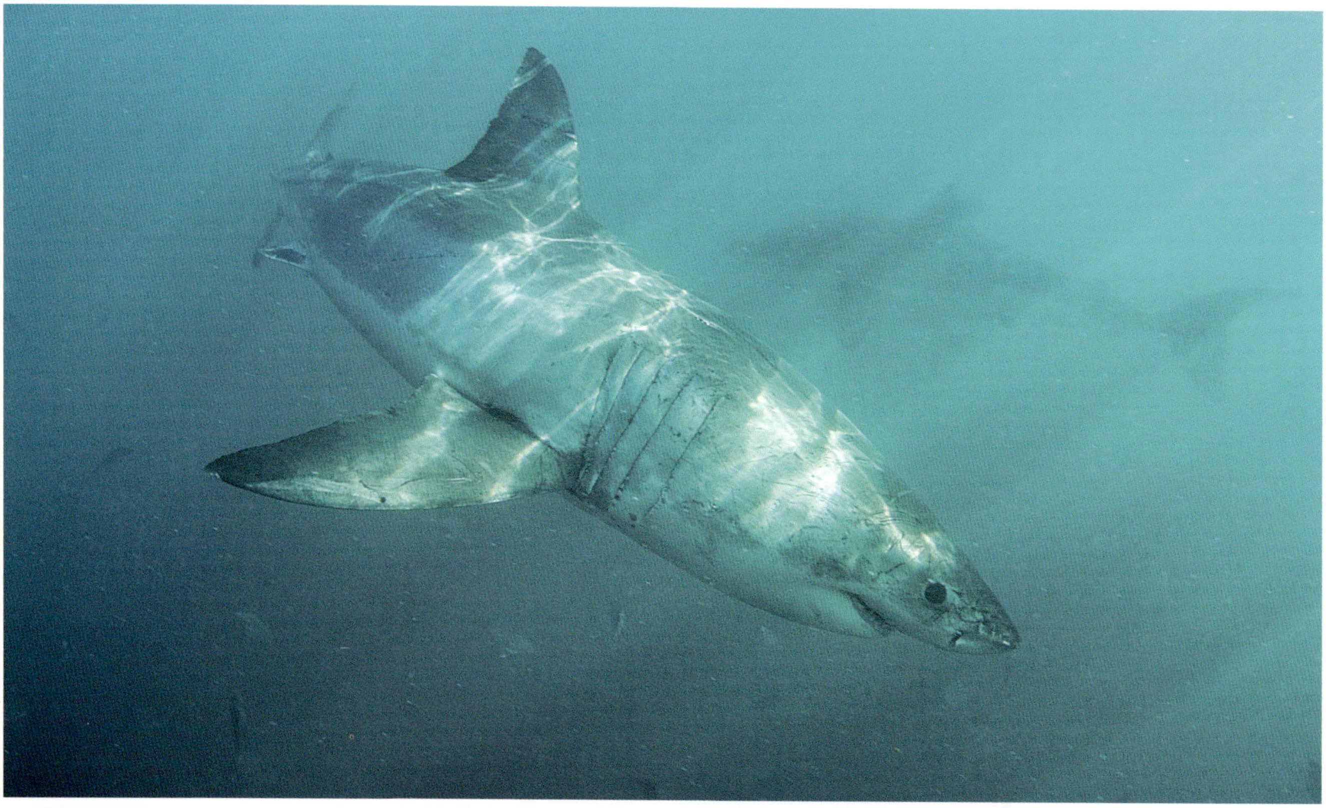

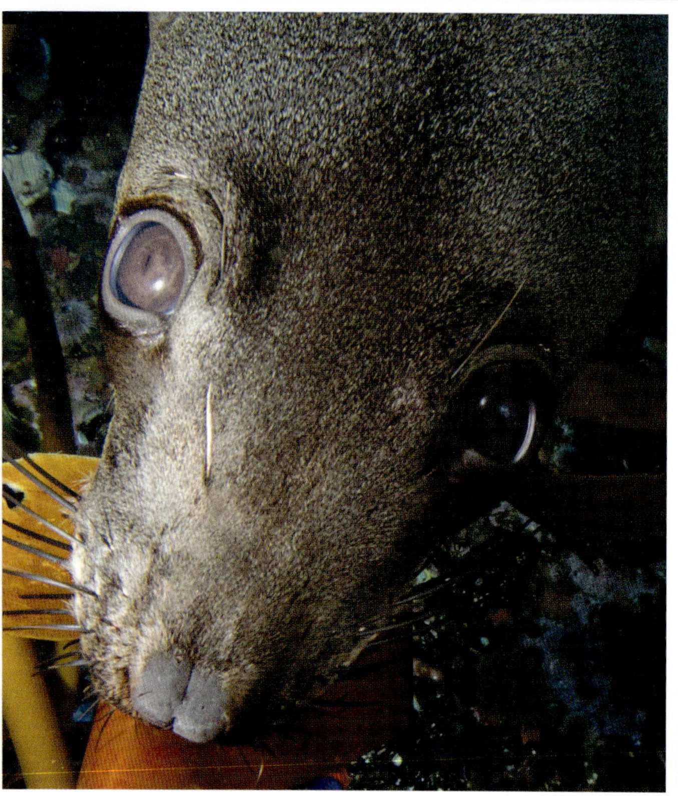

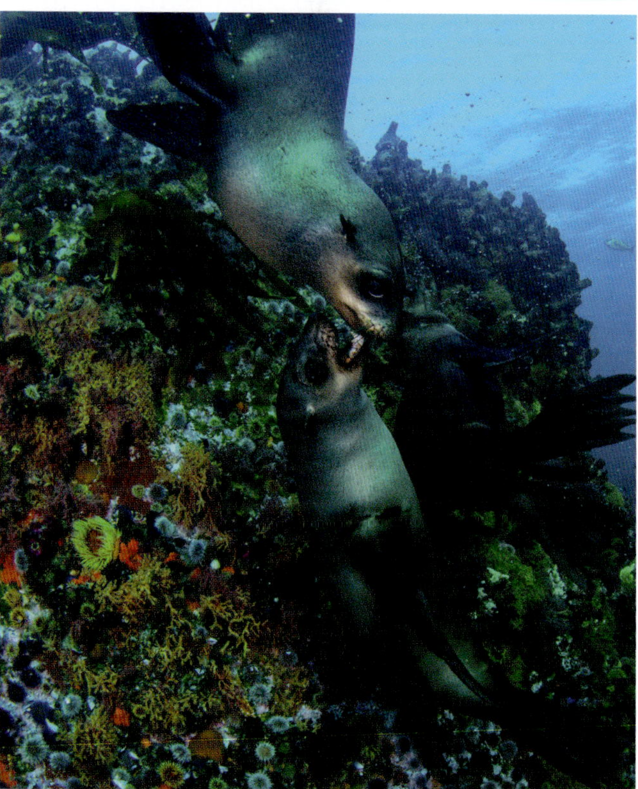

image. One seal joined me and so I hovered close to the bottom waiting for the games to begin again. However, I was not expecting what happened next. It put its nose down on the sea floor and I edged closer to see what it was looking at. It had gently got hold of one arm of a yellow starfish and was pulling backwards, easing it off the rock it was attached to. Once it had freed the starfish it swam up to me, slightly above my head, and dropped the starfish down onto me. What do you do in this circumstance? What is the correct etiquette? I decided to err on the side of caution and stayed still, waiting for the seal to make its next move. It darted back down, picked up the starfish again and stopped in front of me, looking right into my eyes and then gently dropped the starfish onto my hand. I let it fall once again to the seabed and the seal seemed to be confused that I did not understand the game. To be honest it was exactly like playing with my Golden Retriever puppy and their expressions were incredibly similar. With its big sad-looking eyes it implored me to join in the fun, but I held my ground and all too soon it got bored of this one-sided game and with a flick of its fins headed back up to the kelp fronds. It was time for me to also head back to the surface, from a dive that I would never forget – playing fetch (well sort of) with a wild Cape Fur Seal.

OPPOSITE TOP Great White Shark cage diving is popular in this area.

OPPOSITE BOTTOM LEFT Giving me the big puppy dog eyes.

OPPOSITE BOTTOM RIGHT Cape Fur Seals playing together.

BELOW The fur seal dropped the starfish near me several times.

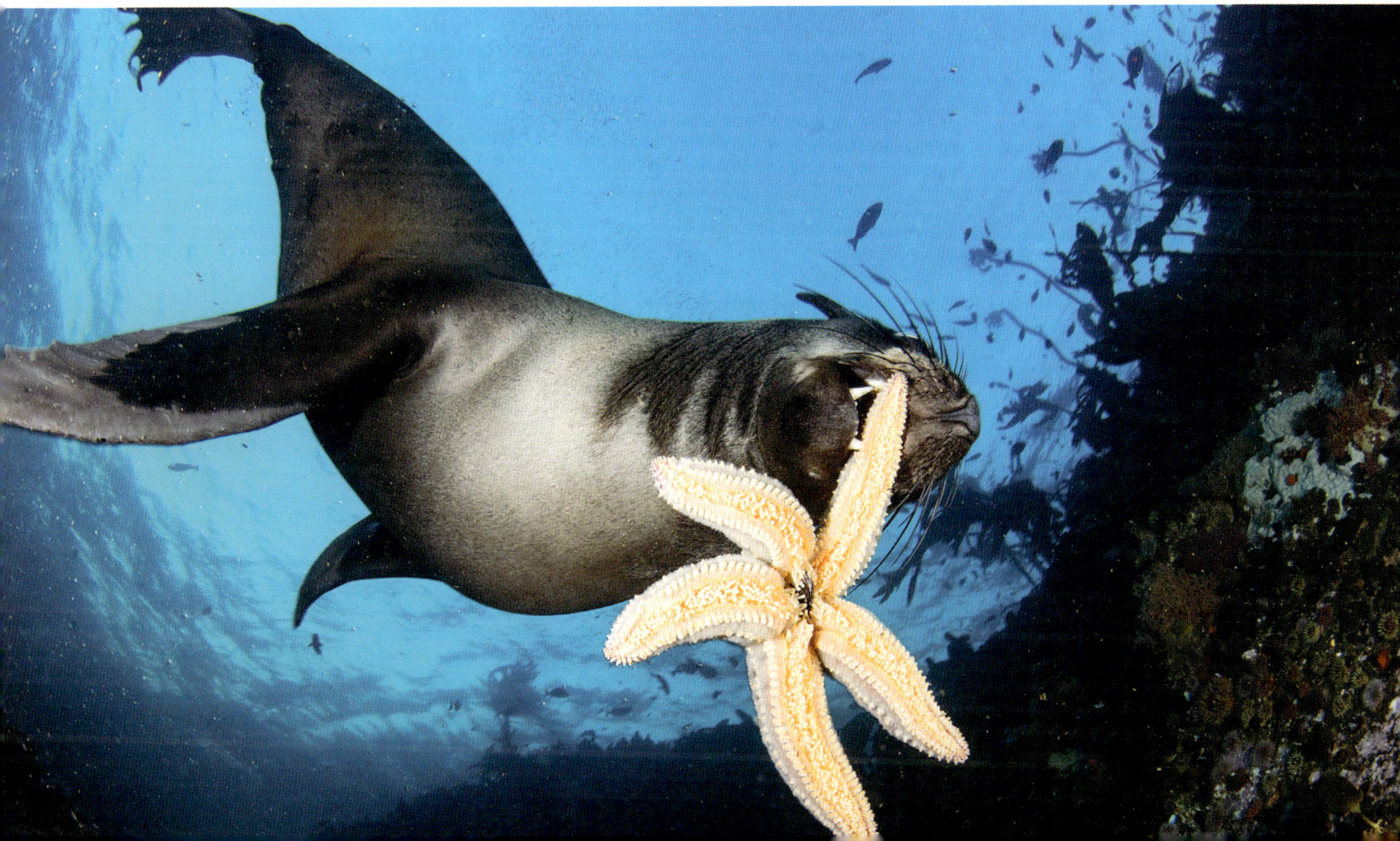

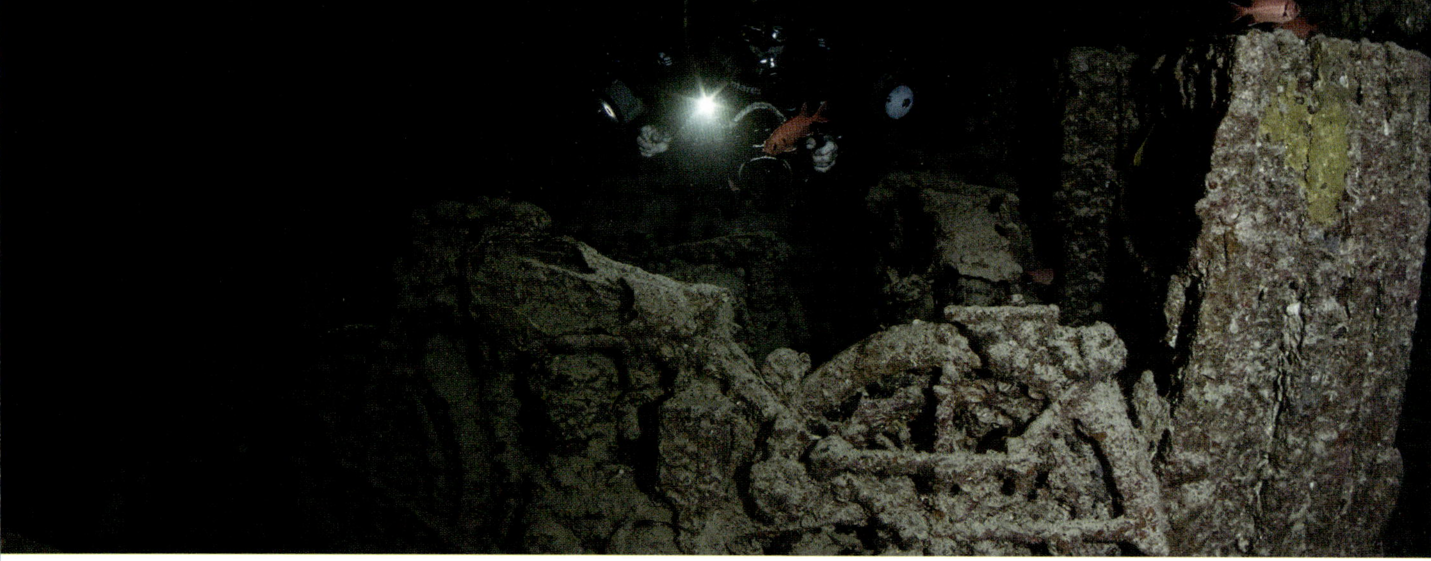

03 THE RED SEA'S *THISTLEGORM* WRECK

EGYPT

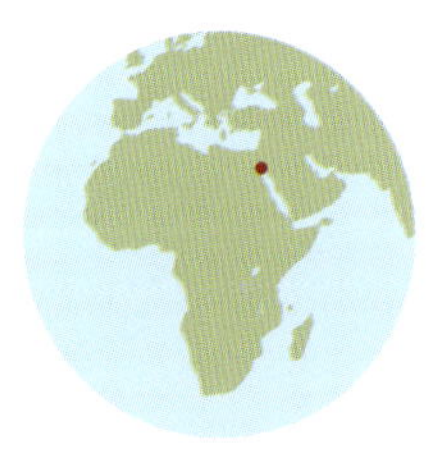

OPPOSITE Nearby reefs are vibrant and healthy.

SHARM EL SHEIKH IS A RED SEA RESORT TOWN on the Sinai Peninsula of Egypt. It had been a popular scuba diving resort for decades due to its sandy beaches and stunning reefs, until its popularity waned in the wake of a terrorist attack in 2015. While many people still flock here for the 365 days a year sunshine, tourism has dwindled in comparison to the heady heights of the early part of the 21st century. Those who do venture here are rewarded as the reefs, which were always spectacular, are now less busy but a whole lot more vibrant than ever.

GENERAL DIVING

The diving in this part of the Red Sea can be stunning, and there is a huge variety of dive sites to visit. Shore dives in Naama Bay offer sheltered and easy dives for those starting out or divers wanting a more relaxed experience, although juvenile eagle rays and other bucket-list creatures can be regularly seen here. However, a short boat ride can take you to some of the best reefs in the world. Huge pink sea fans,

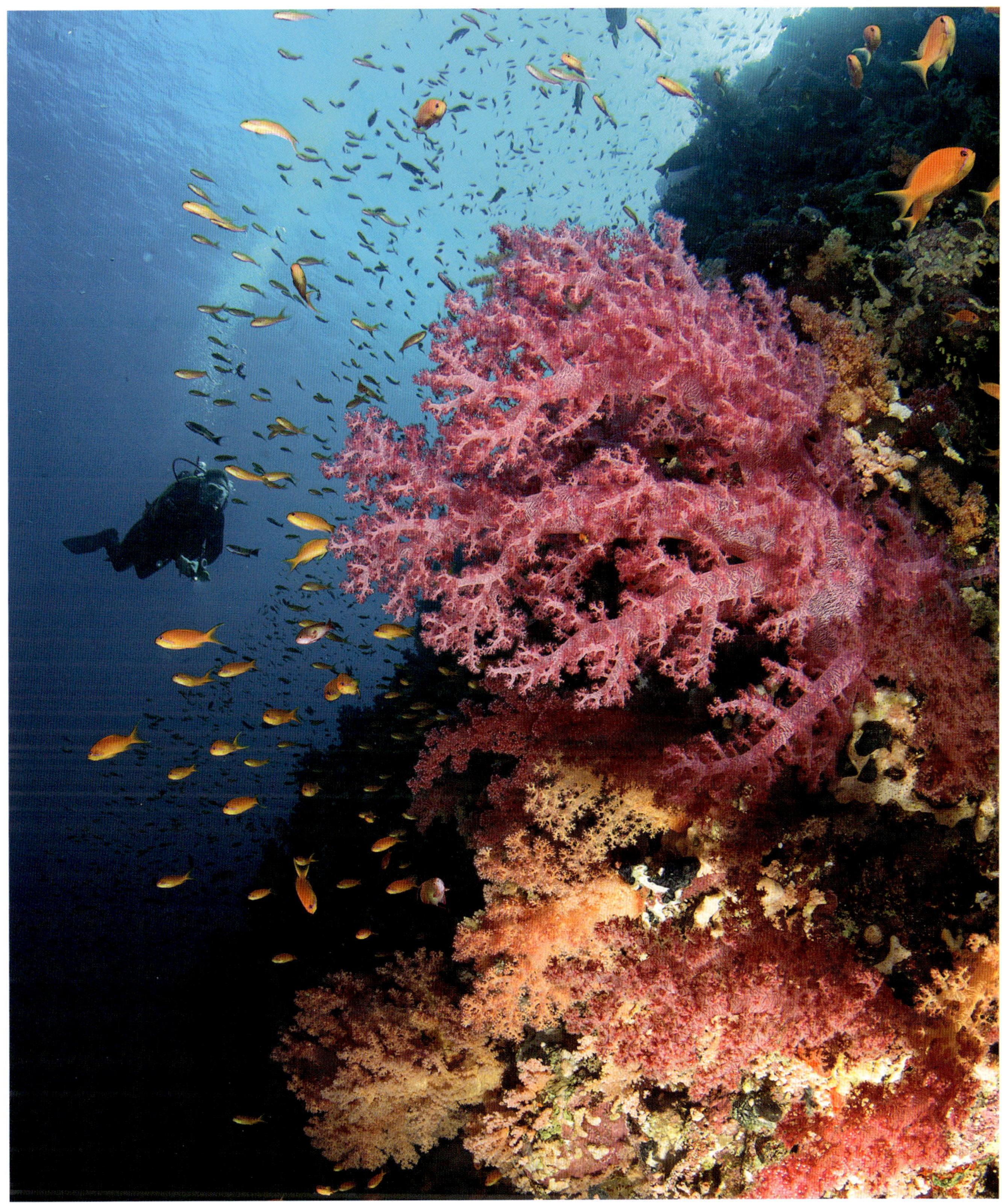

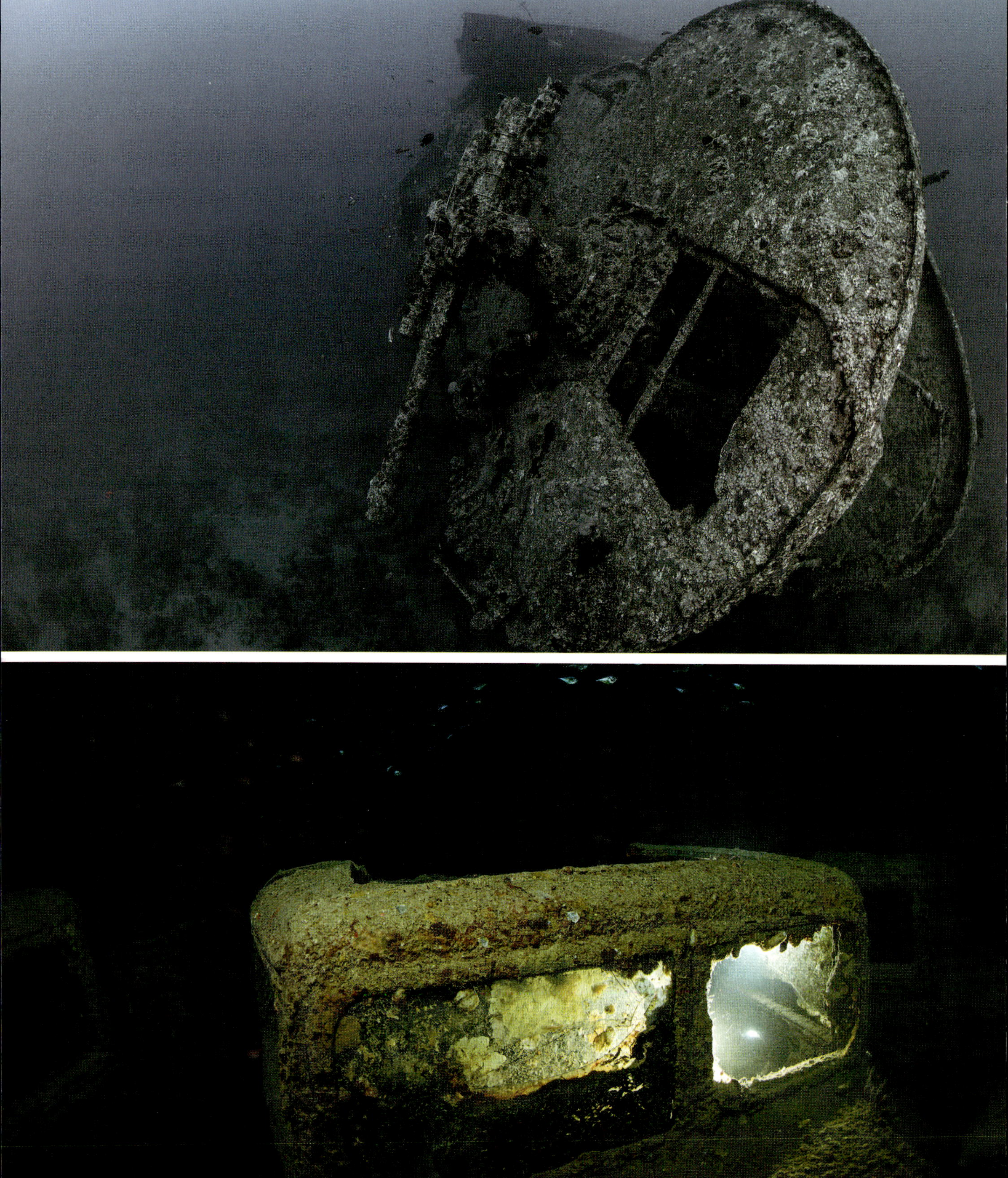

each as big as a car, grow together in deeper water and create beautiful curtains across the reef. Drifting along the wall at Shark Reef the soft corals sway with the moving water as you are pushed round before you find yourself regarding the bathroom fitting from the wreck of the *Yolanda*. Here a massive Napoleon Wrasse may come to inspect you and check out if you have brought any food. On Jackson's Reef in the summer months huge schools of fish come to mate, forming impossibly large cones of swirling barracuda and snapper. Large manta rays and even Whale Sharks cruise these waters in the warmer months too. Turtles graze, oblivious to the divers so excited to see them. On the seabed Blue-spotted Stingrays hide in the sand and anemones have their protective clownfish on guard, while on the reef lionfish patrol the overhangs looking for small prey. The reefs here are something every diver should experience.

THE WILD DIVE

In the past diving the SS *Thistlegorm* could be a crowded experience with several day boats as well as liveaboards moored to her throughout the day and overnight. The demand for divers to experience diving what must be one of the most famous wrecks in the world was simply too great. However, on our recent return we were

OPPOSITE TOP The iconic gun on the *Thistlegorm* wreck.

OPPOSITE BOTTOM The artefacts within the holds of the *Thistlegorm* are now home to a host of marine life.

BELOW Huge pink sea fans stretch over the reef.

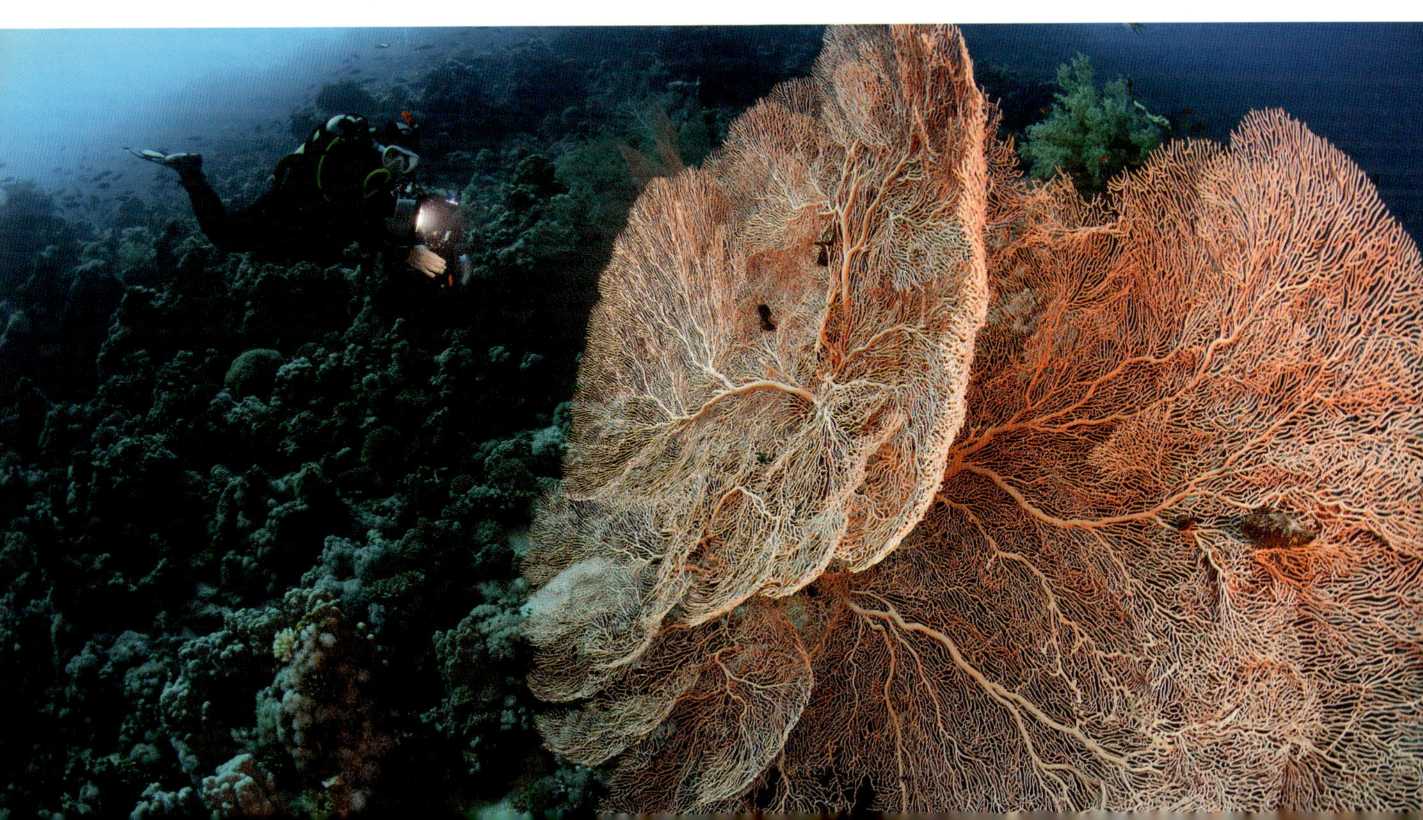

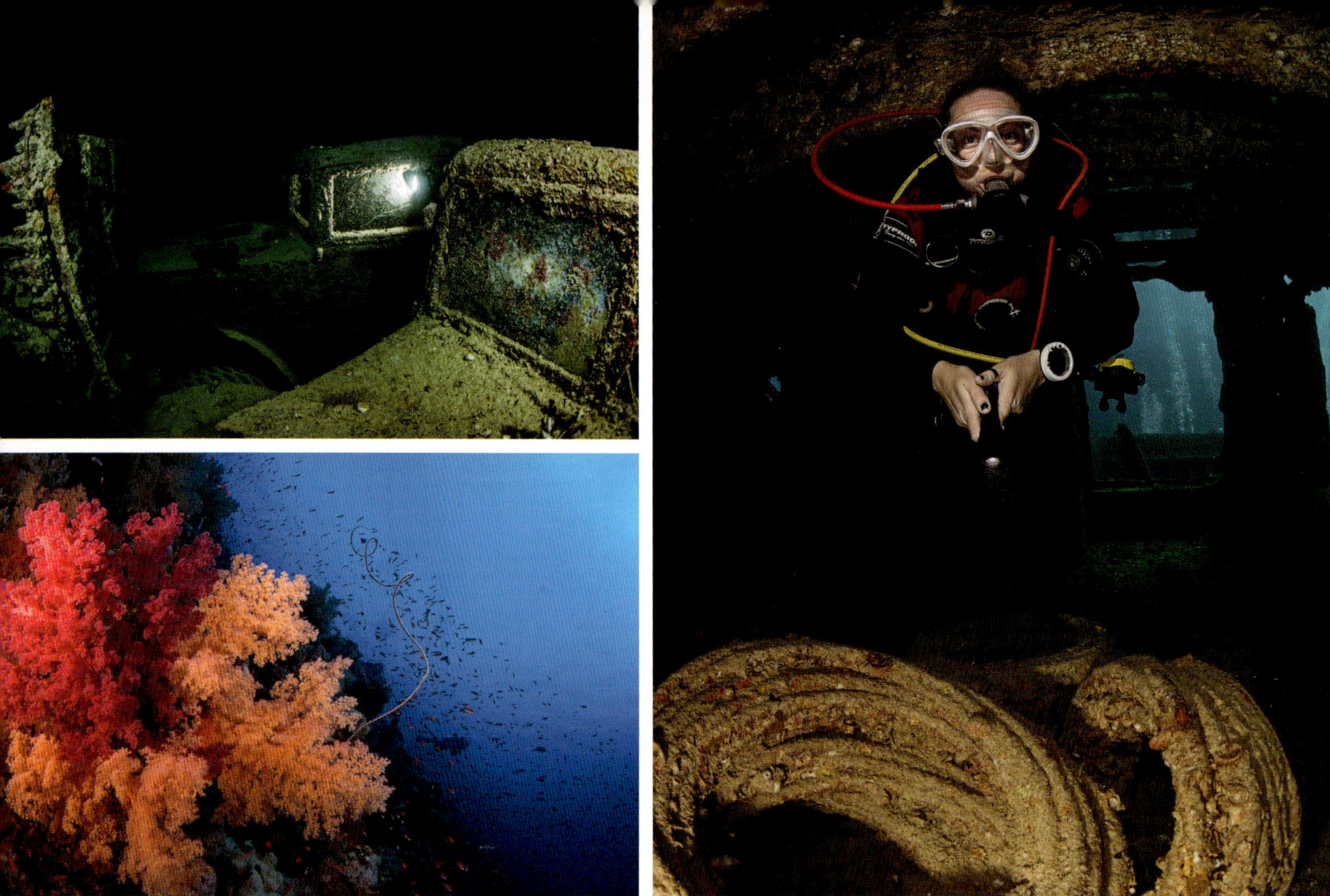

astounded to be the only boat there. We had set off from Sharm very early in the morning, watching the sun rise as we motored around the peninsular. With only a handful of people on the boat, arriving to find that we were going to be the only divers in the water was a very pleasant surprise indeed.

TOP In the hold of the *Thistlegorm* trucks are lined up ready for action.

TOP RIGHT A diver explores the rope room.

ABOVE The reefs around Sharm el Sheikh have recovered well.

The SS *Thistlegorm* was a British Merchant Navy ship that was sunk near Ras Mohammad in 1941, while in the process of taking supplies to the Allied Forces that included munitions, military vehicles, motorcycles, trucks and even wellington boots. She was hit by a German bomber looking for troop carriers. Two bombs hit the hold and the explosion of the bomb, and then some of the munitions too, caused the ship to tear in two and sink. Nine crew members were lost, but the rest were rescued successfully by another ship in the convoy. The wreck now lies in around 30m (100ft) of clear blue water.

On our journey to get to the *Thistlegorm* we sat down with our guide to plan the two dives we would make that day. We had dived this wreck on several occasions before and so were familiar with the general layout, but we had never previously

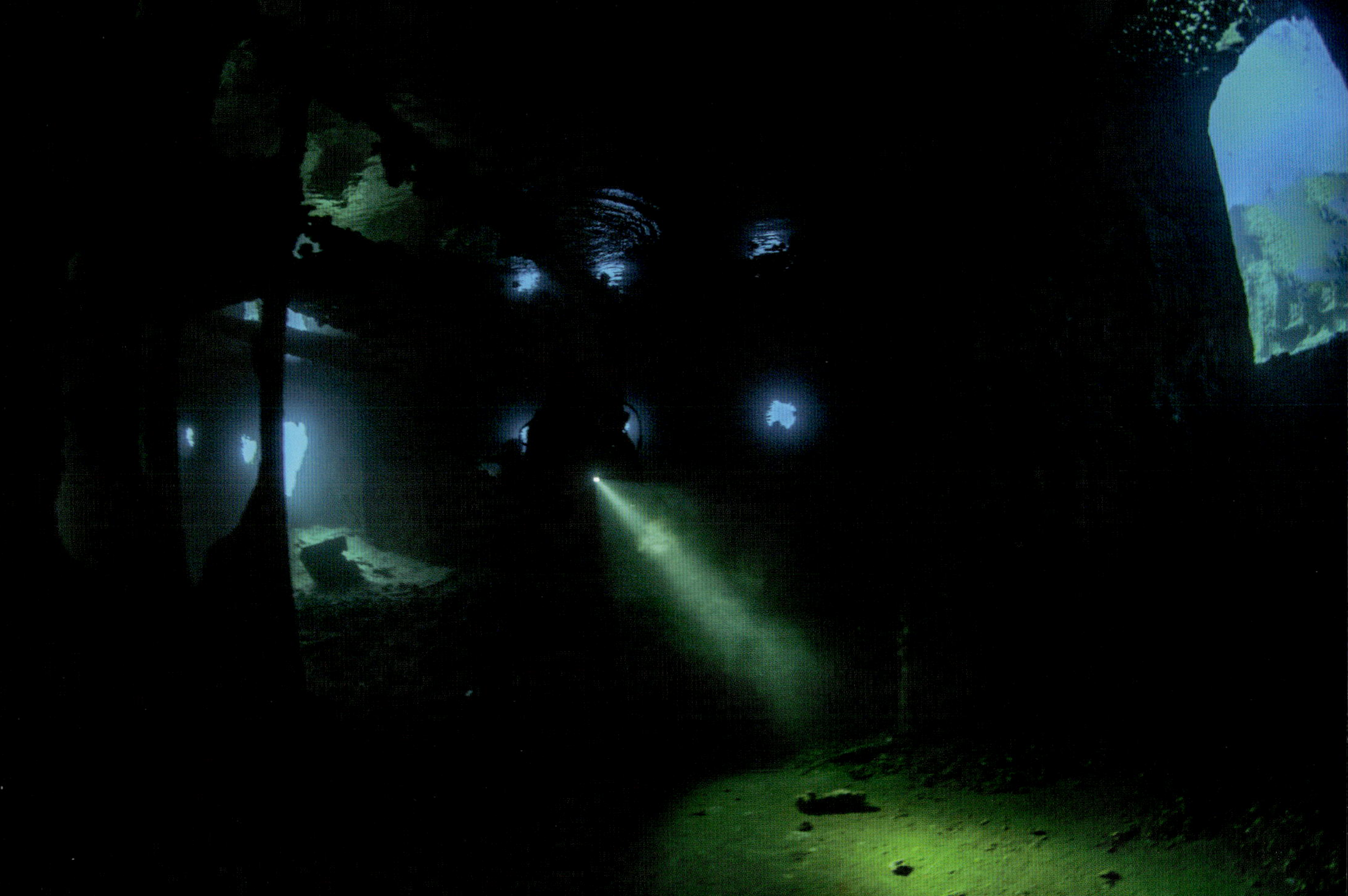

explored the holds and this was what we wanted to do on our first dive. We pored over a map giving us more detail on the internal layout of the ship and agreed a plan. Once we were all happy, we quickly got our equipment together so that we could be the only three divers in the water, leaving the other small group to follow us in later as our dive was ending.

ABOVE A diver inside the wreck.

We hit the water and started to make our way down the line. Although the shipwreck rests at a depth of 30m (100ft), you can make it out from the surface. Even though we had seen the *Thistlegorm* before, resting on the bottom, reacquainting ourselves with the huge tear blasted into the hull was still a sobering moment. A couple of tuna shot past our masks and through our bubbles and were out of sight again in a silvery flash. We continued down the line until our guide signalled for us to follow. As she headed into the structure, I switched on my dive light and followed closely behind. Soon we were in a hold that contained a series of trucks lined up next to each other. They are in incredible condition considering they have been underwater for decades, with thousands of divers exploring them

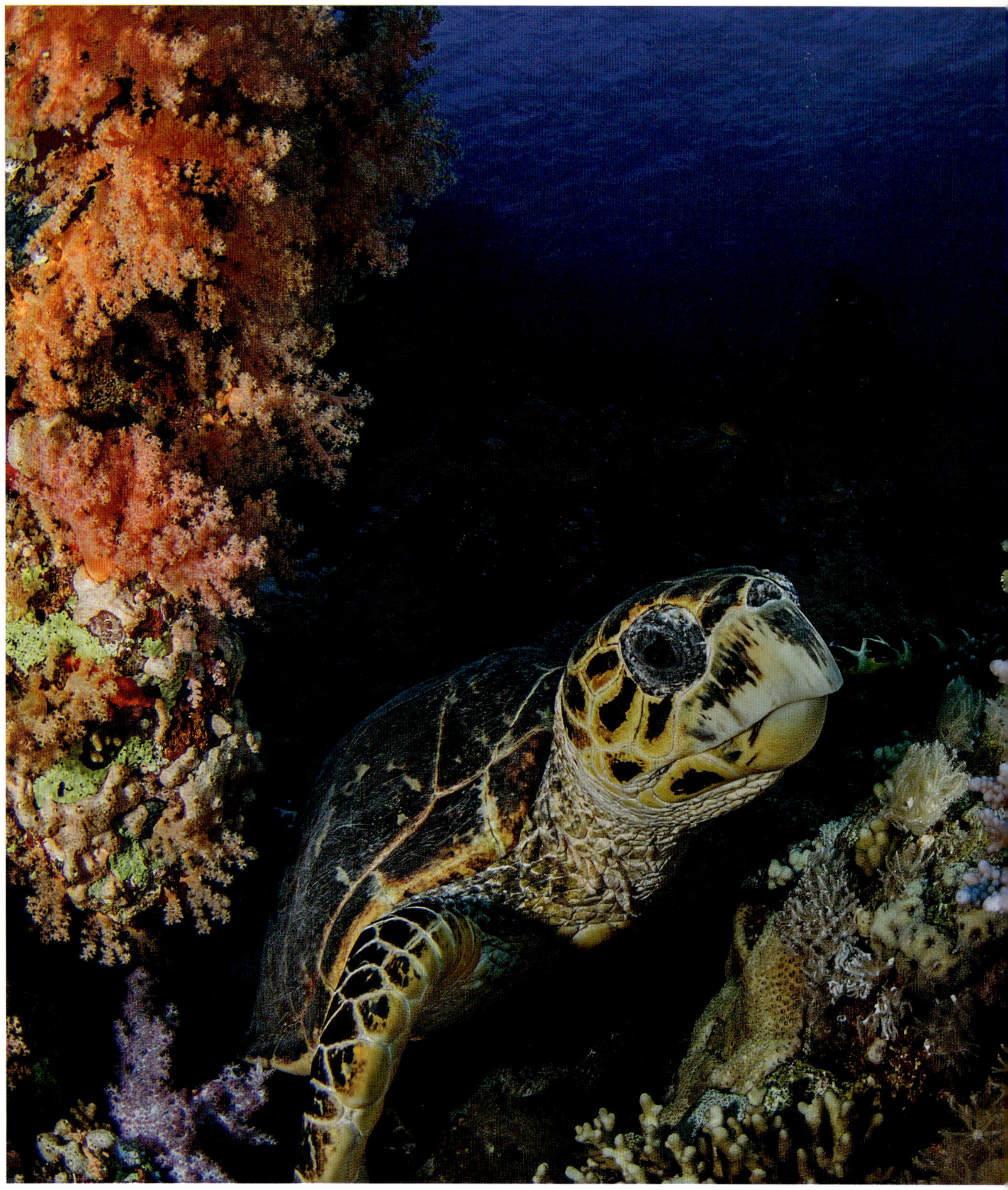

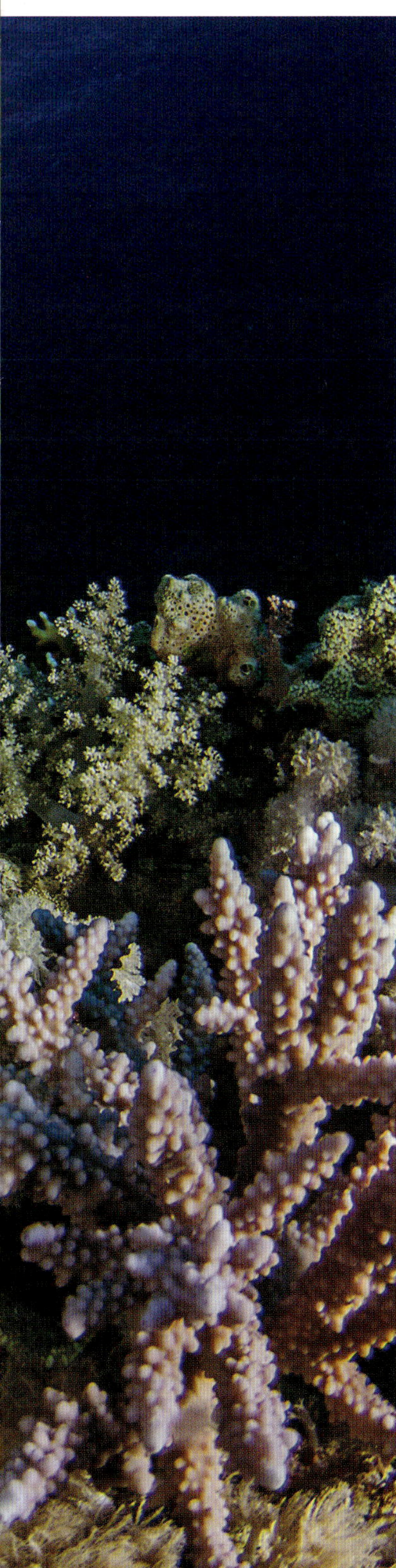

over the years. We picked out motorcycles still lined up in crates and there were rifles, as well as ammunition.

I marvelled at the history preserved inside this wreck, and smiled at the thought that at least some marine life had made this place home. Glassfish and snapper joined us in the holds, illuminated in our torch light and then dashing away for cover. Our guide led us through a series of more open rooms, pointing out further artefacts as we slowly explored, finning really gently and carefully so we did not stir up silt and ruin the visibility inside the holds. Then I found myself once again in brighter waters on the outside of the wreck. I checked my air and no-decompression time and decided there was enough time to quickly see the anti-aircraft gun on the stern. It is an iconic sight for any diver.

All too soon it was time for us to head back to the surface. Our companions from our boat were making their way down as we headed up the line, and a pang of jealousy shot through me that they were just starting out on their adventure exploring this incredible wreck. Then again, there was always the next dive...

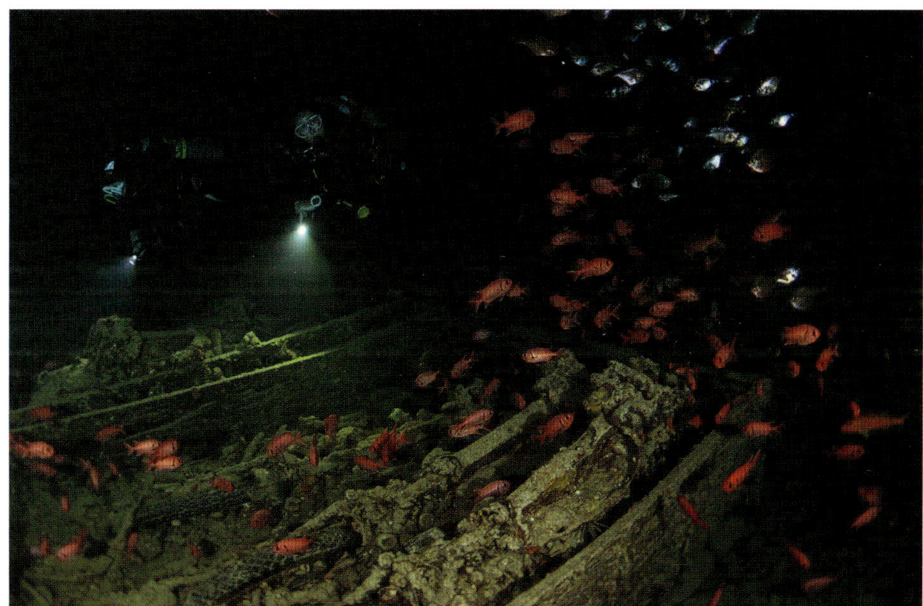

OPPOSITE A turtle forages for food on the reef.

ABOVE Fish gather over the motorbikes still in their crates.

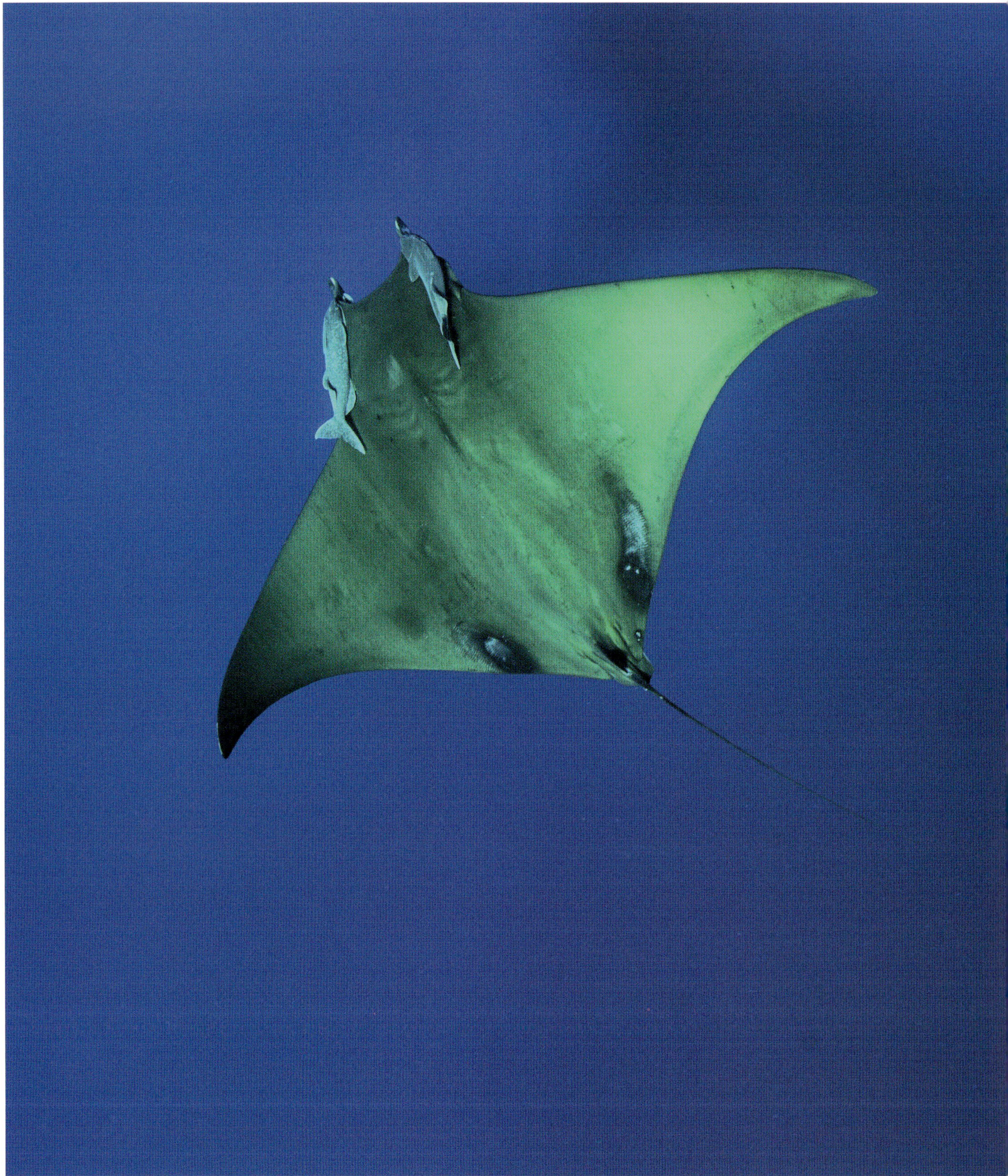

EUROPE

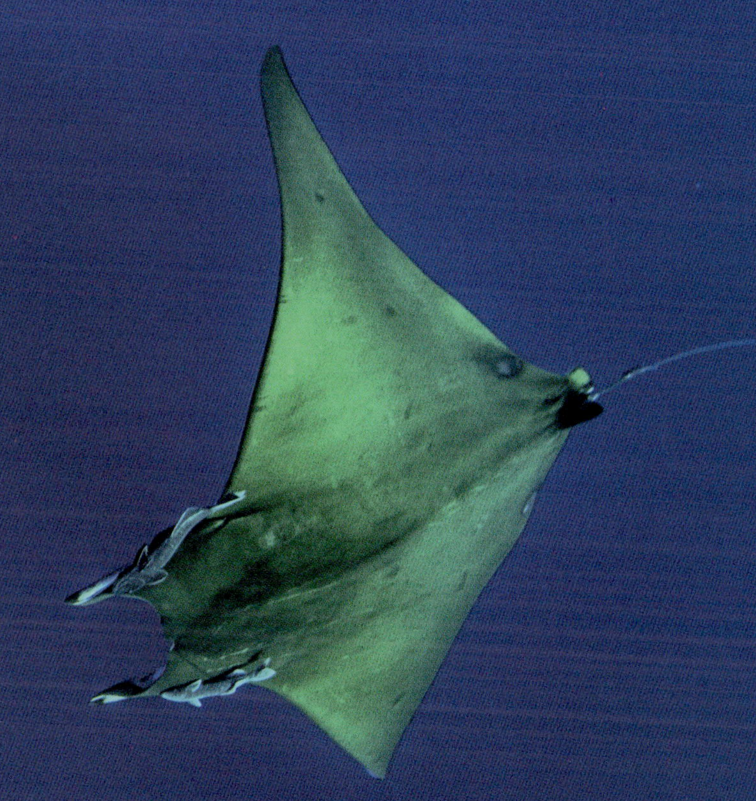

04 SAILING TO SEE MOBULA RAYS

AZORES

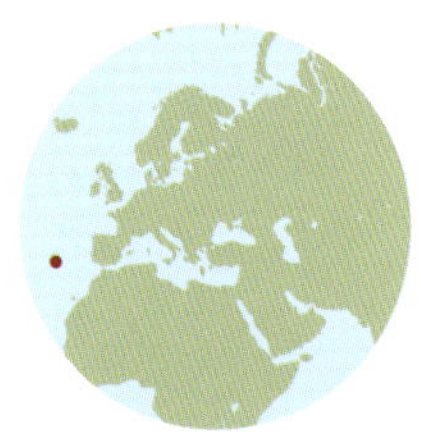

BELOW Rays circle us in shallow water.

OPPOSITE A mobula ray dances in the current.

THE ISLANDS OF THE AZORES form an archipelago in the mid-Atlantic, and they are an autonomous region of Portugal. All the islands are volcanic with dramatic landscapes and underwater vistas. People travel here to hike the mountains, partake in the local wines and enjoy the local history and culture. Sitting nearly 1,600km (1,000 miles) west of the mainland of Europe, these islands act as a refuge for a host of marine life and are famous for Blue Sharks and cetaceans, with huge pods of dolphins and several species of whales regularly visiting the waters. An incredible 26 different species of dolphin and whale have been recorded off the Azores. At night, on the island of Pico, visitors can witness the return of Cory's Shearwaters to their nests, which are often situated in crevices in the dark volcanic stone. These birds have an astonishing call that sailors used to believe was the cry of drowned souls – it certainly is a haunting sound.

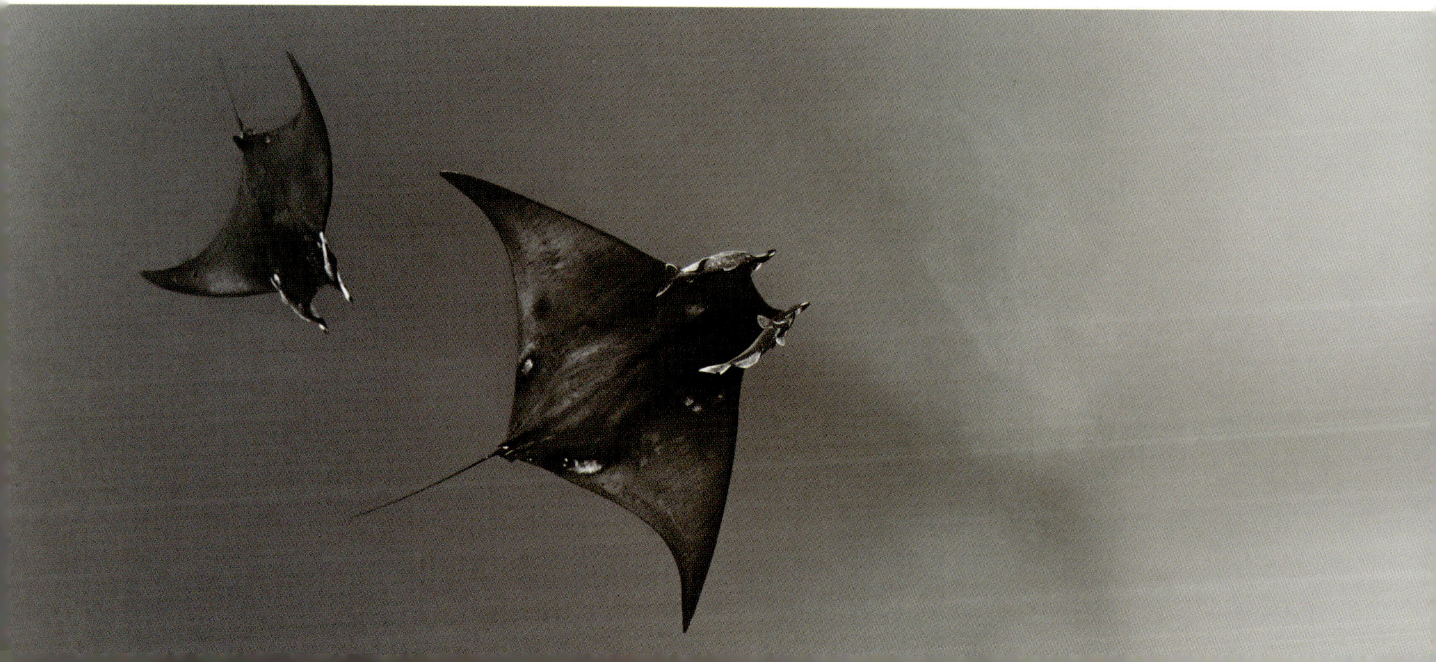

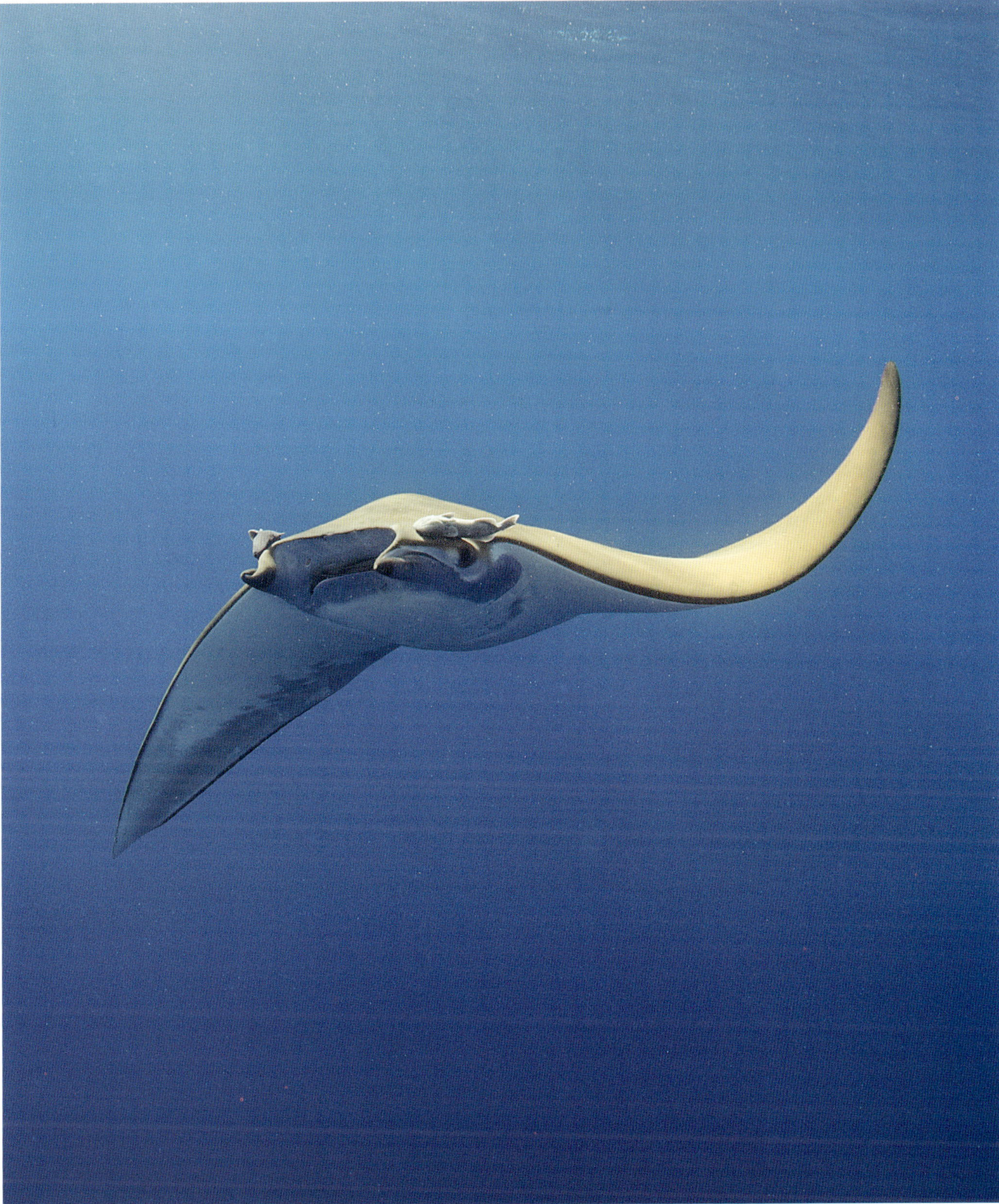

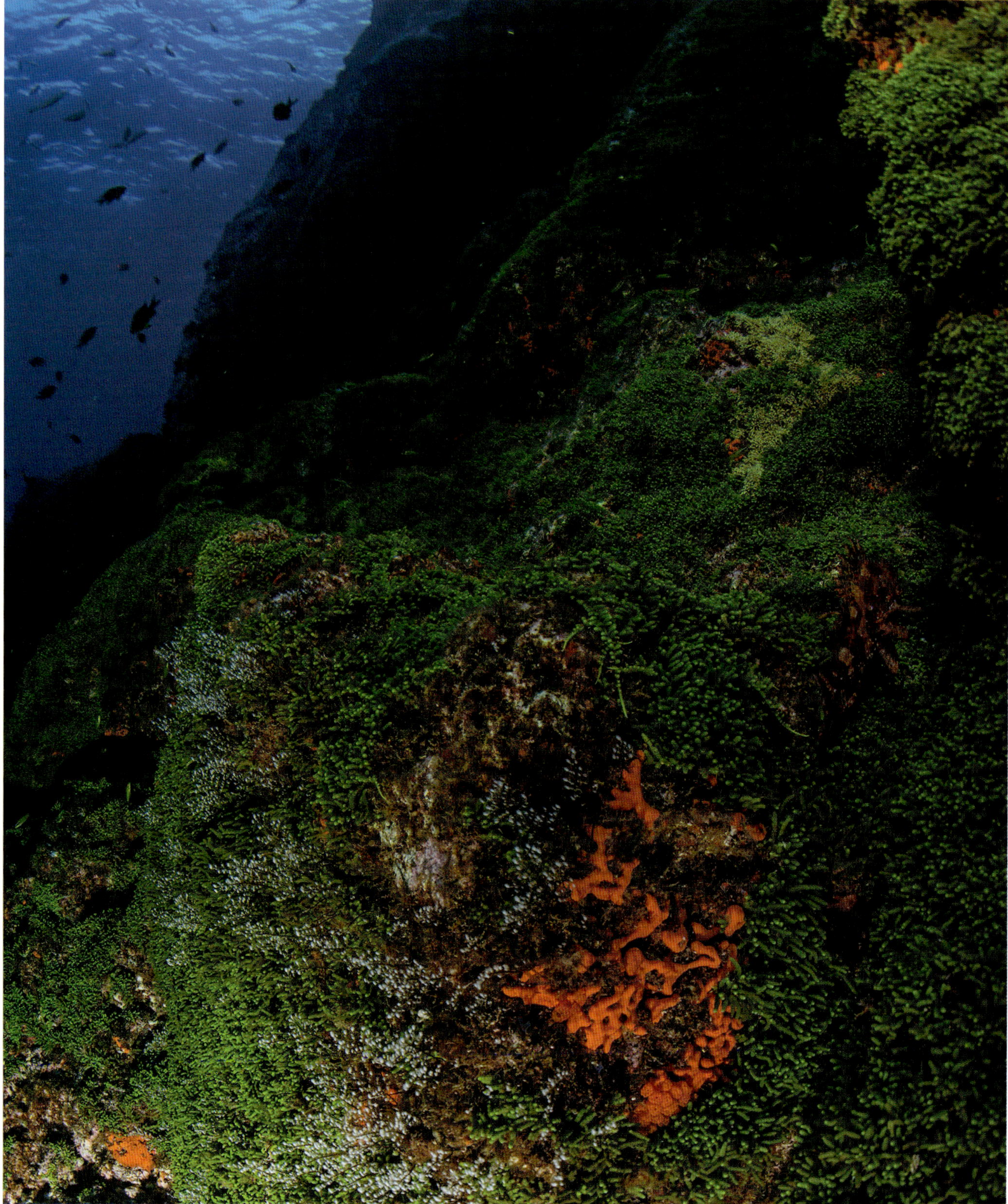

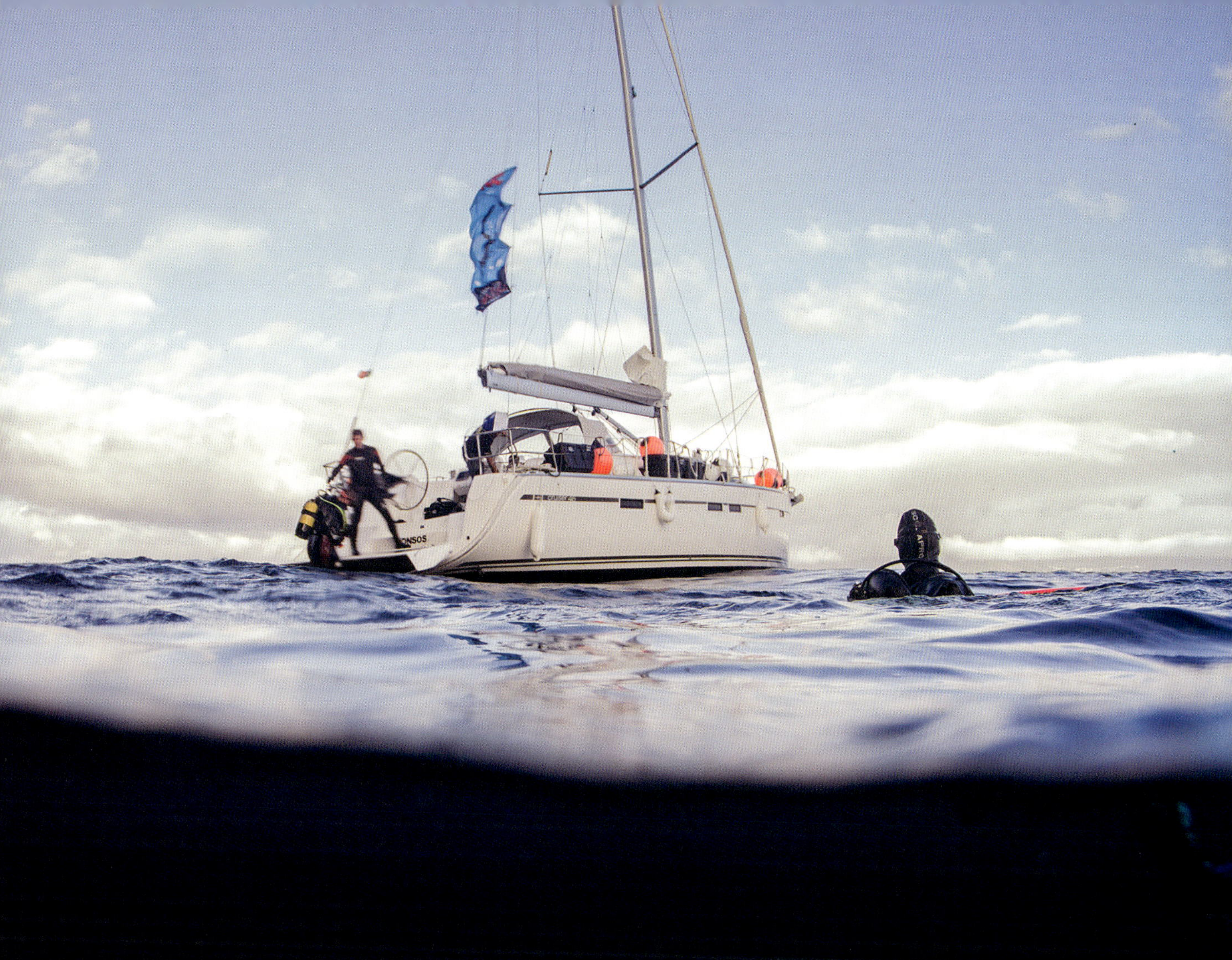

GENERAL DIVING

The seascape is dominated by dramatic volcanic boulders, cliffs and caverns carved out and moulded by the sea over millennia. Divers venturing into caverns can find rays and small schools of fish enjoying some shelter. On the shallow reefs, rocks are encrusted with algae amongst which small critters such as crustaceans and nudibranchs feed. In the open ocean, dive organisers bait for Blue Sharks – one of the most beautiful and elegant sharks in the ocean – and divers can join them in the water for a close encounter beneath the boat. With a rich maritime history, the islands of the Azores have underwater wrecks and artefacts to explore, with anchors from across the ages abandoned on the seabed, as well as several wrecks.

ABOVE Our yacht waiting to pick us up after a fantastic dive.

OPPOSITE Volcanic rocks covered in hydroids and algae.

THE WILD DIVE

There is a very special place called Princess Alice Bank, which is situated 93km (50 nautical miles) off Pico. It is a submerged seamount that is of huge importance to marine life. It is most famous for the majestic mobula (or devil) rays that aggregate here in large numbers. Because the banks are so far out in the Atlantic Ocean, the weather can change in an instant and large waves, or choppy waters, can develop quickly. Rather than take the usual teeth-shattering three-hour RIB ride each way, we decided to take a more sedate and easy-going crossing method, under sail by yacht. We chartered the boat with a few friends and the plan was to set off in the evening, and for our crew to sail us to the Princess Alice Bank overnight so that we would arrive before any other dive boats and have the whole site to ourselves. We loaded our dive and camera gear on board and set sail. Once underway and out on the open sea, the crew laid on a simple dinner for us and then we shared a bottle of the local wine beneath the stars. We headed to our beds full of ideas and thoughts about how our dives may pan out the next morning.

We were woken by the sound of our anchor being dropped – a procedure that required a great deal of precision to hook onto the submerged seamount below. We had a quick breakfast and grabbed a coffee out on deck to take in our surroundings. Nothing – that is what we saw – absolutely nothing but water and sky in every direction. I had read many accounts about the diving here though and was really keen to kit up and get into the water. Our guides warned us that despite the fact that the sea looked particularly calm, with only a few gentle swells, the current would be strong and so we should keep hold of the mooring line rather than drifting away from the boat. I took a giant stride off the back of the yacht and plunged into the water. I reached up to grab my camera and then hauled myself along the line to get to the descent line. The current was, indeed, very strong and it took a while to get my breath back under control after the effort of getting underwater. A few feet down I stopped to get my camera ready ,and out of the corner of my eye I caught a glimpse of something big and moving. A ray, gliding below me, rose up as if to check me out and then just as quickly disappeared too deep for me to follow. I waited for a few moments and then three rays appeared and completed a few turns around the line, each time getting a little closer. I could hear my buddy breathing hard just above me. We were both hanging on with the

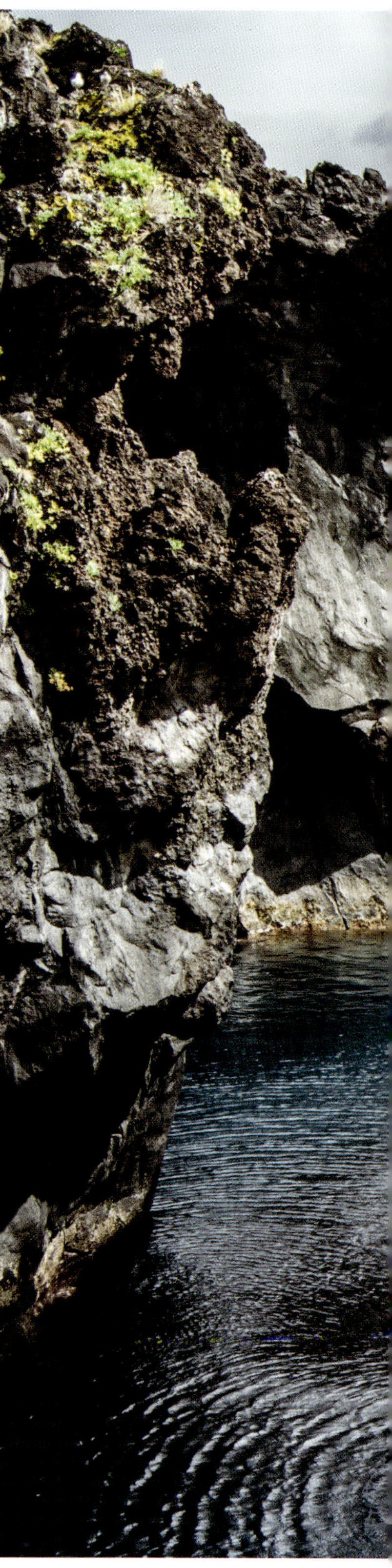

RIGHT Spectacular rock formations line the islands of the Azores.

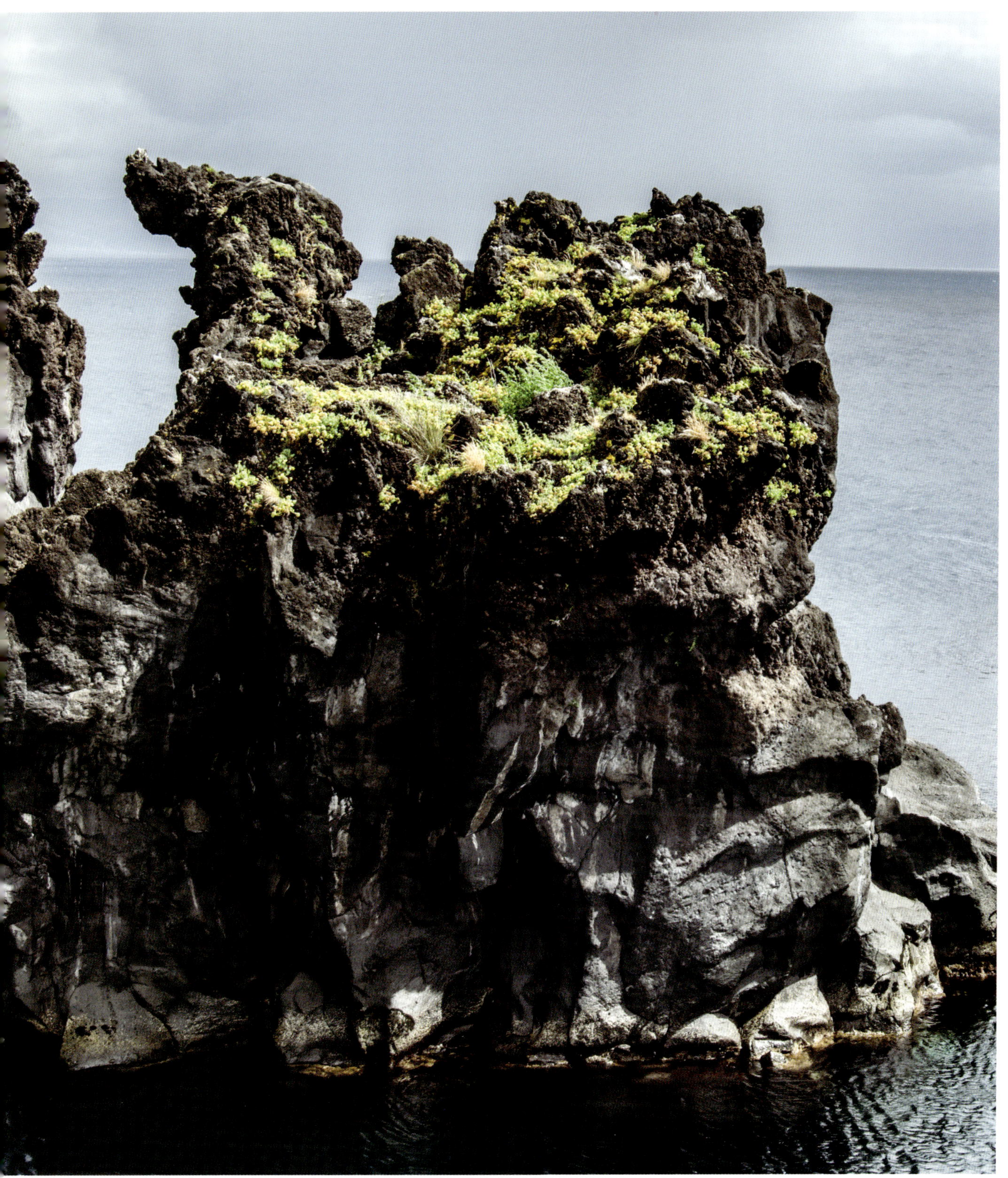

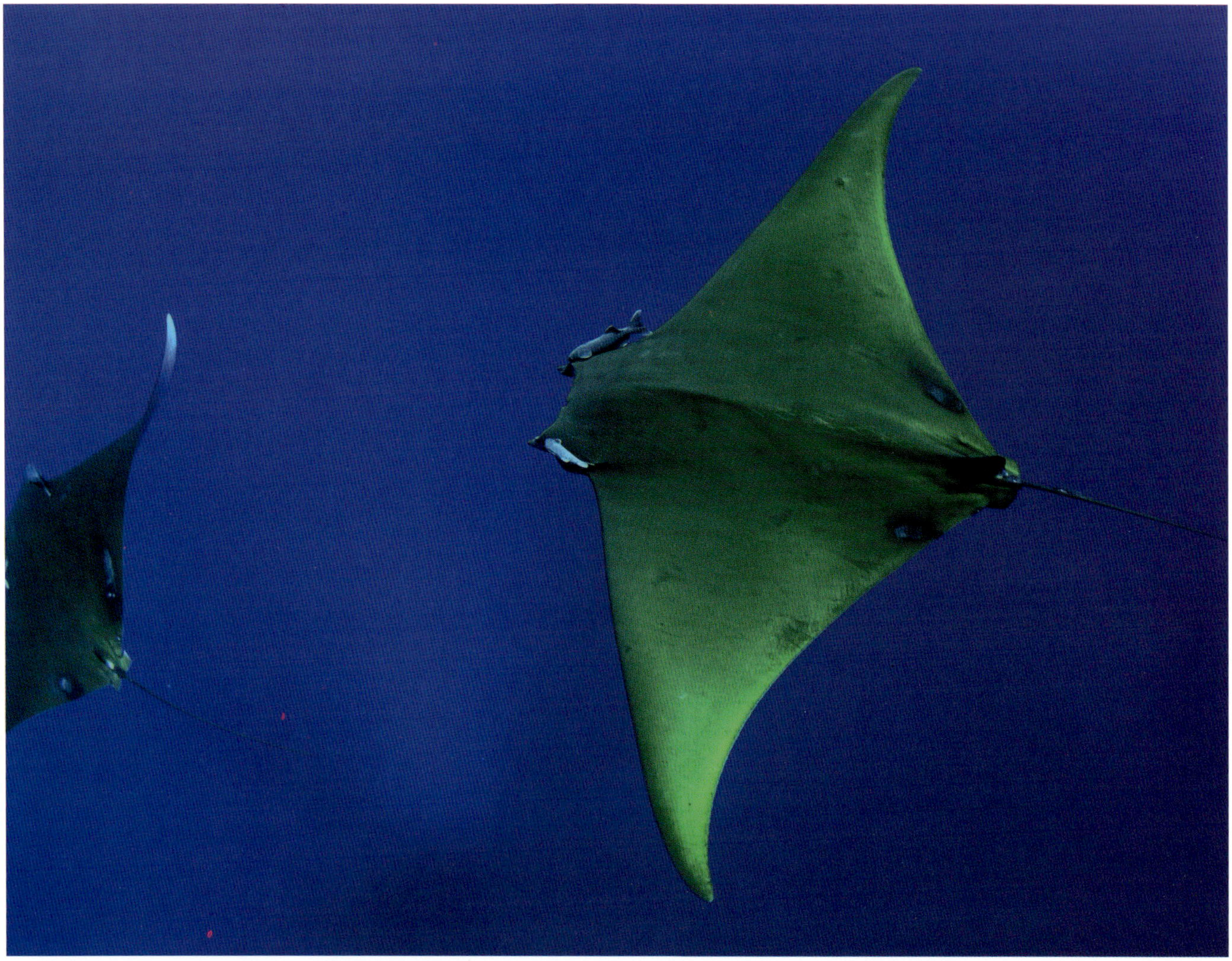

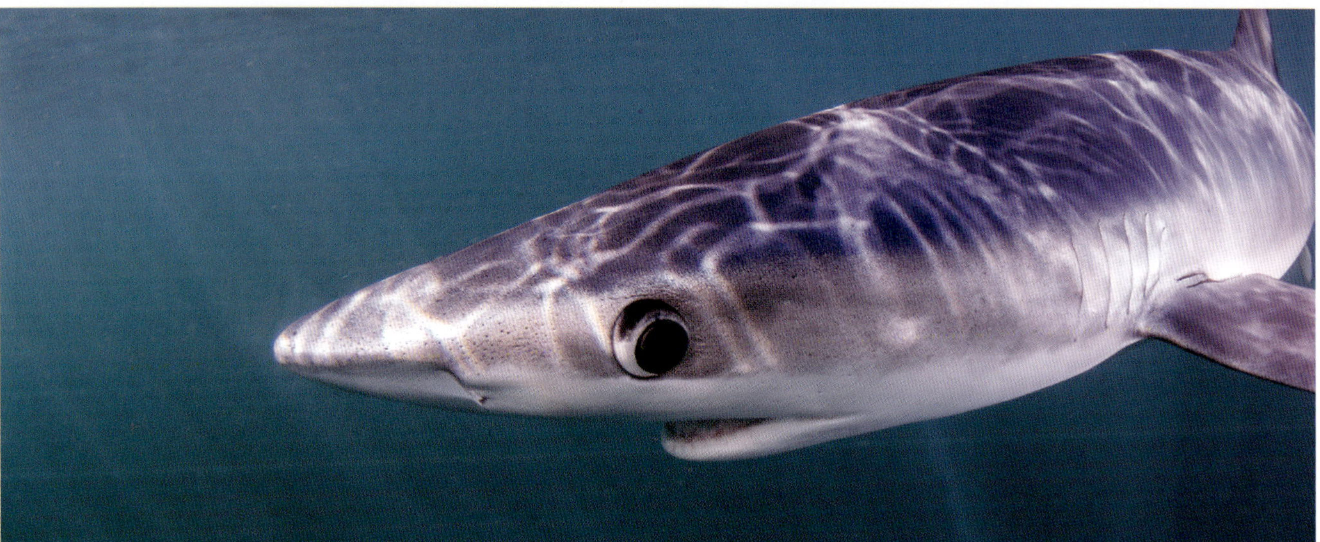

current still pushing at us, making it hard to hold the line and take photos. I was quite content just to hang there for a while, watching the rays dancing effortlessly in the current and moving around us, as if they were trying to show us how easy it was. Then I felt a tap on my shoulder. My buddy signalled that we should drift with the current to try to get some better shots. As soon as we let go of the line the plan did not seem quite so brilliant, as we were quickly swept away from the line. The rays, however, seemed to think this was a great idea, picking up on this and joining us on our rapid ride. The next few minutes were magical, as the rays performed and danced in the current, but all too soon we had to surface and start a lung-busting swim back to the boat before we were lost at sea!

OPPOSITE TOP Looking down on mantas from above.

OPPOSITE BOTTOM Blue Sharks are a wonderful treat for divers to see in the Azores.

BELOW The rays follow us as we drift.

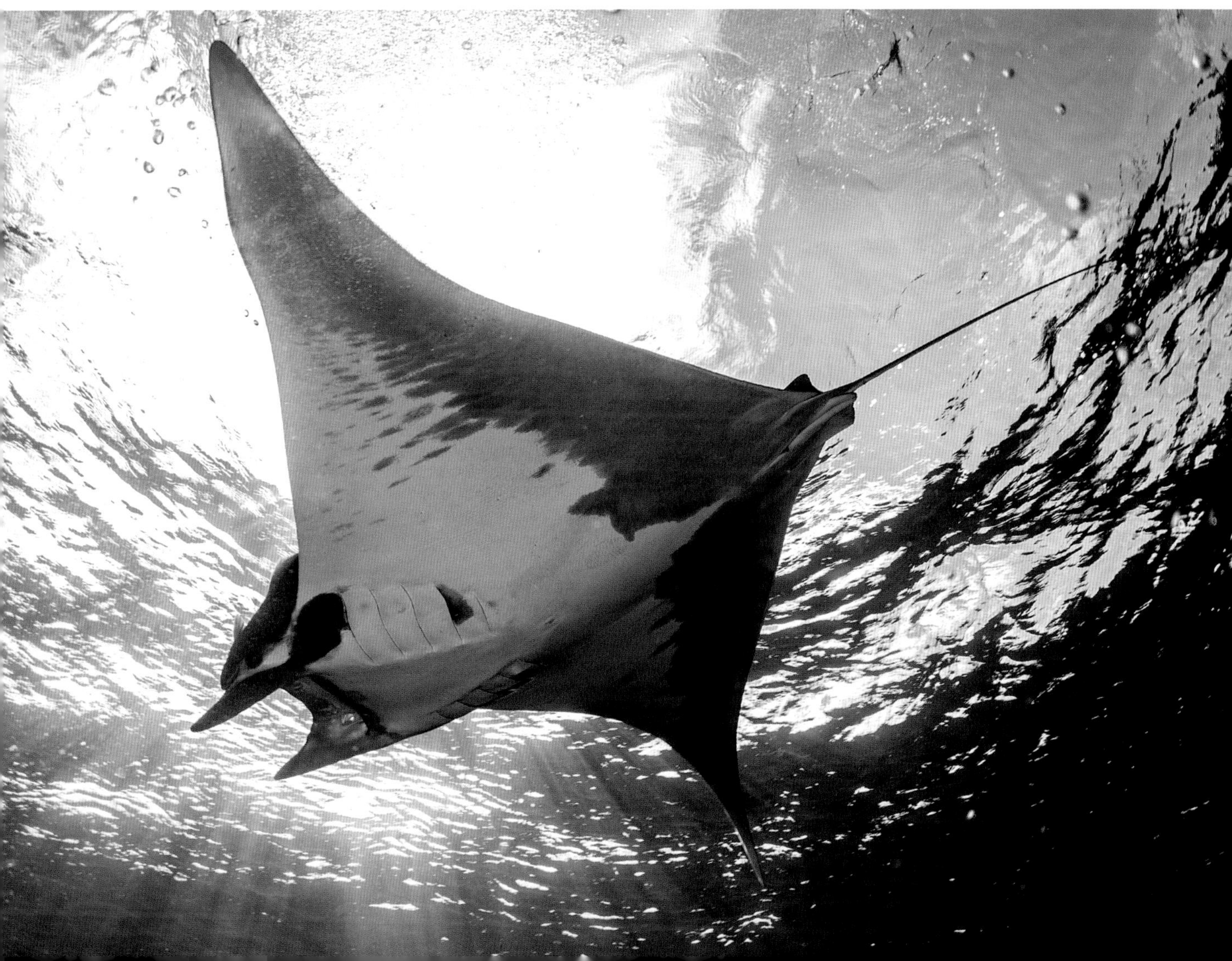

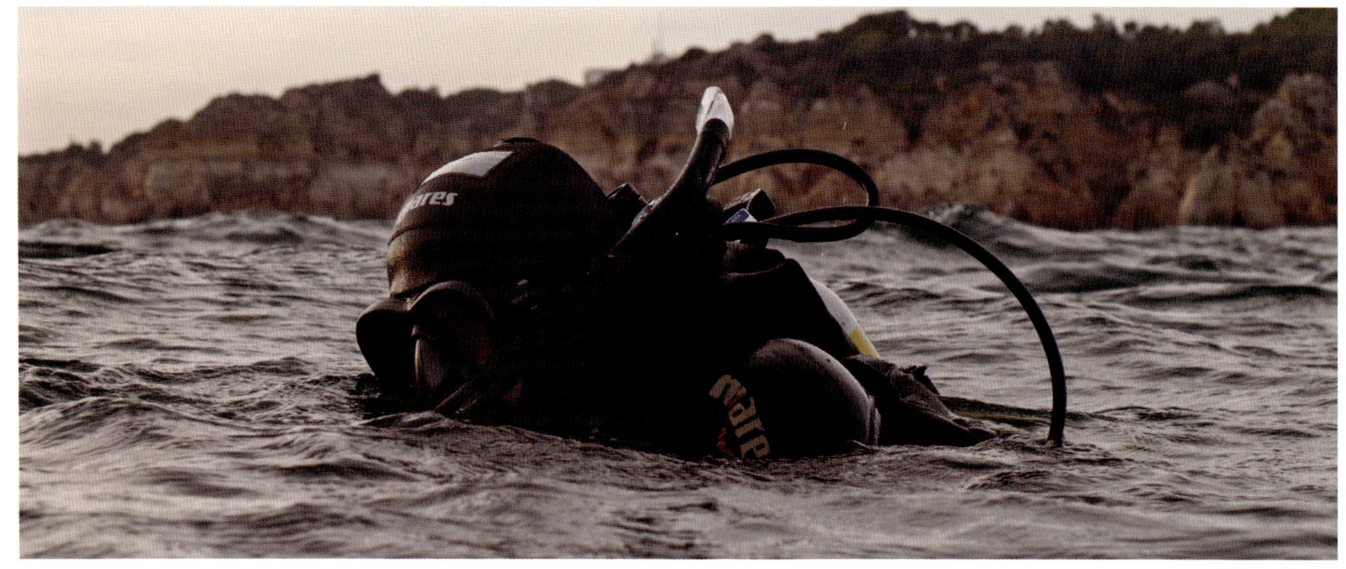

05 OCEAN REVIVAL WARSHIP EXPLORATION

PORTUGAL

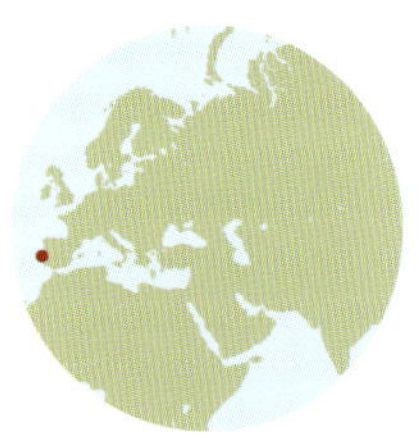

OPPOSITE Posing behind the wheel inside the wreck.

ABOVE Returning to the boat after the final dive of the day.

THE ALGARVE IS THE MOST SOUTHERN REGION of mainland Portugal, which itself is situated along the western seaboard of the Iberian Peninsula. It is particularly famous for its rugged Atlantic coastline and is a very popular tourist destination. People flock to the region for the warm and sunny weather and its wonderful beaches. The Algarve has a lot to offer, especially for family holidays, as not only are there quiet and secluded beaches, but there are also numerous resorts which offer just about every watersport in buzzing locations. If culture is on your horizon, then the inland villages and quaint old towns will surely capture your interest.

GENERAL DIVING

Divers have a wide range of diving on offer when they travel to this region of Portugal. This is where the Mediterranean meets the Atlantic and the area usually offers calm waters in which to explore the underwater world. The area has a rich

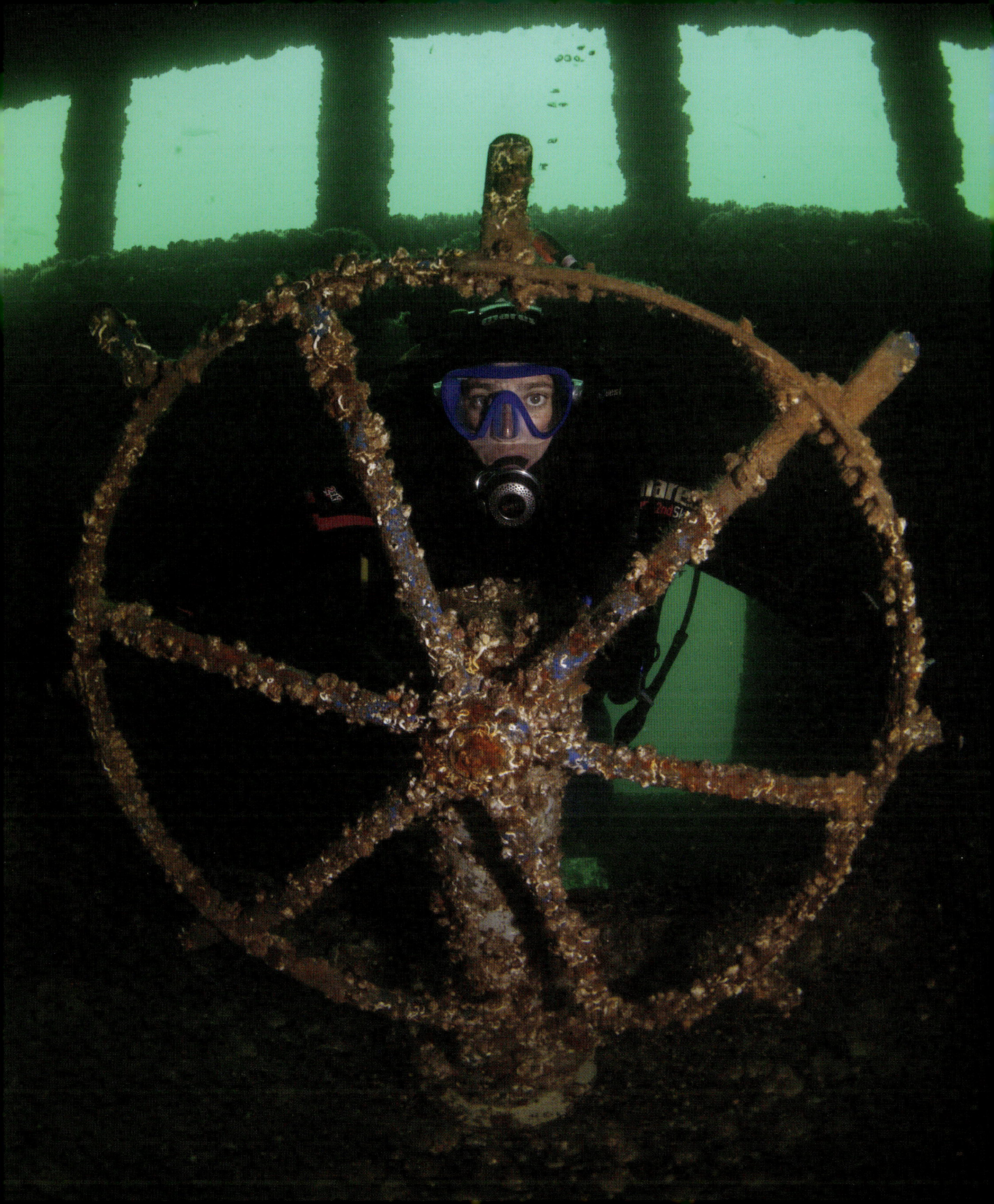

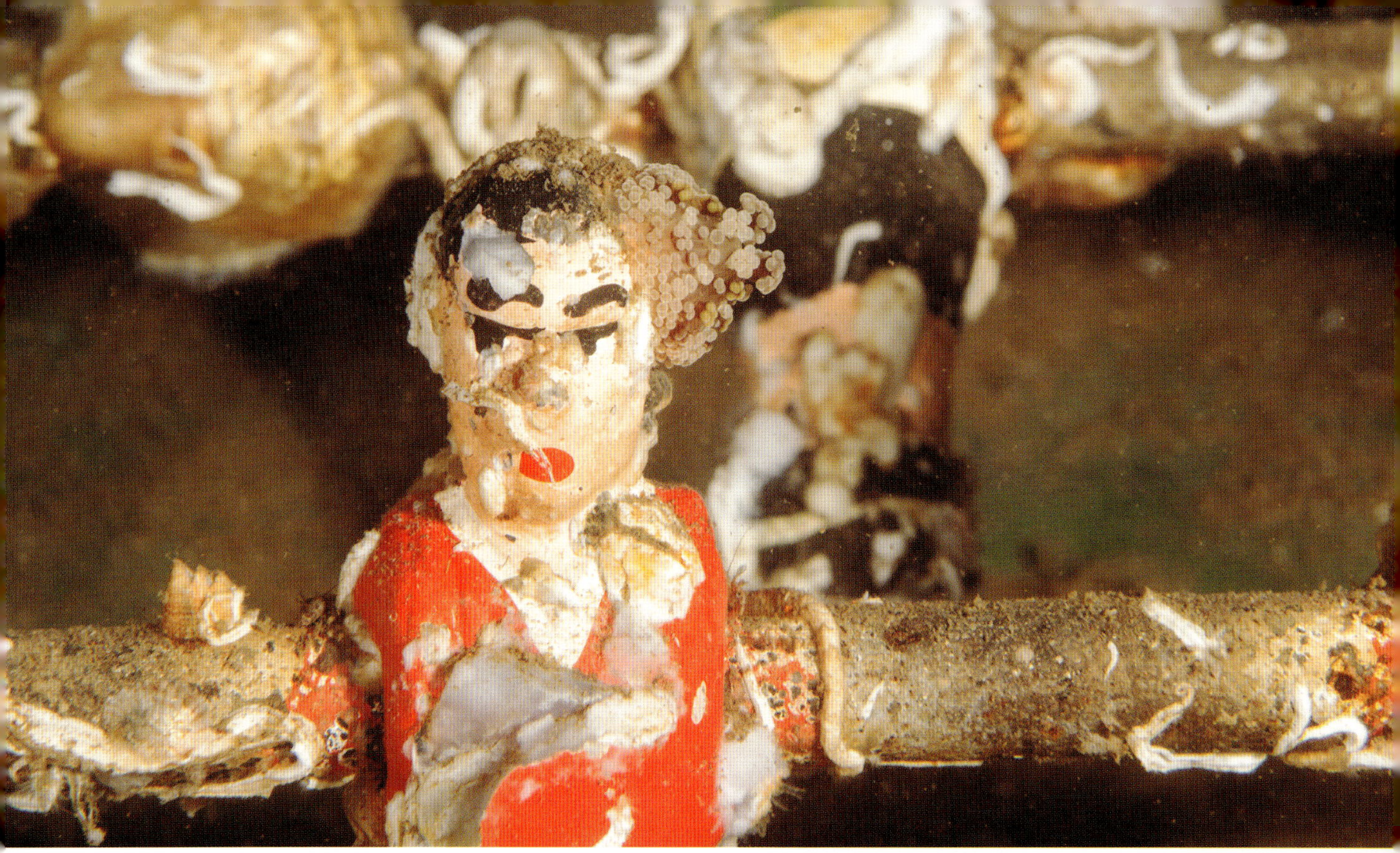

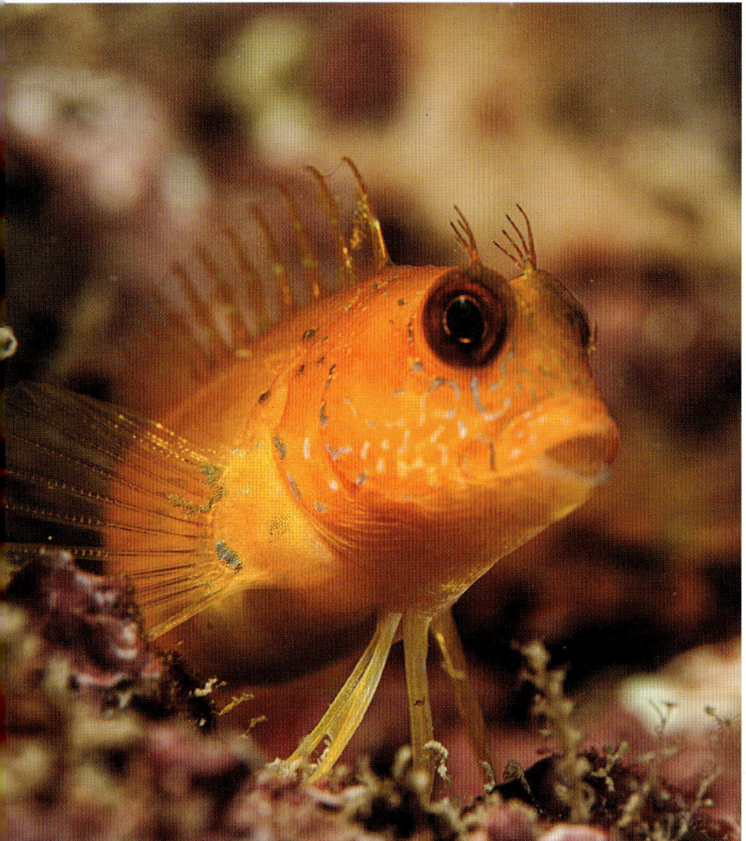

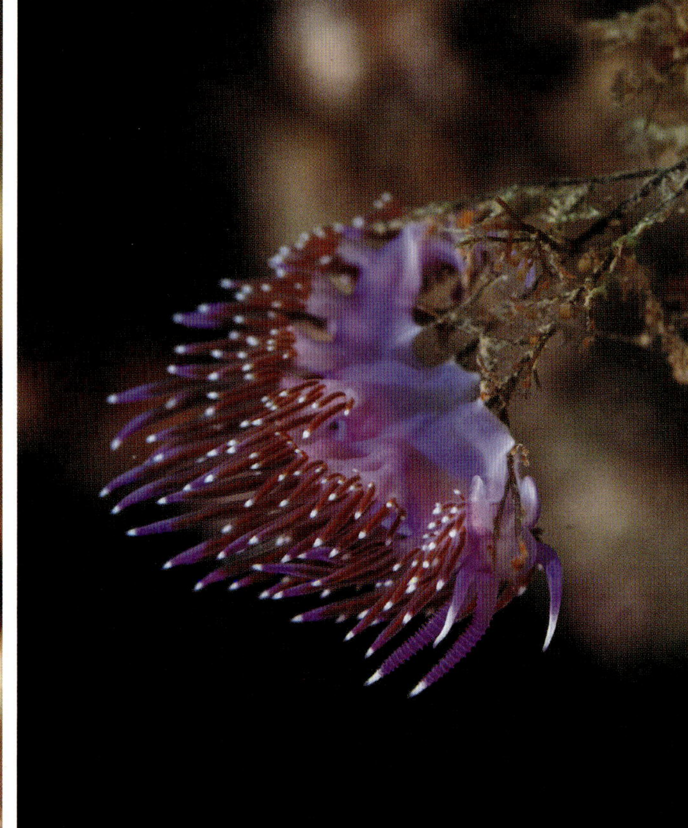

maritime history as the Portuguese have a tradition of producing great travellers and explorers, who opened up large parts of the 'New World' during the age of discovery. There are wrecks dating from the 18th century until the Second World War littering these waters, and should you prefer to look for wildlife there is no shortage of rocky reefs – these provide the perfect hunting ground for octopus and grazing grounds for a host of weird and wonderful nudibranchs. In 2012 a project called Ocean Revival saw four naval warships sunk to make a huge artificial reef. Another little-known fact about this region is that seahorses can be found among the seagrass beds in sheltered bays. The best place to find the two species of seahorse that live here is at the Ria Formosa Nature Reserve, which is claimed to have the highest concentration of seahorses anywhere in the world.

THE WILD DIVE

We were really looking forward to diving what is one of the largest artificial reefs in the world. Ocean Revival has seen the sinking of four naval vessels to create this new playground for divers and home for an abundance of local marine life.

OPPOSITE TOP Table football piece with marine growth.

OPPOSITE BOTTOM LEFT Bright orange blennies were everywhere on the reef.

OPPOSITE BOTTOM RIGHT Colourful sea slugs are found on the reef.

BELOW A dual destination trip could see you treading grapes in the Douro Valley.

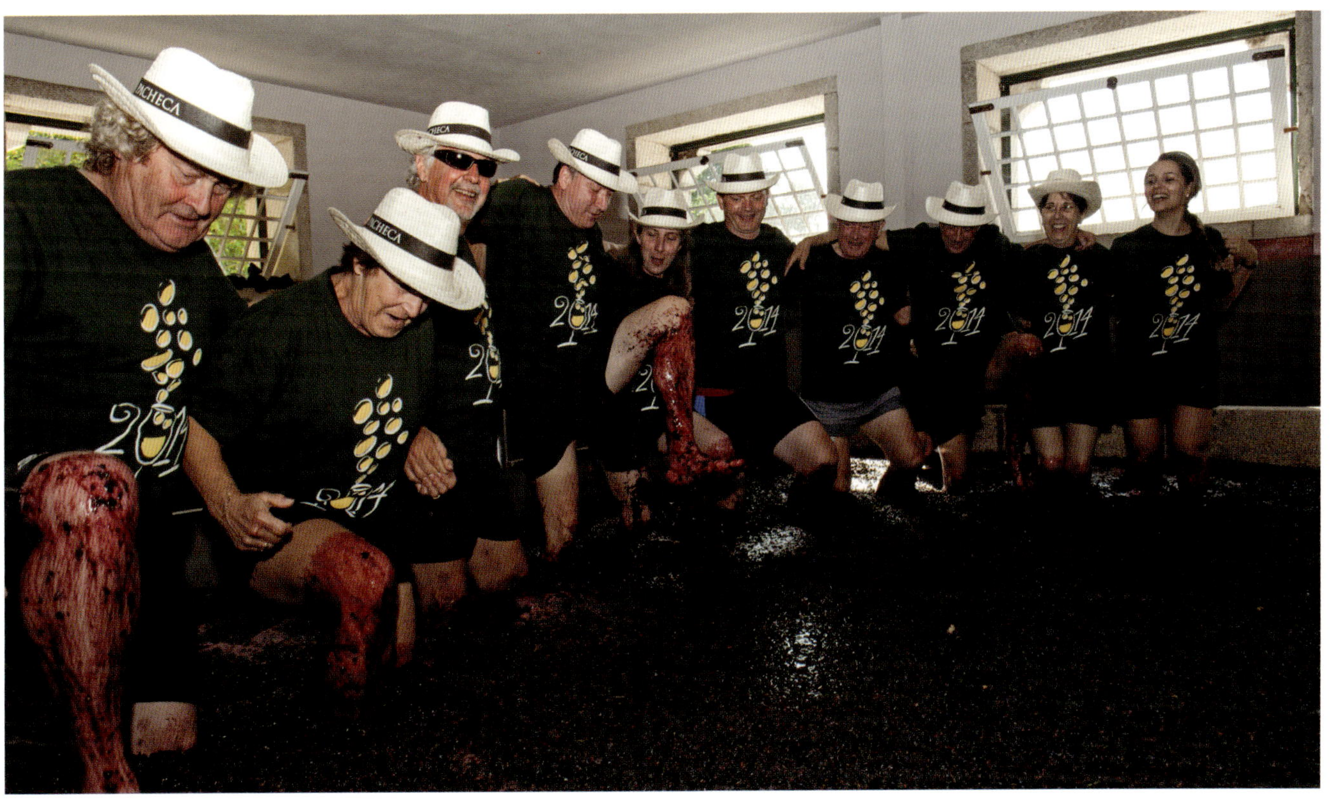

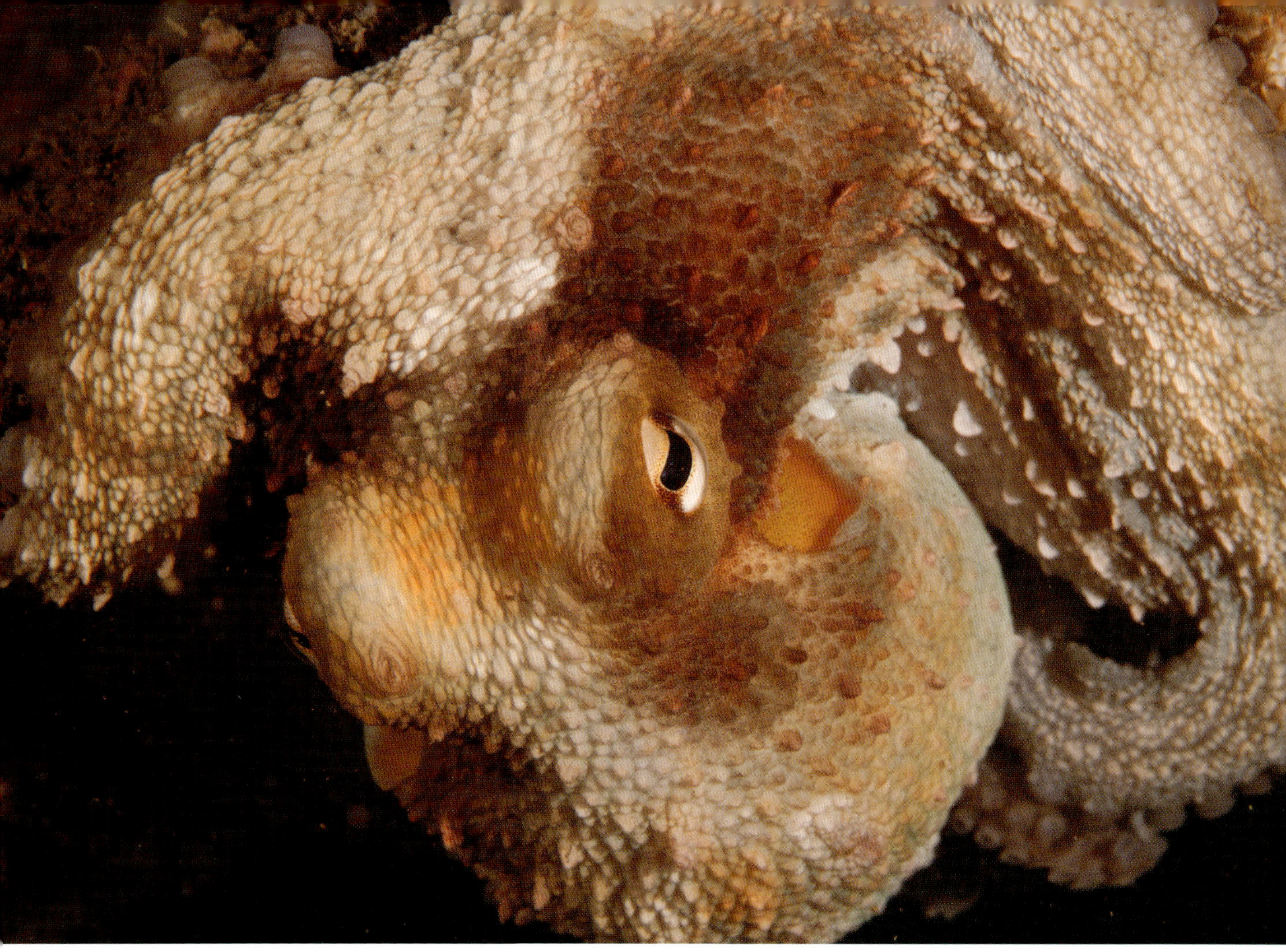

ABOVE Octopus were seen on every dive we did.

OPPOSITE The Ocean Revival wrecks have been made safe to explore inside.

The weather had been against us and sizeable waves had churned up the water, making the visibility less than ideal, but still we were keen to see these huge wrecks for ourselves. Ocean Revival is just a couple of miles out to sea from our base at Portimão, which meant a short boat ride was all it took to reach this site. The four ships all rest in more than 20m (65ft) of water and, as they are so big, each of the wrecks should easily take up more than one dive in order that you can truly feel you have explored them fully.

My buddy and I had plans for a shot in the wheelhouse and our guide explained to us that the wreck of the NRP *Hermenegildo Capelo* F481 frigate would make the best subject and so this is where we moored up. We descended into the gloomy green water, which is typical of the North Atlantic, and made our way down the line, watching in awe as the wreck loomed out of the murky water. As we got closer we could just about take in the sheer size of this ship, and for a frigate her figures

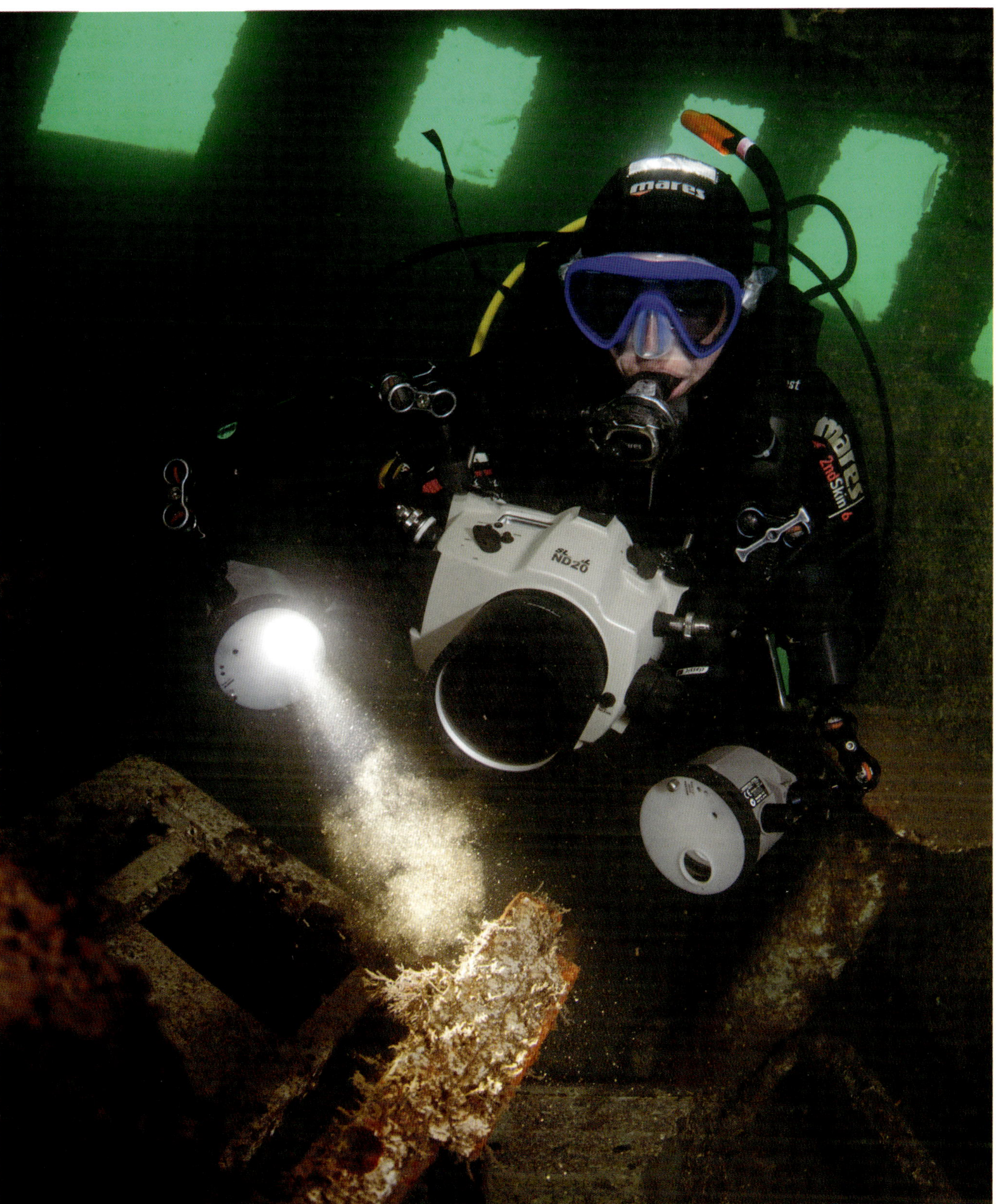
mares
ND20
mares
2ndSkin 6

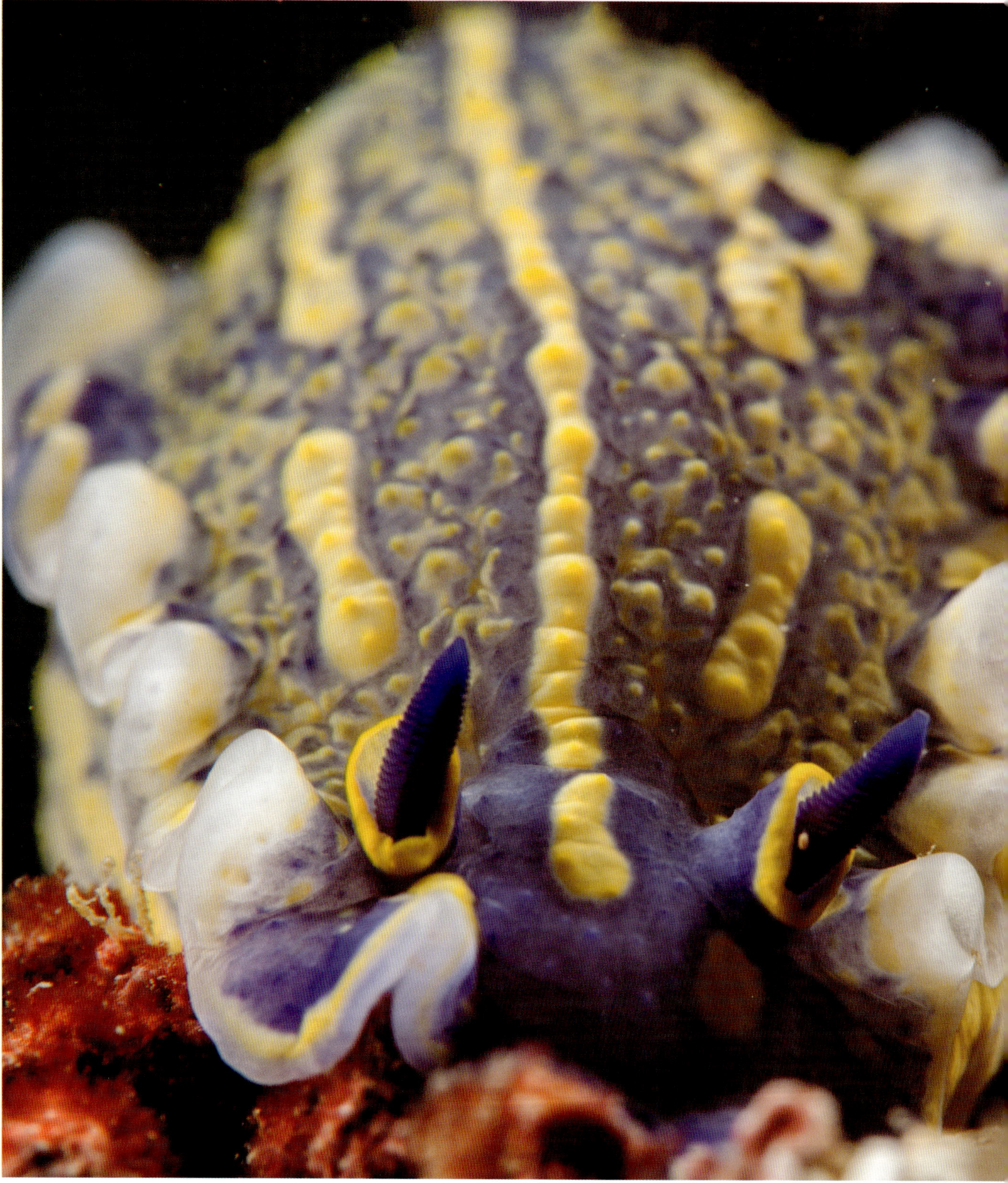

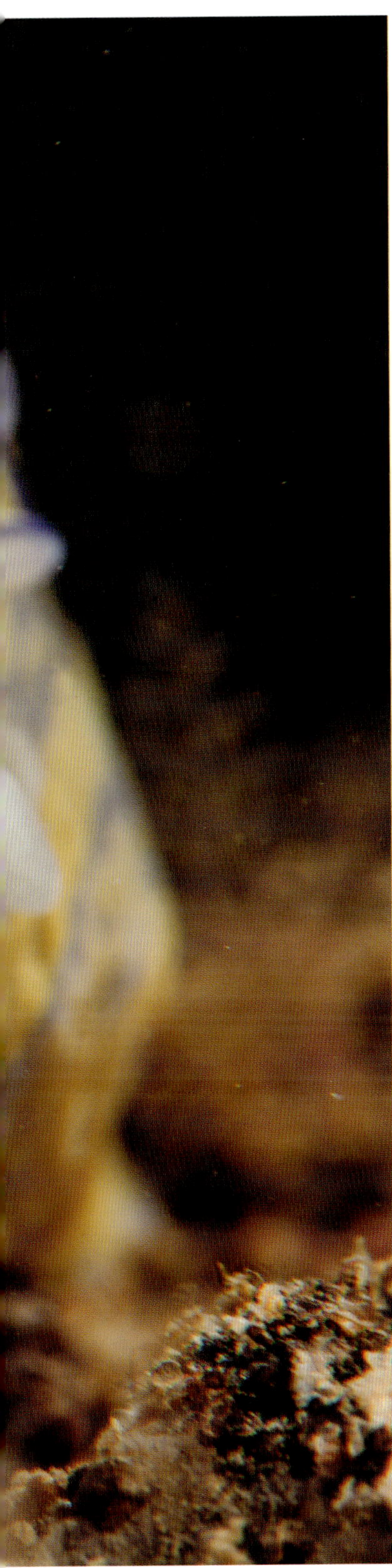

are an impressive 12m (39ft) breadth and 103m (338ft) in length. She sits in 35m (115ft) of water at the deepest point, although you will reach the highest point of the wreck at just over 15m (50ft). A small group of Oceanic Triggerfish greeted us as we arrived at the deck, and we spotted several octopus hiding in small holes in the superstructure too. But on this dive we were keen to get into the wheelhouse and use our limited no-decompression time to set up and capture the shot we wanted. I led the way, manoeuvring through the metal doorframe into the wheelhouse. The view from inside the wheelhouse was spectacular, and I found myself looking out into the ocean through several small windows, while hovering behind the already encrusted wheel. However, it was my job on this dive to provide the additional lighting and so I made my way behind the wheel, trying desperately not to disturb any debris or sediment as I manoeuvred myself into position. My buddy was right behind me and soon he too was looking at the spectacular view I had just left behind. I lit up the wheel from behind, and as he shot the main images from his position we imagined the view that the captain of this frigate would have had as he commanded his ship.

On our second dive on these wrecks we encountered a table football (foosball) table complete with its players, but with the playing surface now encrusted with marine life. In another room we found a phone still hanging from the wall. As is the case with all artificial reefs, Ocean Revival has attracted a host of marine life, but it was the wheelhouse with its beautiful spoked wheel that really took our breath away. As we headed out of the wheelhouse, our time to ascend rapidly approaching, we saw a beautiful stripy nudibranch making its way across the deck. Alas, my buddy had a fisheye lens on his cameras so it was down to me to attempt to capture a decent image of the tiny creature. However, it is clear that the local life is making itself at home here.

The Ocean Revival wrecks are a treat because whilst any polluting or dangerous artefacts have been removed before the sinking, many of the original features remain. The three other wrecks which make up this project are all similar in size to the frigate. There is the hydrographic ship NRP *Almeida Carvalho*, the ocean patrol boat NRP *Zambese* and the corvette NRP *Oliveira E Carmo*. I would say that these wrecks alone are a good enough reason to travel to the region, and for those with a true lust for rust this dive site is a must.

LEFT Wonderful nudibranchs can be found on the wrecks and reefs.

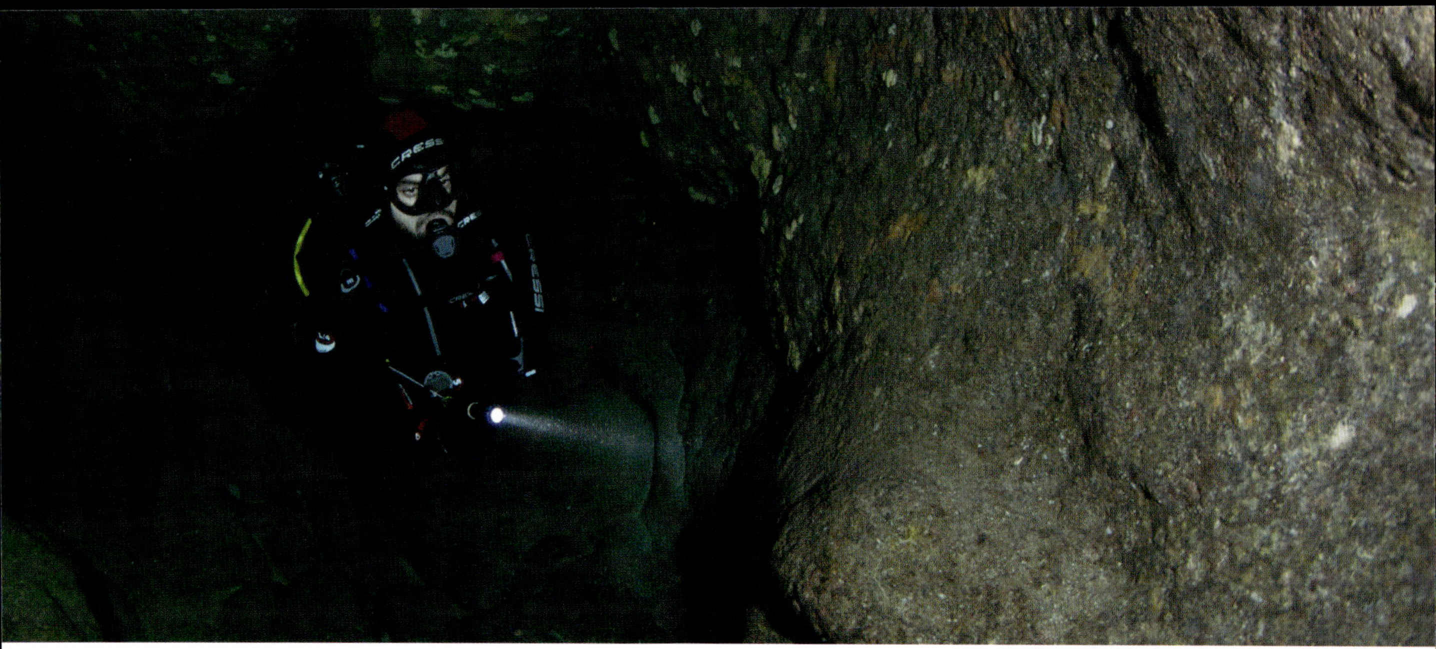

06 AN UNEXPECTED TRIP TO DIVE KARAVOMILOS CAVE

GREECE

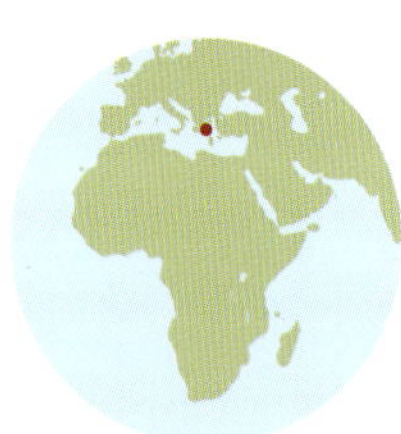

ABOVE A diver heading out of one of the caves that line the coast.

OPPOSITE The light fades as you enter the cave system.

KEFALONIA IS THE LARGEST of the five main islands of the Ionian region of south-western Greece. This region is steeped in history and has been invaded or occupied by numerous empire builders. These invaders have included the Romans, Byzantines, Normans, Venetians, Turks, Russians, French and British. Most recently the Germans and Italians occupied the region during the Second World War, as was depicted in Louis de Bernières' novel *Captain Corelli's Mandolin*.

Today Kefalonia is considered to be one of the best holiday destinations in the Mediterranean thanks to its fabulous beaches and crystal clear turquoise water. The coastline is one of the island's key attractions and diving is becoming more popular, with the region's government keen to promote it.

GENERAL DIVING

Marine wildlife in the Mediterranean is scarce, largely due to overfishing during the past 20 years or so, but wrecks, caverns and caves are abundant in this region.

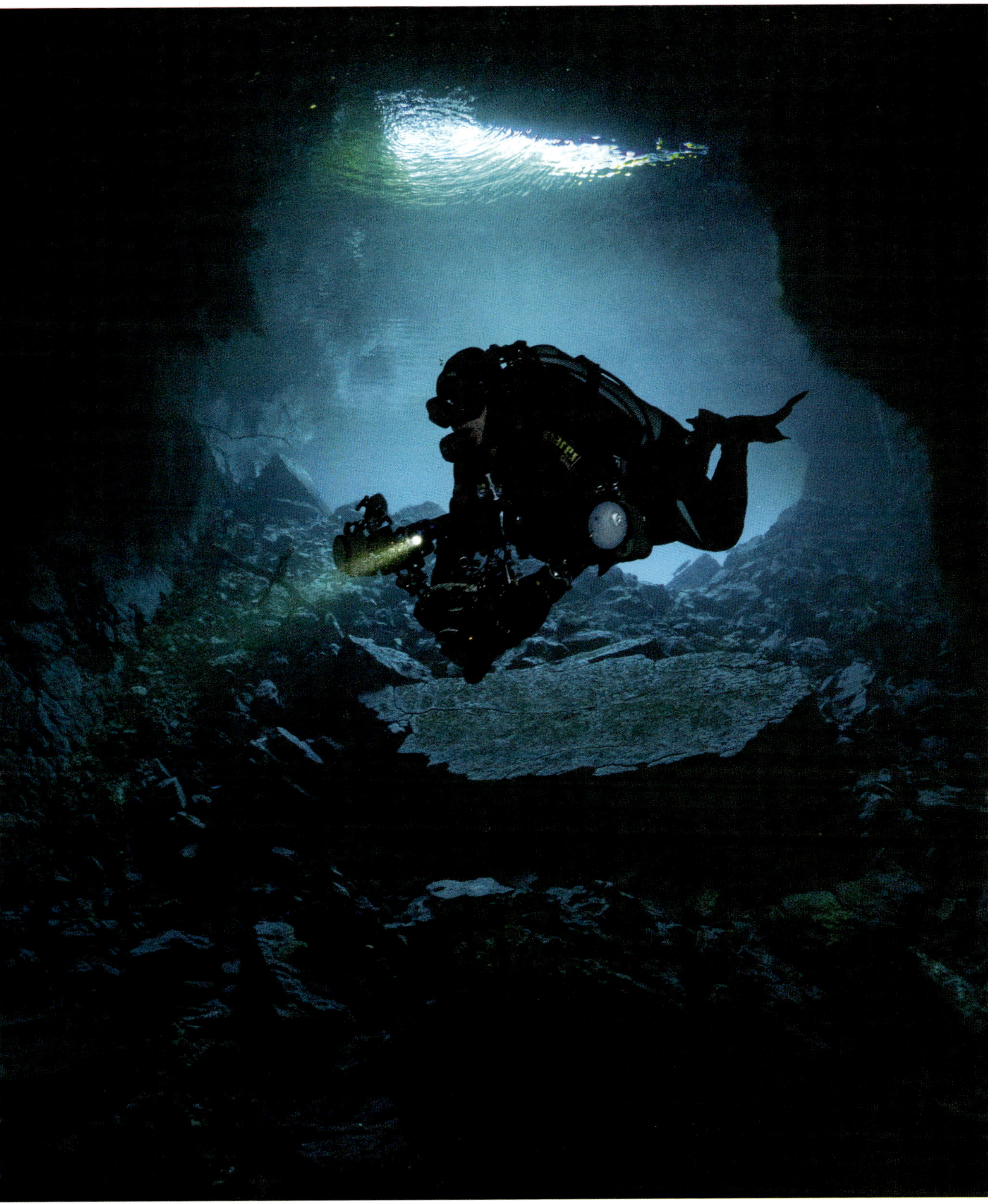

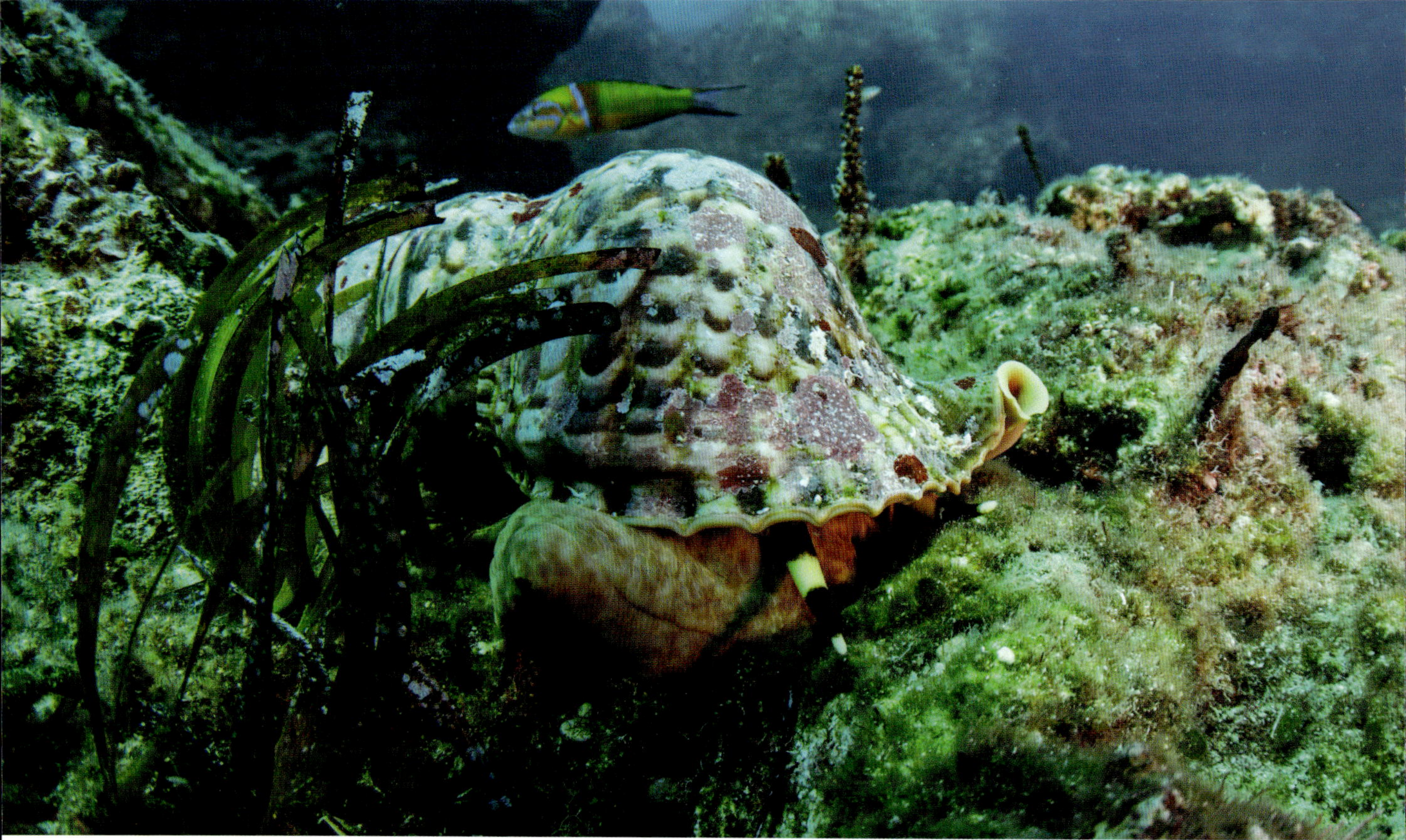

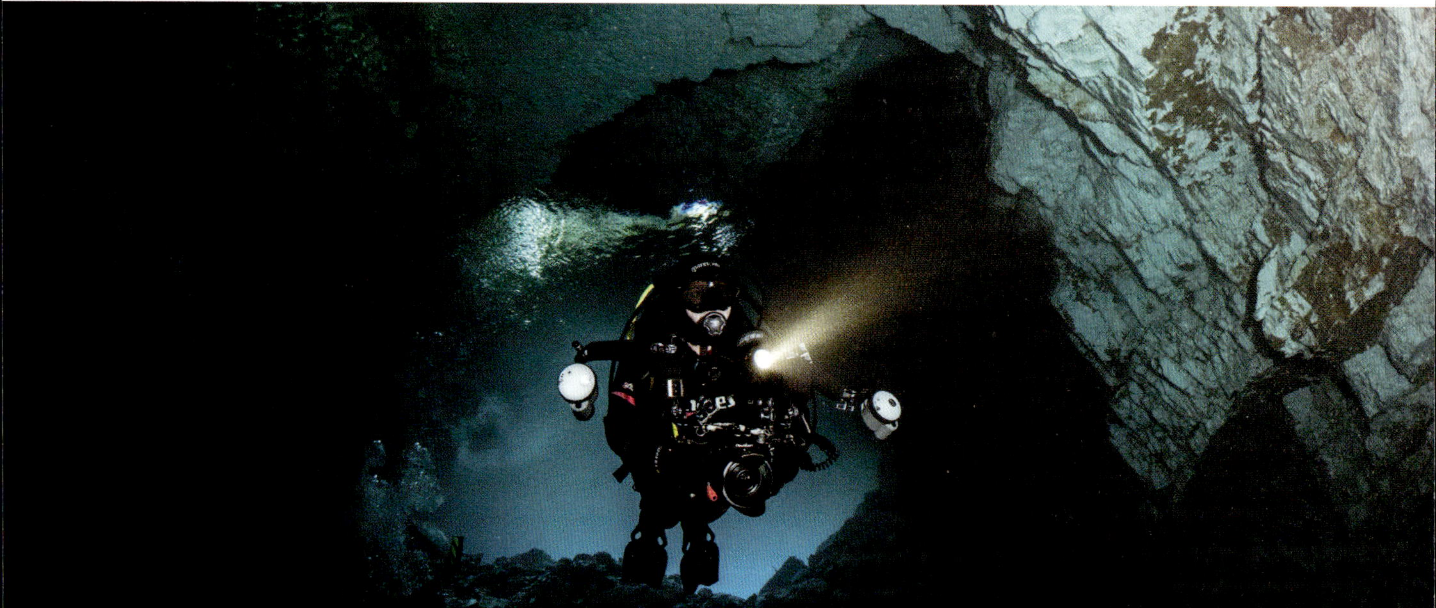

However it is the historical dives that have attracted the most interest in recent times. The waters off Kefalonia have seen plenty of action, dating back to Roman times. At one dive site amphoras lie exposed in the water, shallow enough for divers to explore. More recent wrecks are from the Second World War, the most famous being the HMS *Perseus*. This British patrol submarine lies in deep water, which is perfect for technical divers, but with the top section still accessible to more advanced recreational divers. This wreck has an amazing story. The submarine was on combat patrol in December 1941, and while cruising at the surface at night it hit a mine and sank. From the crew of 59 only one person, the Royal Navy leading stoker John Capes, managed a daredevil escape from a depth that no one had attempted before. With the assistance of local people he escaped capture by occupation forces. While his feat was legendary in Royal Navy circles, almost nobody believed his story until 1997, when a team of Greek divers located the submarine and verified the details of the escape he described.

OPPOSITE TOP
A Giant Triton combs the rocks looking for food.

OPPOSITE BOTTOM
The cave entrance.

THE WILD DIVE

Sometimes circumstances beyond your control, such as the weather, can be serendipitous, and this turned out to be the case for us on a diving trip to Kefalonia. We were supposed to be diving some of the many Second World War wrecks in the area, and of course we were excited to have been asked to join a group of journalists to write about these unusual dives. However, Mother Nature had different ideas and as the wind picked up and the waves got bigger and bigger we eventually but reluctantly realised that making a coastal dive was going to be impossible for the ensuing few days that we were on this beautiful island. Whilst every experienced diver knows that every now and again the weather can turn against you, and that this will happen at some point, it is still difficult to brush aside the disappointment right away. However, in this case the back-up plan turned out to be one to lift the spirits of even the most hardened pessimists – we were going to explore a famous freshwater cave.

Now I am not a qualified cave diver, and so actually I was only going to explore the opening of the Karavomilos cave system. It takes specialist training and lots of planning to actually explore underwater cave systems, but at least I would be able to reconnoitre and experience the entrance, and my buddy and I agreed that we would only explore the area still in view of the exit and go no further into the complex cave system. We loaded our dive gear and camera rigs onto the minibus and with some excitement headed off to the small fishing village of Karavomilos on the central east coast of Kefalonia.

ABOVE Rain on our return from the dive into the cool shallow lake.

OPPOSITE Some of the cave formations are huge.

The dive starts in a beautiful shallow brackish lake that is home to vivid green vegetation, European Seabass and Grey Mullet. The first thing you notice is the incredible visibility which must extend further than 40m (130ft). The water, however, is a chilly 14°C (57°F) due to a freshwater spring that runs up through the cave system and out into the lake. As you head towards the entrance of the cave system you gradually find it getting deeper, although the maximum depth for the whole dive was just 10m (33ft). Only a few minutes into the dive we reached the mouth of the cave and we were completely amazed at what we saw.

The arched entrance to the cave system, with the shallow lake behind, allows plenty of light to penetrate into the first chamber. After just a few fin-kicks into the cavern we encountered our first stalagmites and stalactites, respectively rising up

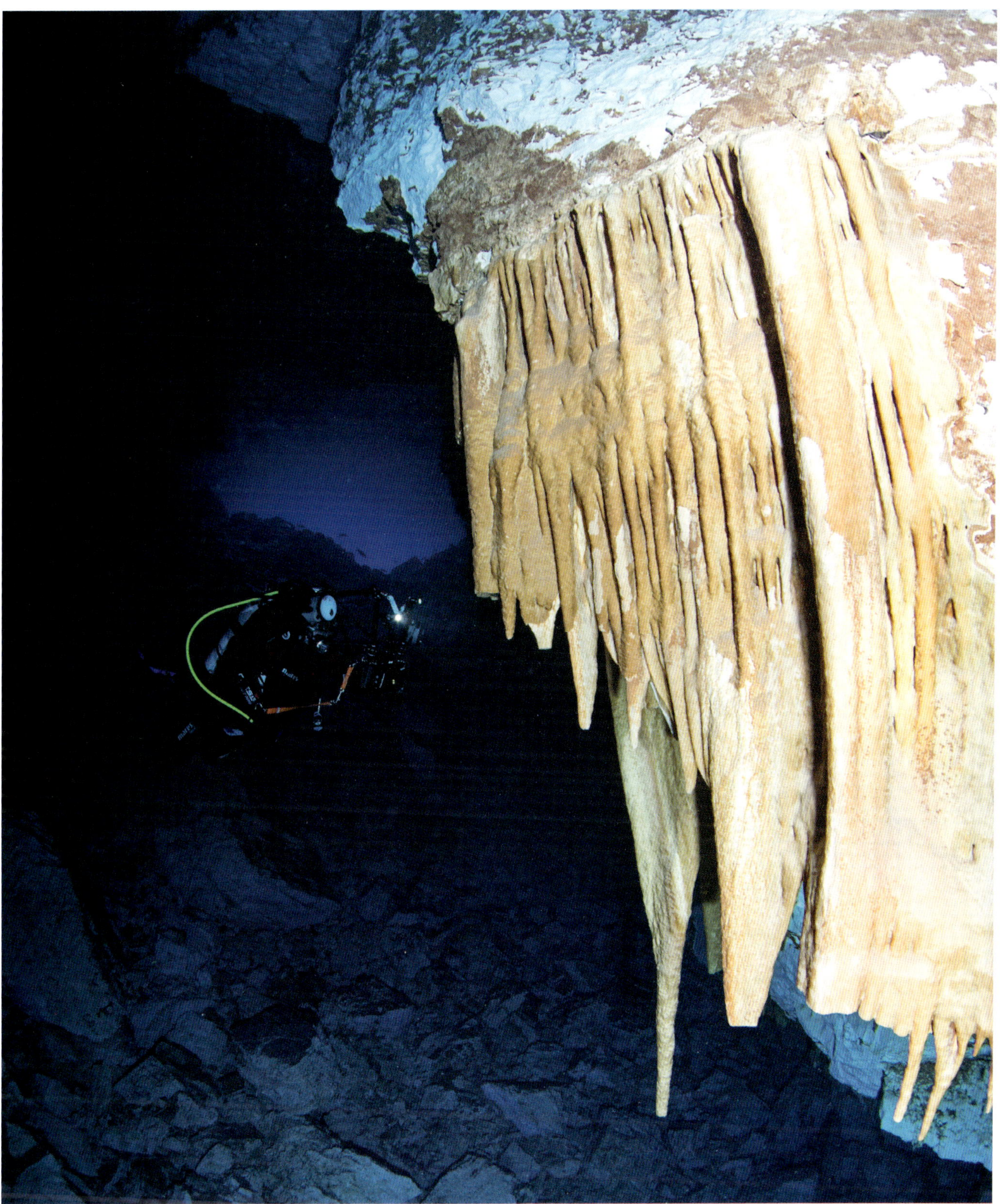

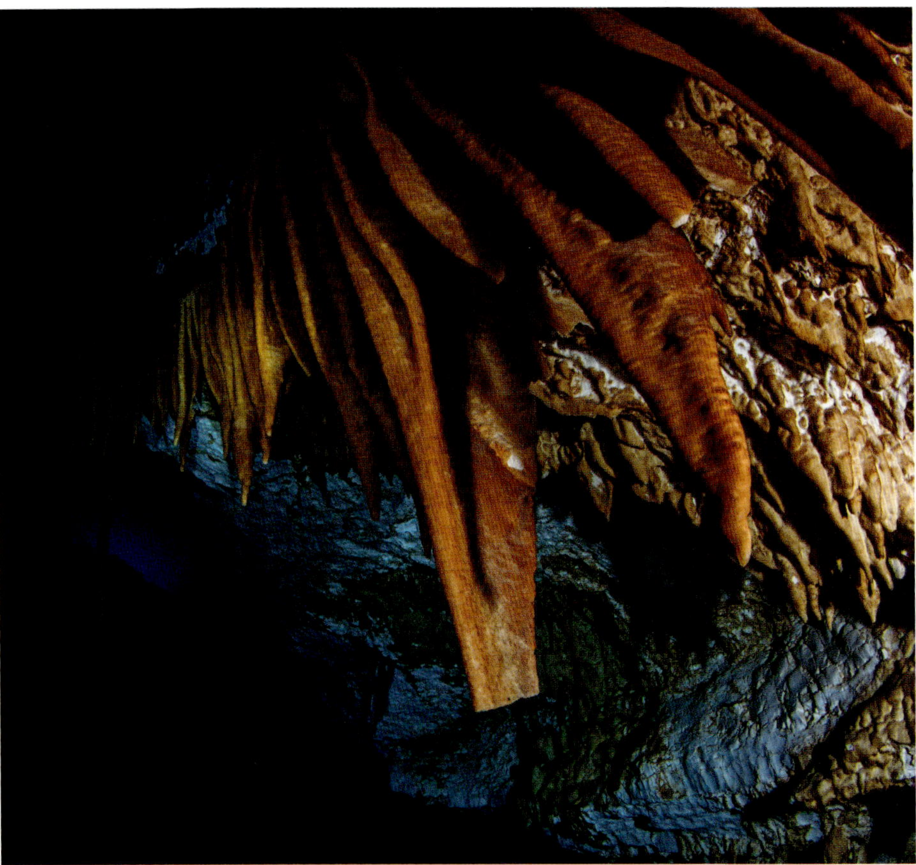

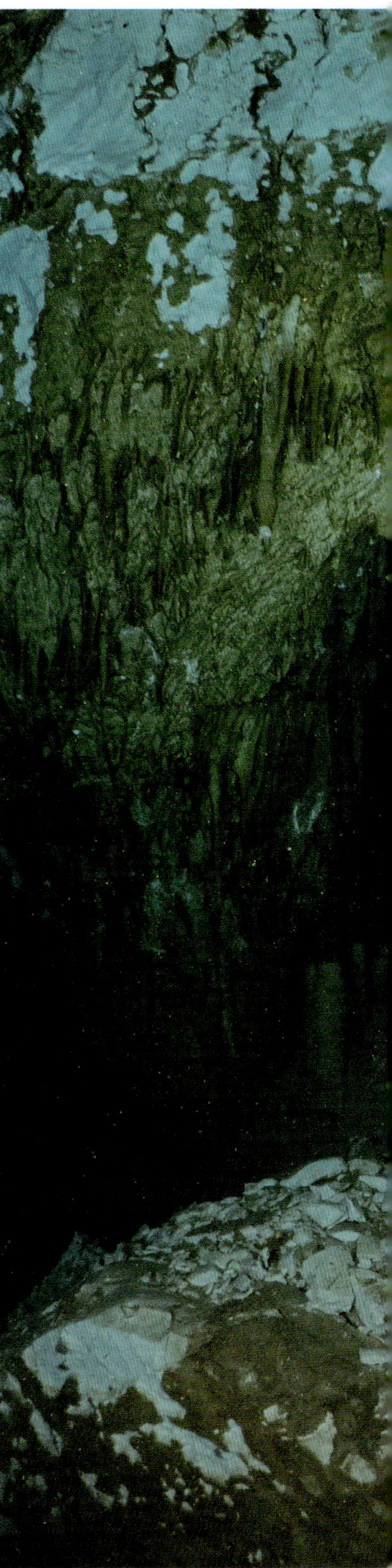

from the cave floor and hanging down from the ceiling. They were a beautiful sight, and as you turned around, either beneath the water or at the surface, you realised that they were everywhere and that even the solid calcification on the walls creates a truly incredible sight. Some of them appeared to sparkle as my torch swept across the structures, with others seeming almost to be made of gold. I wished I had completed the training which would have enabled me to access further along into the cave system, but all too soon it was time to turn around and head back to the surface. The glorious sunshine had turned into wild rain by the time we returned, but my spirits could not be dampened after such an incredible underwater experience.

ABOVE Stalactites in gold cling to the roof of the cave.

RIGHT A diver exploring the cave system.

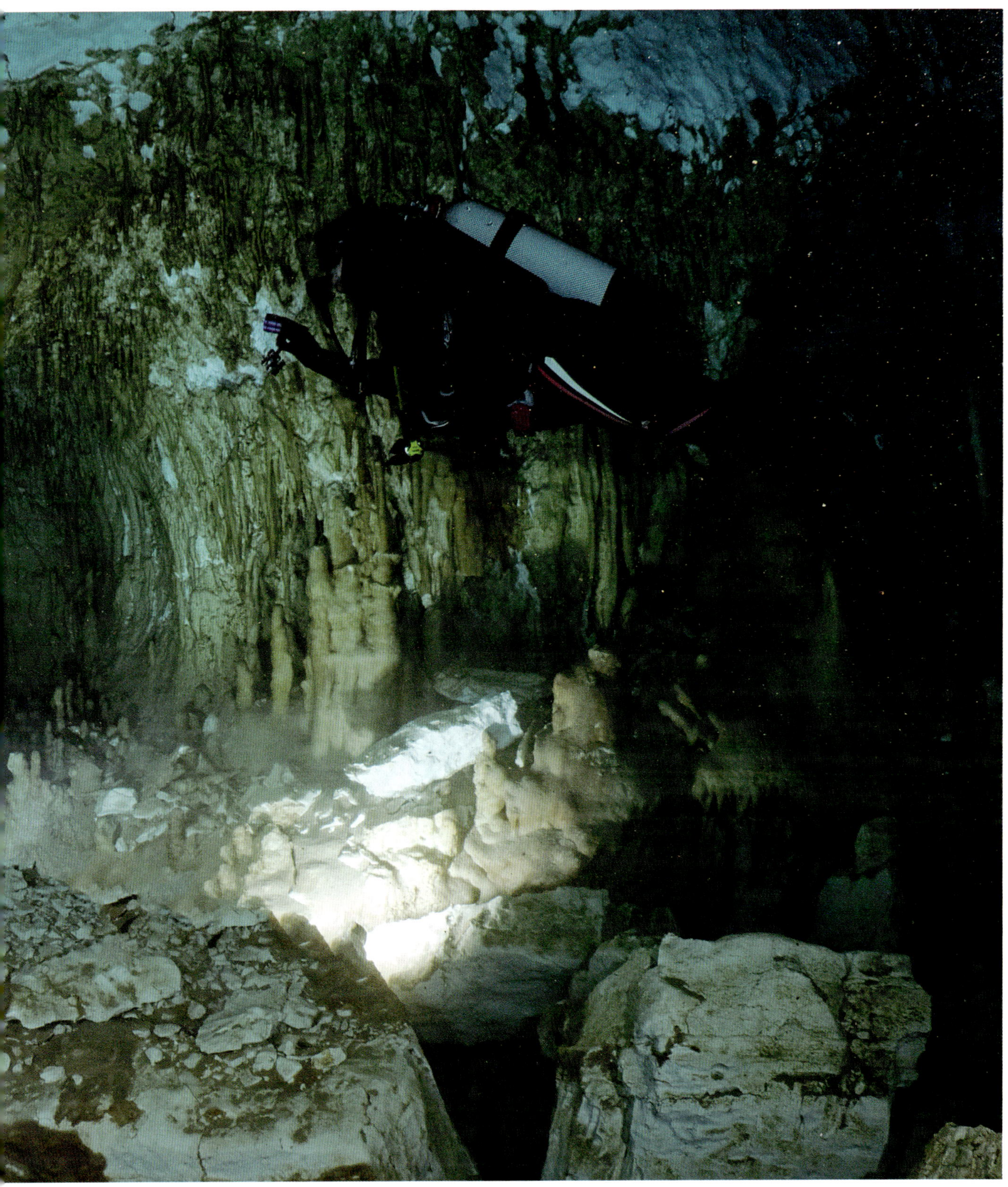

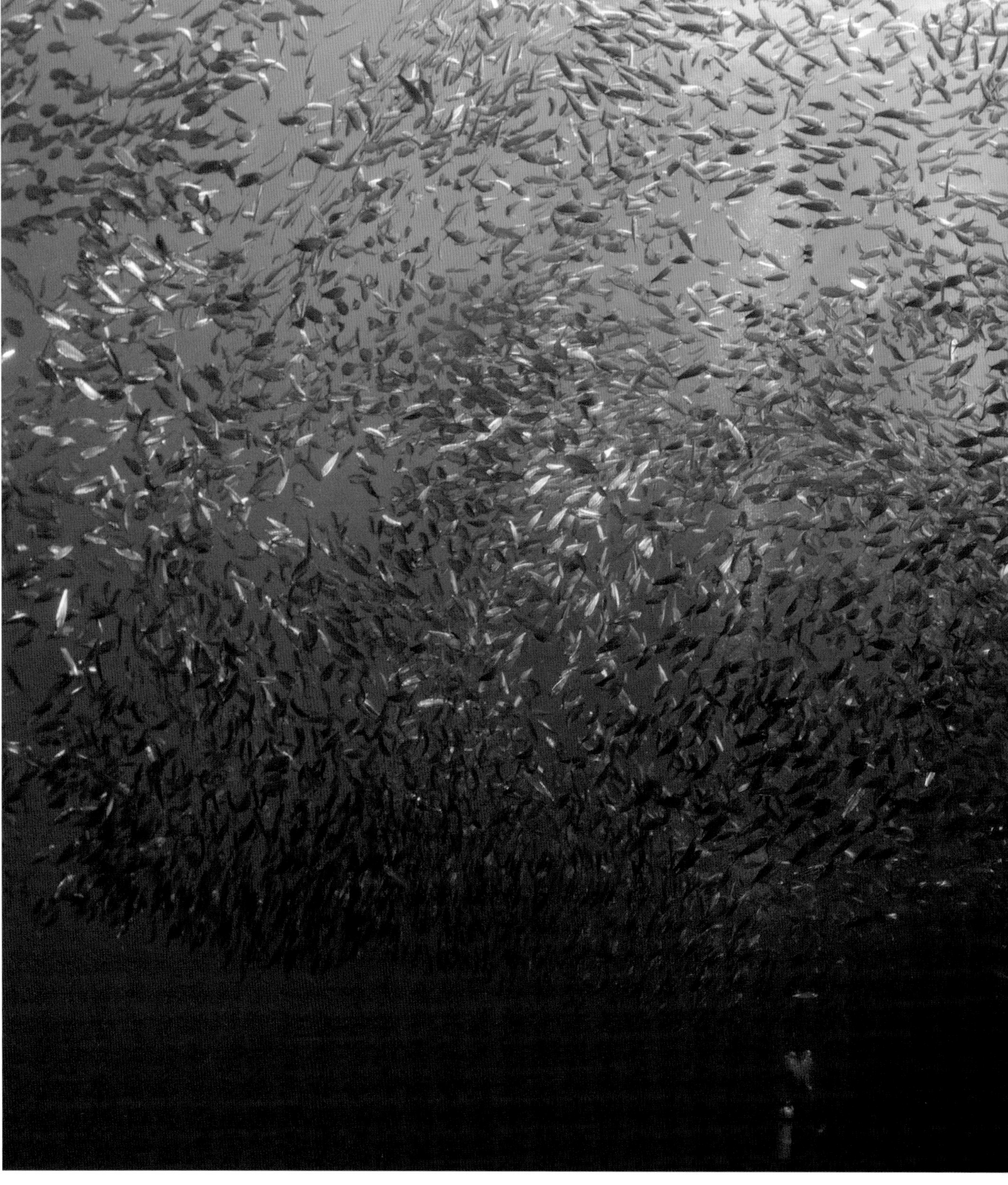

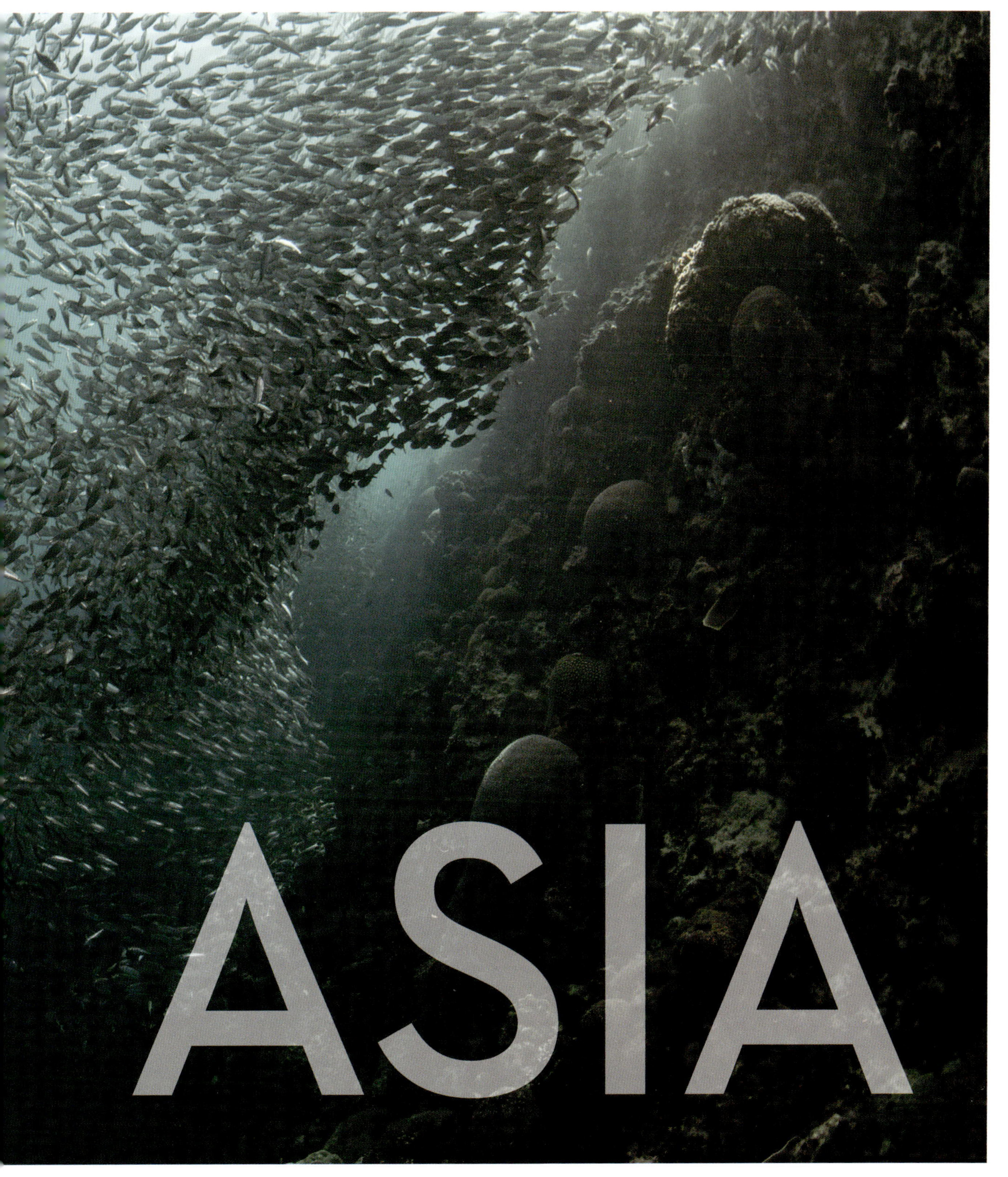

ASIA

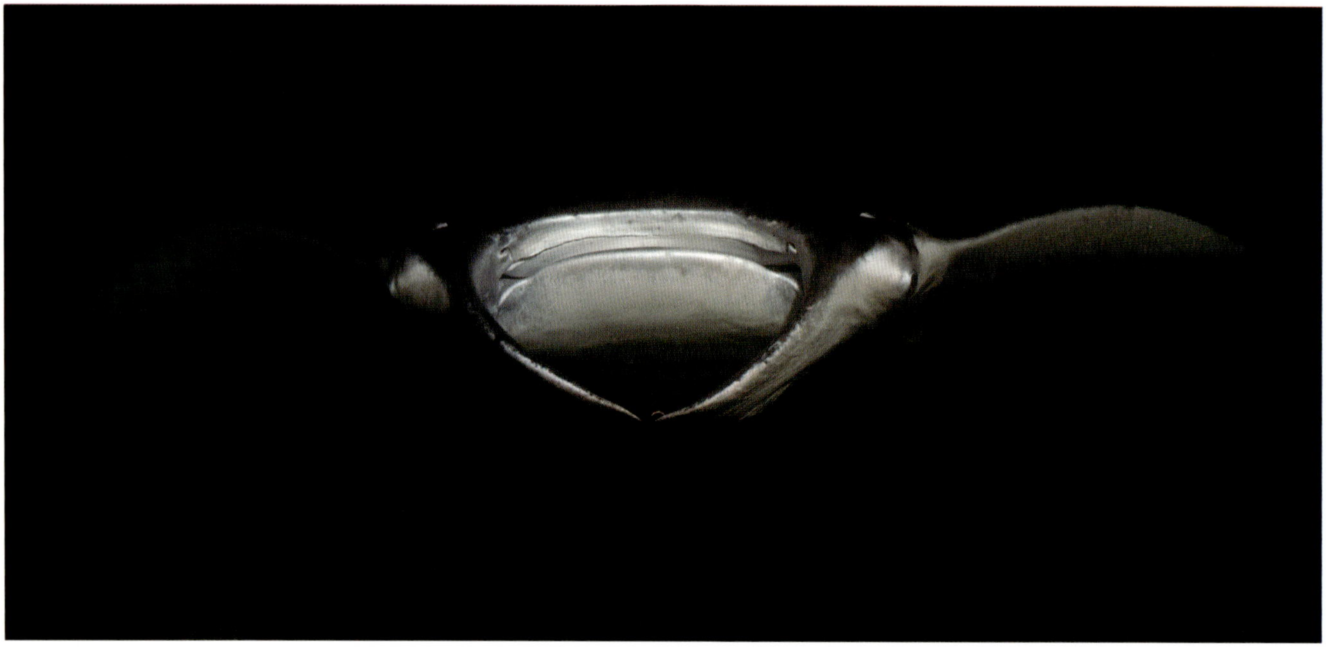

07 MANTAS AT NIGHT

MALDIVES

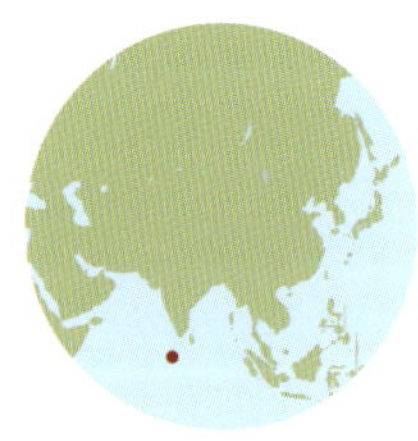

ABOVE It is incredible to see mantas swimming through the water at night.

OPPOSITE Large numbers of fish live inside a wreck.

THE MALDIVES ARCHIPELAGO conjures up thoughts of small isolated islands with white sandy beaches and a handful of palm trees, surrounded by turquoise waters. In fact there are more than 1,000 coral islands in the Maldives, spread over 26 atolls set in the Indian Ocean. They are part of a vast underwater mountain range that breaks the surface at these islands. Each one is a tropical paradise and many have visitors flocking here to soak up some sun, pamper themselves in luxurious hotels and explore the underwater world. Only 185 of the islands are occupied, so many of them are completely untouched by humans. There is another way to visit these islands, and that is to live on a boat – called, somewhat unsurprisingly, a liveaboard – for the duration of your stay. It might not have all the creature comforts, but it can allow you to visit the reefs of multiple atolls in a single trip, following the best underwater action in the prevailing conditions.

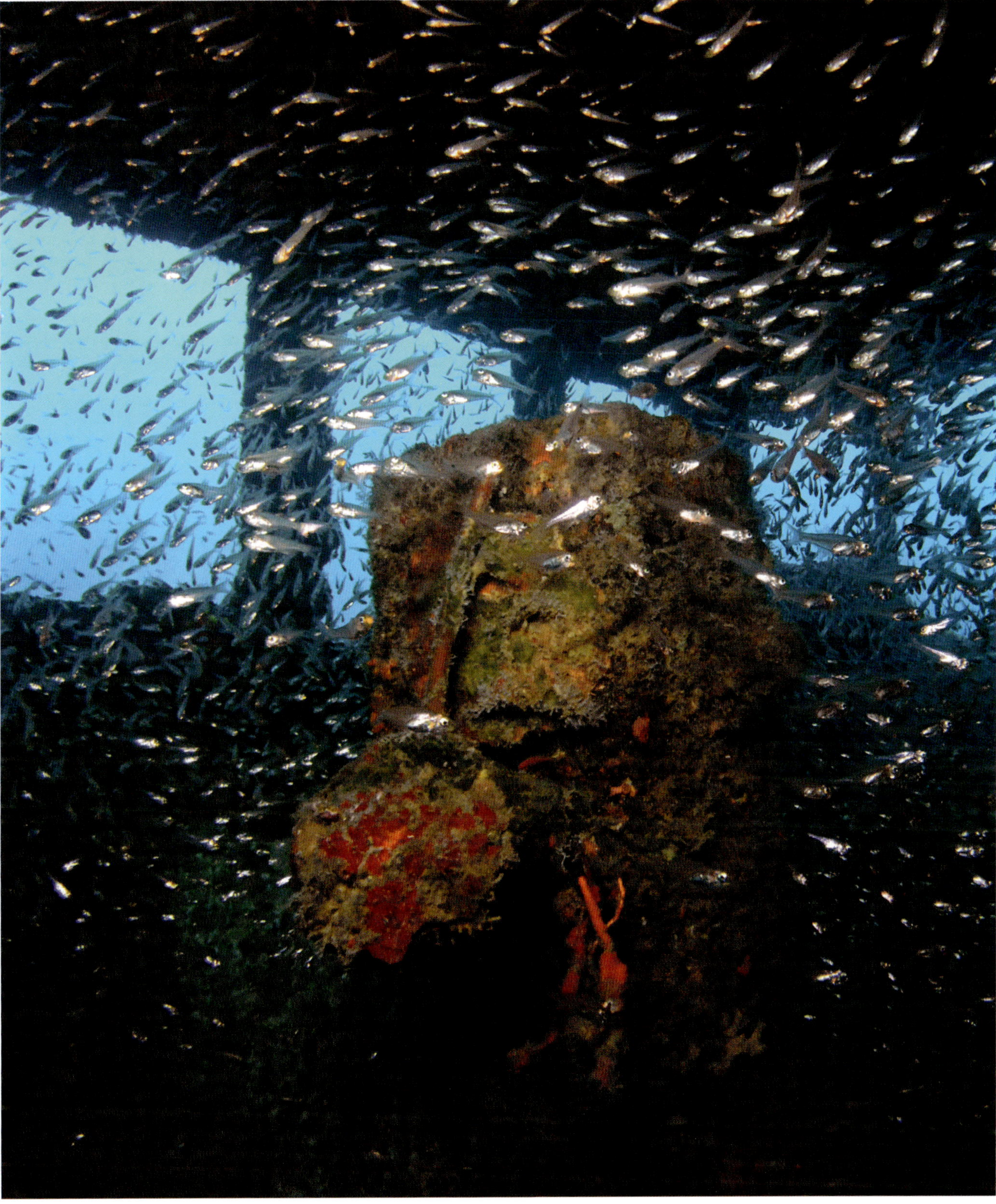

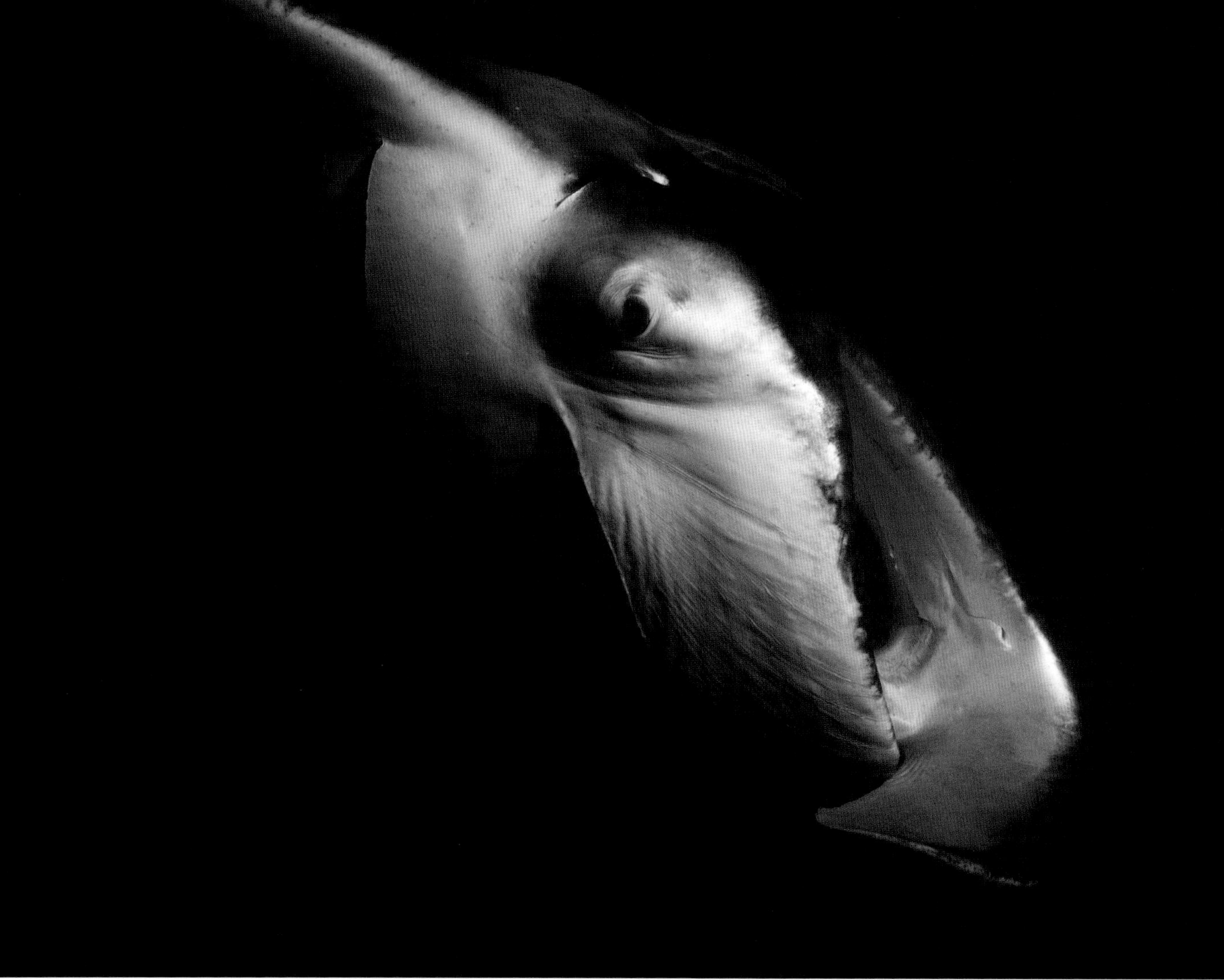

ABOVE A manta eyeballs me as it makes a turn.

GENERAL DIVING

The Maldives are famous for their big marine animal action, with manta rays and Whale Sharks being regular visitors to these waters. Sometimes these huge animals gather together to form large feeding aggregations, while during other periods they are solitary, hovering over a favourite coral reef to get cleaned of parasites by the specialist fish that reside there. Smaller sharks are found hunting in the strong currents and these same currents bring nutrients to the rest of the reef too. In recent times the Maldives has suffered from some coral bleaching as water temperatures rise, but this does not seem to have affected the sheer numbers of

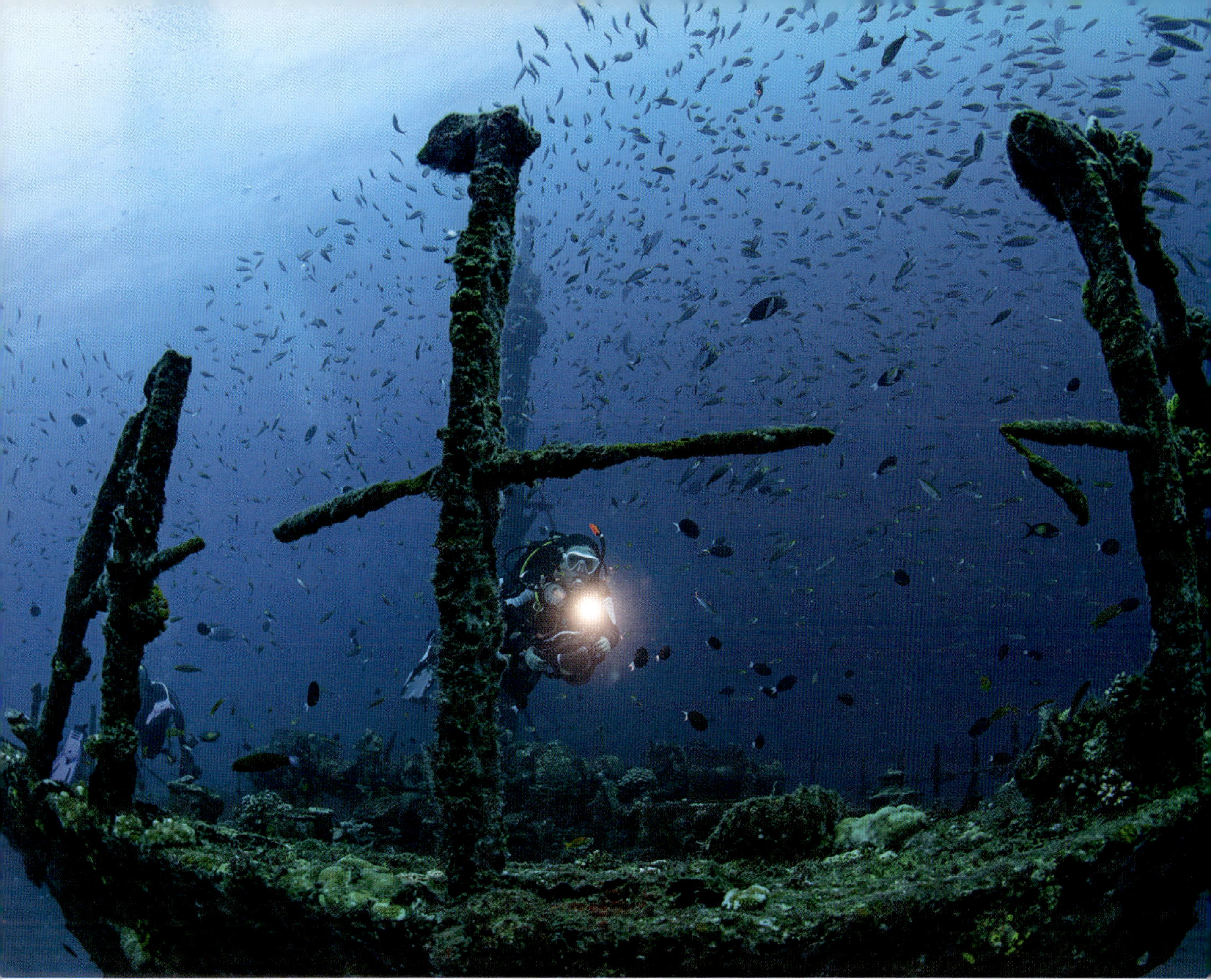

fish that live on the reefs. More sheltered reefs have vibrant soft corals and pretty anemones gently toing and froing in time with the water. In areas with stronger currents, hard corals prevail, providing shelter for a host of smaller fish. Some of the islands have wrecks to explore and they are usually also home to several schools of fish, either on the outside of the structure, or even hiding inside. Divers can also take part in conservation or 'citizen science' projects in the Maldives, with several organisations and charities offering courses on reef surveying, animal behaviour and manta ID programmes. The Maldives is a diving destination that truly has something to offer everyone.

ABOVE Exploring a wreck covered in marine life.

THE WILD DIVE

We were on a liveaboard trip to the Maldives covering a couple of the more northerly atolls in a search for sharks and rays. Earlier in the day our guide had mentioned to us that this particular evening would be a good one to do a night dive, as the boat was going to stay in the sheltered bay where we were moored and not move off to another island just yet. As the sun started to go down we gathered in the lounge to hear his briefing about the dive we were going to do. It was to be on a shallow local reef, and it sounded lovely. However, just as he was wrapping up we heard a shout from the diving deck of the boat – MANTA! The captain of the boat had turned on the diving deck lights and this had attracted a host of tiny critters. In turn this had also drawn in two huge mantas, which were performing loops and barrel rolls in order to scoop up and eat the zooplankton feast. Our guide asked if we still wanted to do the planned night dive, but I was already halfway into my diving equipment and nearly ready to step off the back of the boat with the prospect of watching these majestic animals feeding at night. I had heard about it and seen images and videos but had never experienced it myself and there was no way I was going to miss it. Some of the other divers caught on and started to join me and so the divemaster resigned himself to a long night hovering in shallow water under the boat. We grabbed some additional dive lights and jumped in a short

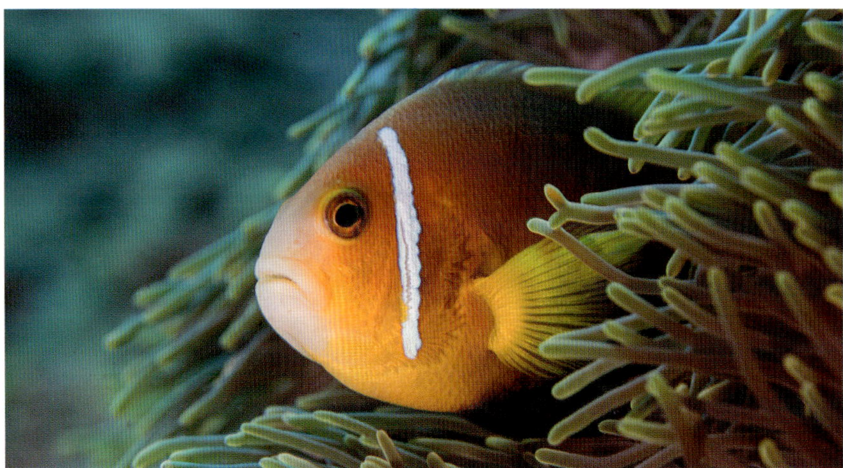

ABOVE Anemonefish are a common and popular reef resident.

OPPOSITE Nurse sharks also join in the night dive fun

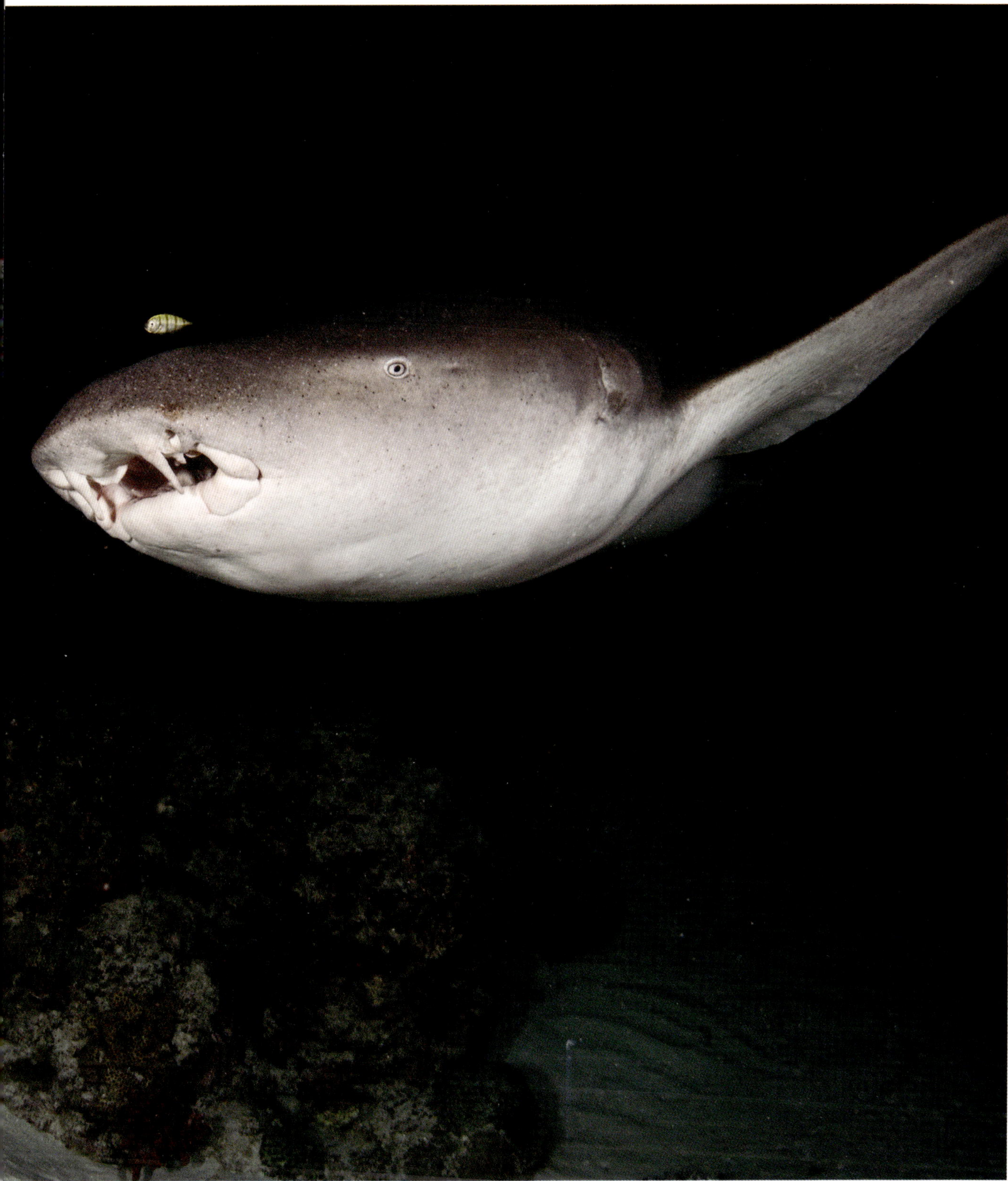

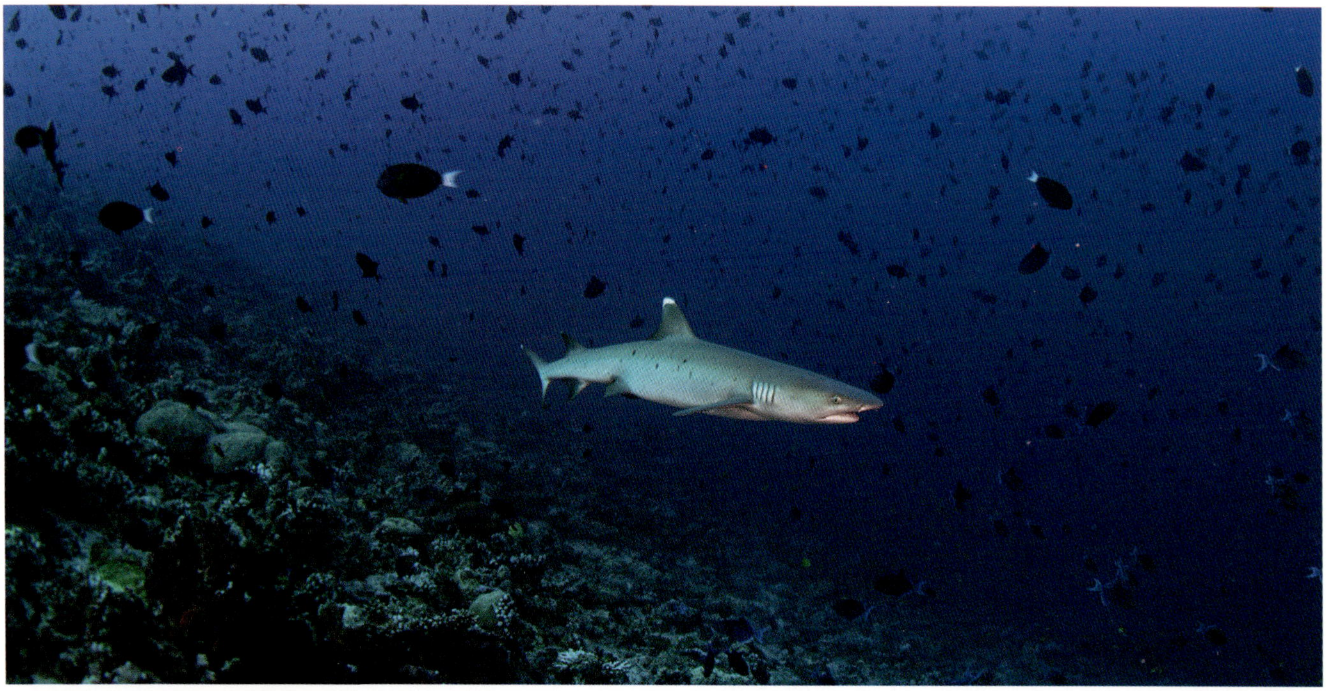

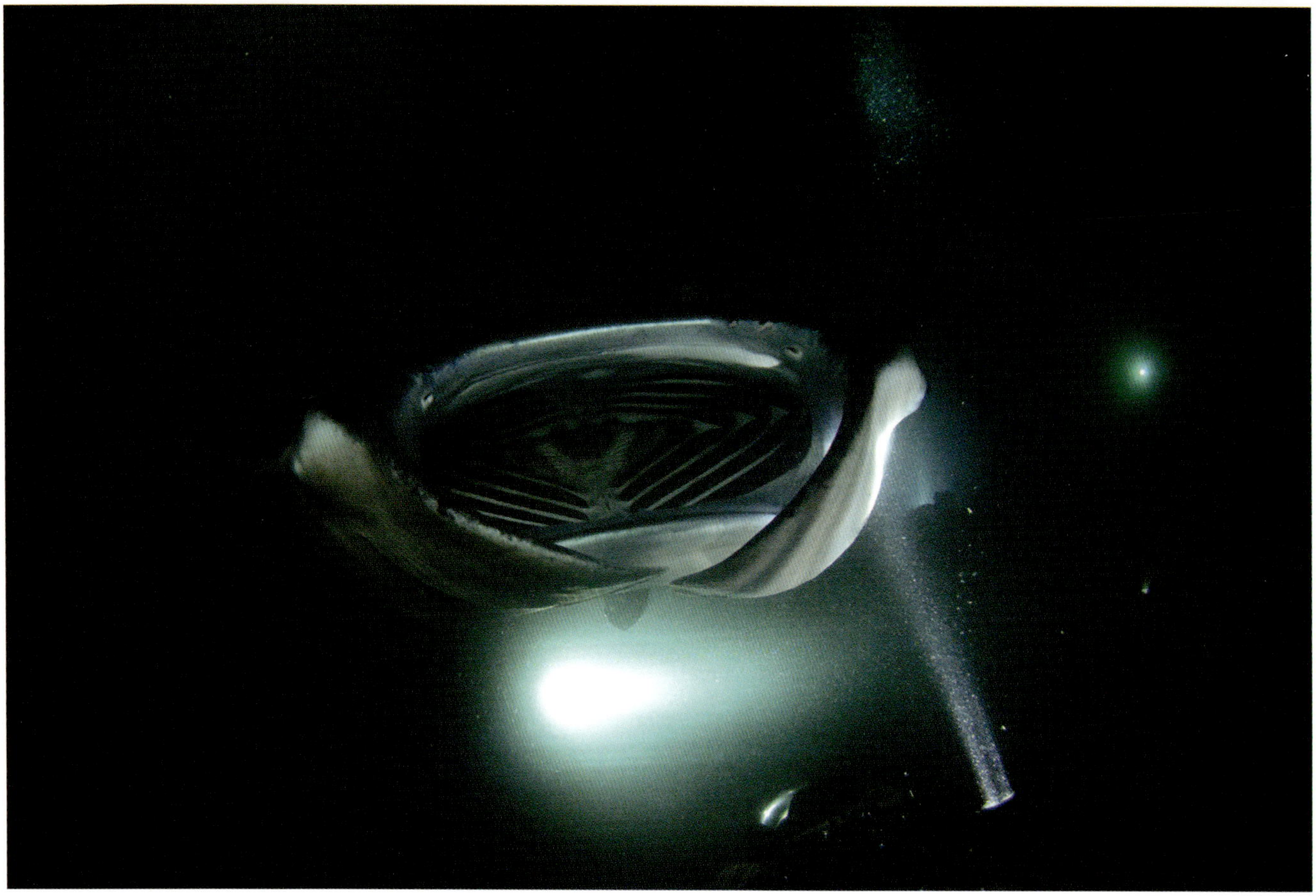

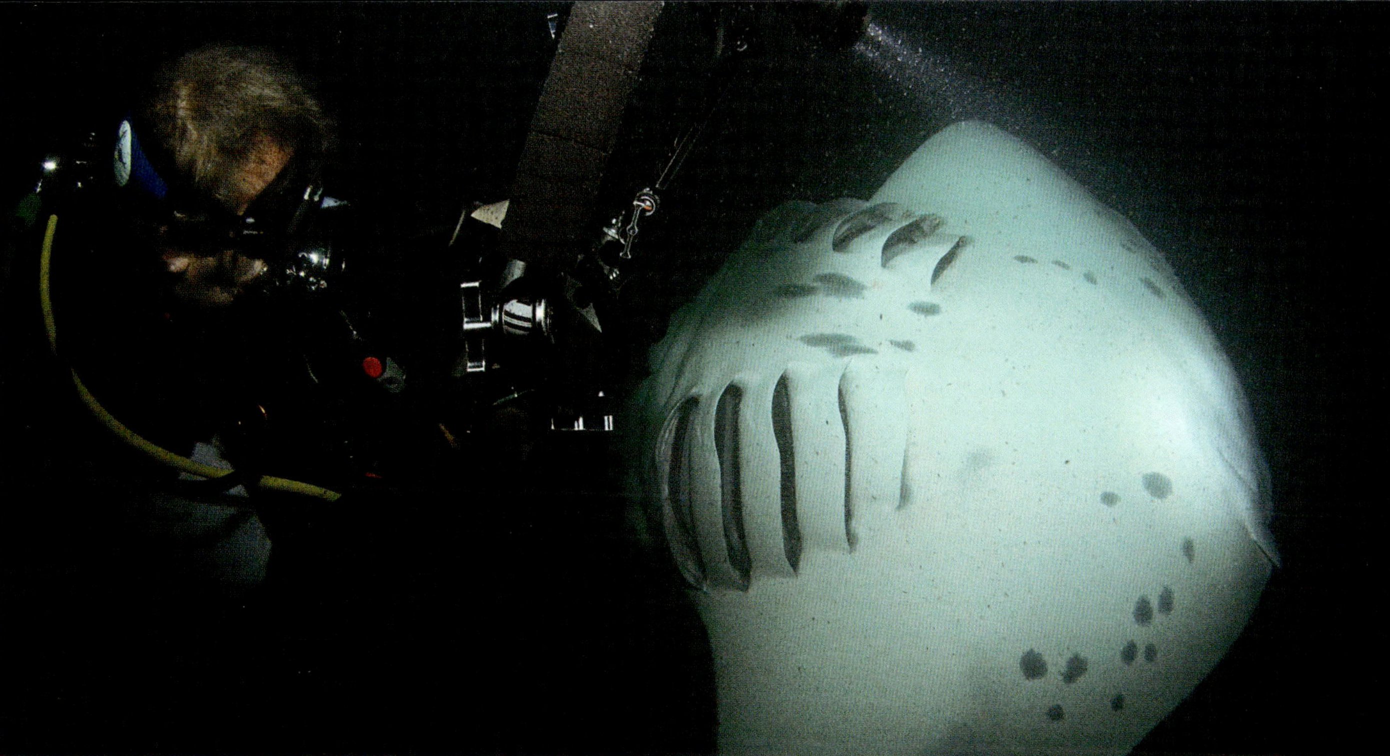

distance away from the mantas so as not to disturb them. Our guide grabbed the extra lights and started setting them up in two straight lines. Before long we had our very own manta runway! These lights attracted more small organisms for the mantas to feed on and created a linear buffet for our mantas. It did not take them long to catch on and soon, while we waited near the lights, they started to swoop towards us.

It is hard to get the camera to focus on the huge black back of a manta ray at night, but as they rolled over backwards and the light caught the white underbelly, the cameras clicked into focus and allowed me to get some shots. I made sure that I did not just focus on getting images, but also took the opportunity to fully experience this incredible spectacle. The two mantas stayed with us for more than two hours, although most of the divers had got cold and hungry by this point and had headed back up to the surface for a well-earned beverage and an incredible Sri Lankan curry made by the boat's chef.

The mantas, with their mouths wide open and their cephalic lobes positioned to scoop up zooplankton as efficiently as possible, flew through the ink black water, gently beating their wings to propel themselves. Looking up I could still see the light from the dive deck of the boat providing a little light to see by. At times, with the huge mantas lit by just a few torches and coming out of the gloom and into view, the dive looked more like a scene from a science fiction movie than a night dive.

OPPOSITE TOP A whitetip shark swims amongst the huge numbers of fish over the reef.

OPPOSITE BOTTOM It is like a scene from a science-fiction movie.

ABOVE A manta barrel-rolls right in front of a diver.

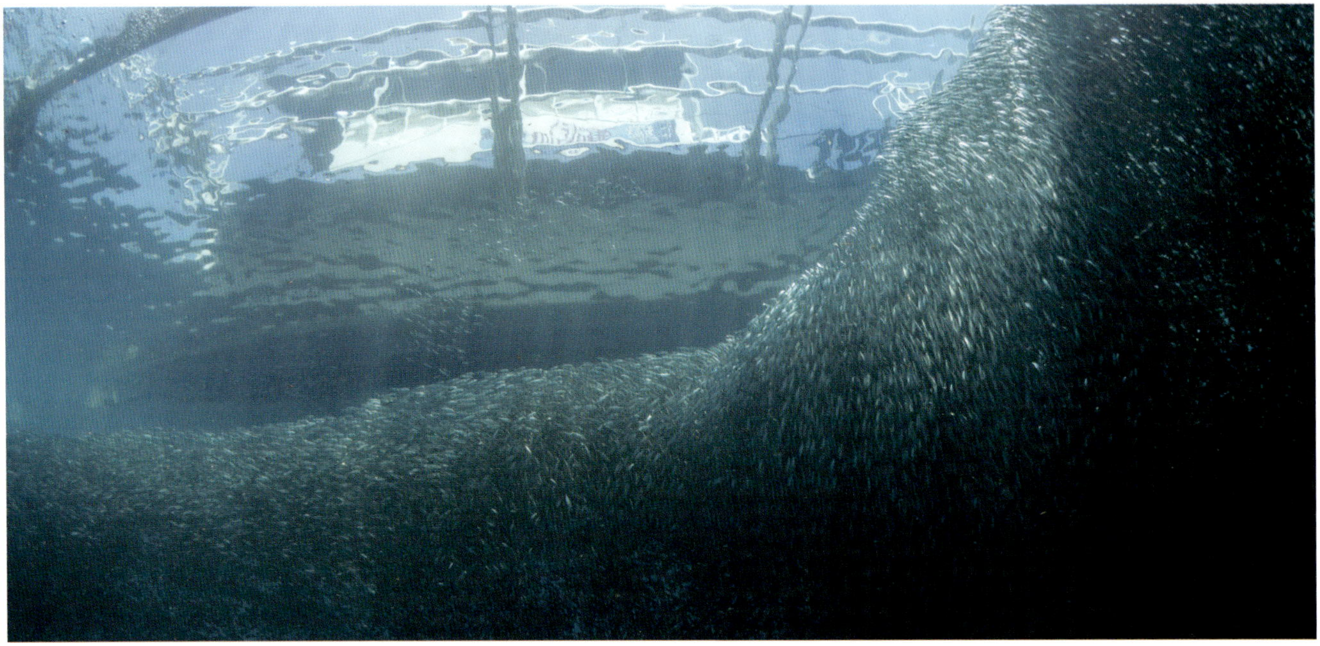

08 THE MOALBOAL SARDINE RUN

PHILIPPINES

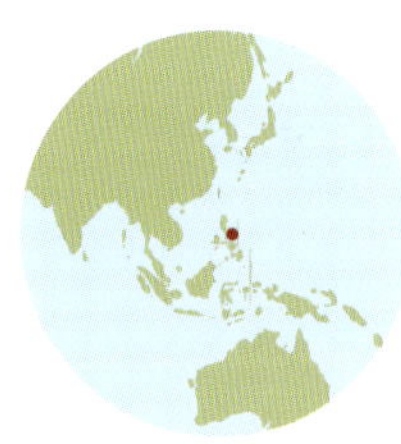

OPPOSITE The reefs here are covered in colourful soft corals.

ABOVE The boat waits overhead.

MOALBOAL IS A PROVINCE on the south-west tip of Cebu Island in the Philippines. If you arrive by plane you can expect a two- to three-hour car or bus ride to get you to your resort, slowly making your way through villages with streetside stalls and winding through impressive countryside. If you like to explore above water as well as below, then this is an ideal location with vibrant markets, beautiful waterfalls, lush landscapes and fantastic opportunities for hiking, biking, kayaking and other action-packed activities, including canyoning (or canyoneering) in some breathtaking locations. Due to its central location in the Philippines, Moalboal is the ideal base for a multi-destination diving trip, offering the opportunity to visit such places as Oslob for Whale Sharks, Malapascua for thresher sharks and Dumaguete for muck diving. Moalboal also offers visitors an active night life with local bars lining the shore and locals and visitors alike coming to watch the sun go down over a few cold drinks.

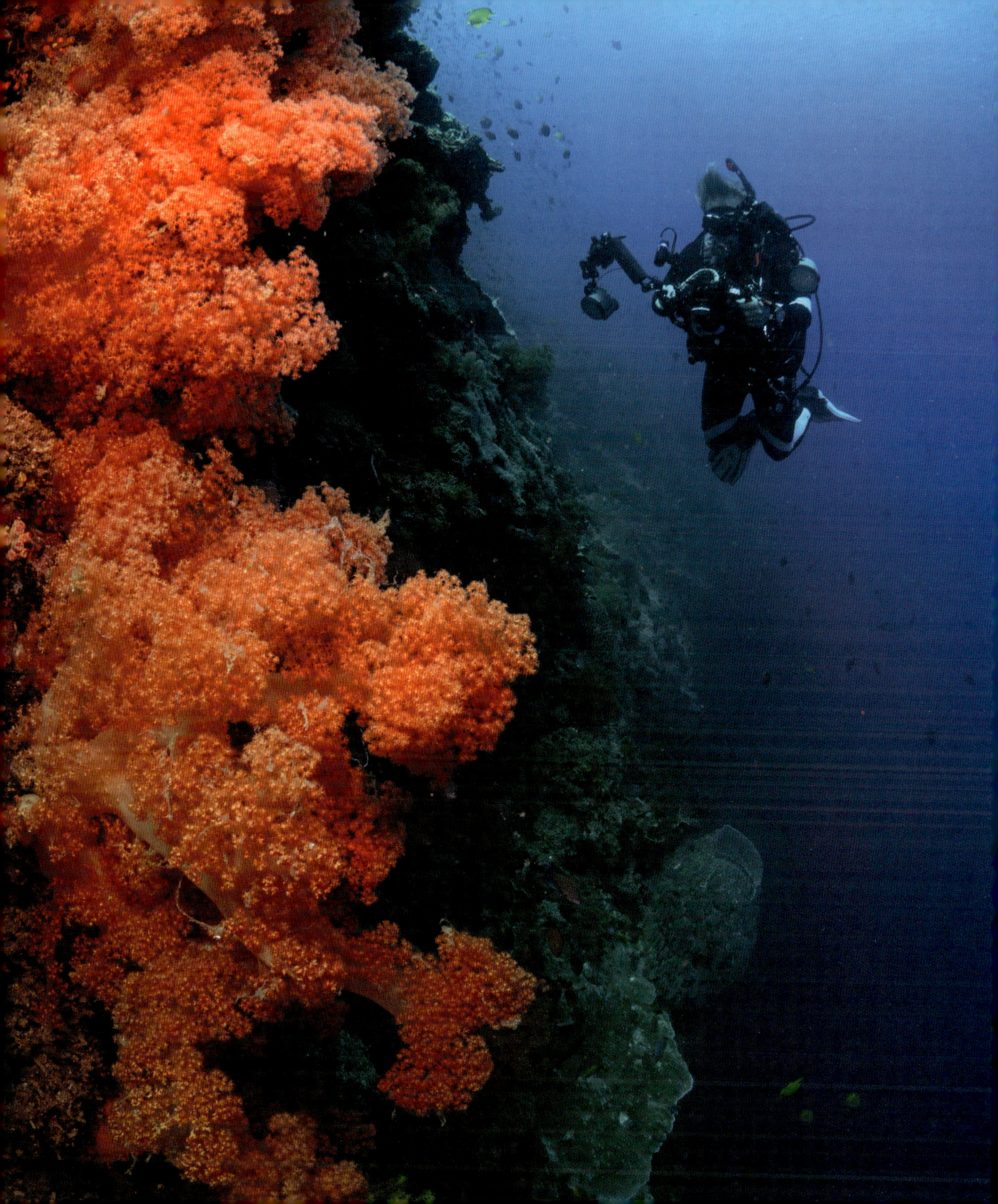

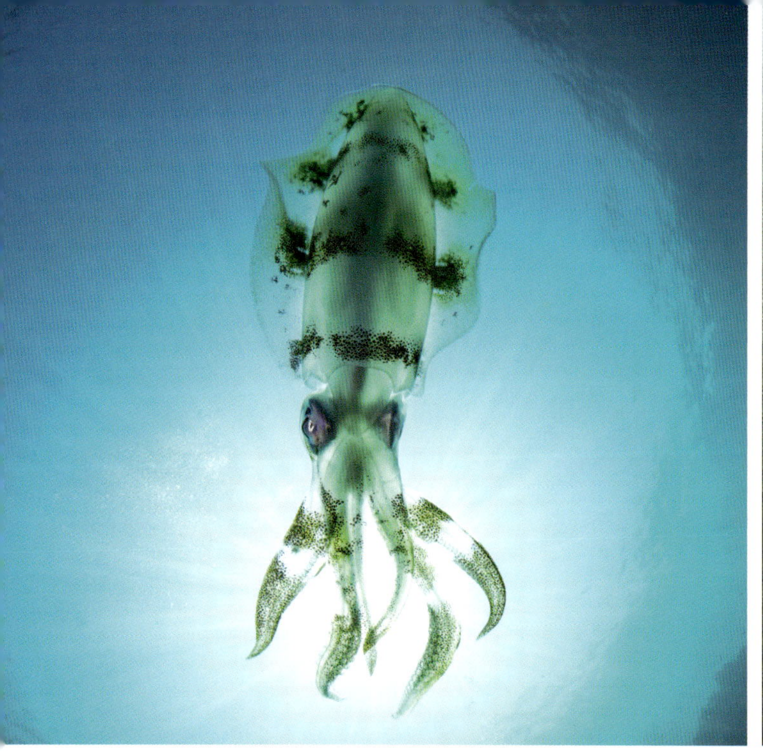

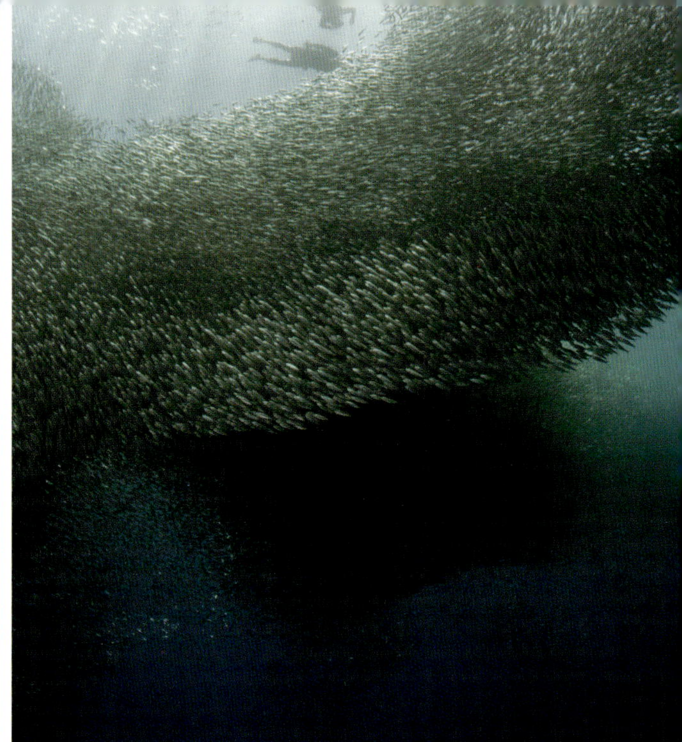

GENERAL DIVING

The diving here is varied and offers visitors a huge range of marine life to discover. The local reefs are teeming with tiny critters that the local dive guides are adept at finding. Pygmy seahorses, frogfish, a plethora of tiny crustaceans, weird and wonderful octopus and tiny many-hued sea slugs are among the favourite critters that divers love to see. You could spend the whole dive here without moving and find an unbelievable array of macro marine life on just a single coral head. Just a few minutes away by boat is Pescador Island, a marine park with amazing walls covered in soft corals, gorgonian sea fans and sponges, as well as caves and caverns to explore. The Cathedral cave has several openings and it is said, if you swim in and look back, that the scene resembles the image of a skull with eye sockets and gaping mouth made from the openings out to the deep blue sea.

ABOVE A squid dances in the light above us.

ABOVE RIGHT This event happens in shallow water, so even snorkelers can enjoy it.

There are several other dive sites that offer something a bit different, including the Cessna plane wreck, which was deliberately sunk as a dive attraction. Venture a little deeper here to find huge sea fans dominating the seascape. There is also 'Sunken Island', which is for experienced divers as the pinnacle starts at 25m (80ft) and you can experience strong currents. You can find all the usual suspects here but there is also a decent chance of seeing sharks and mantas. A huge range of night dives offer a completely different perspective to the marine life that you see by day. On the reef, new creatures come out of their hiding places and give you

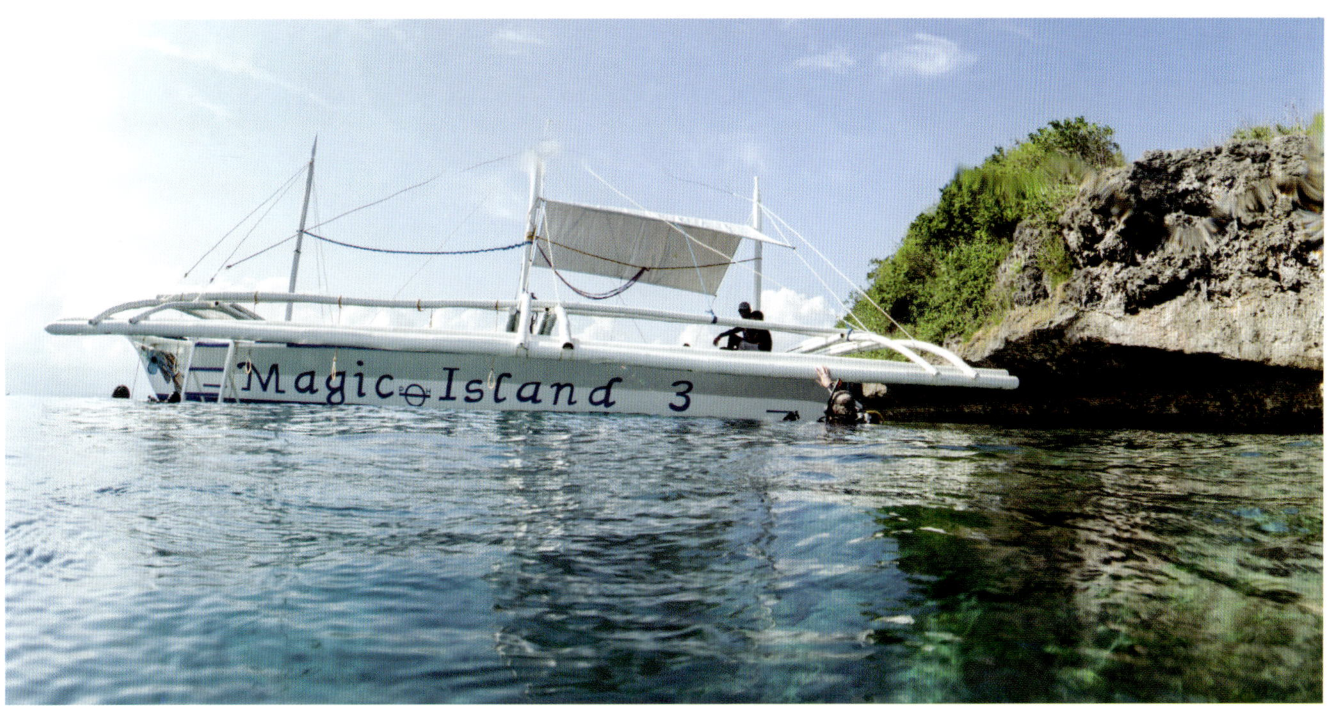

a chance to watch them in the open. Out in the deep water you can drift along and find tiny critters floating with the current. You can even do a night dive here, taking specialist equipment with you to find both corals and animals that fluoresce under UV light, which are usually invisible to the naked human eye.

THE WILD DIVE

Moalboal is probably most famous for its incredible shoal of sardines that reside permanently just off the Panasama Wall. It is so close to the shoreline that snorkelers can swim above this school of fish and wonder at its mesmerising movement, which resembles an underwater murmuration. Our dive started just a short distance away from the shoal, so that we did not manoeuvre the boat too close to swimmers and snorkelers, and also to give us the best experience of approaching the fish. We slipped into the warm water, dropped to around 5m (16ft) and started to fin towards the schooling fish. While I was keen to get to the main event, there were a few distractions as we dived along in the shallows. The first of these occurred as we came across a cuttlefish that seemed totally unfazed by our presence and simply hung above us, showing-off its remarkable skin pattern-changing skills to us. Just a short distance further along, I suddenly made out a huge dark shape in the water. Finning closer, I realised that this was what we had

ABOVE A traditional Filipino dive boat.

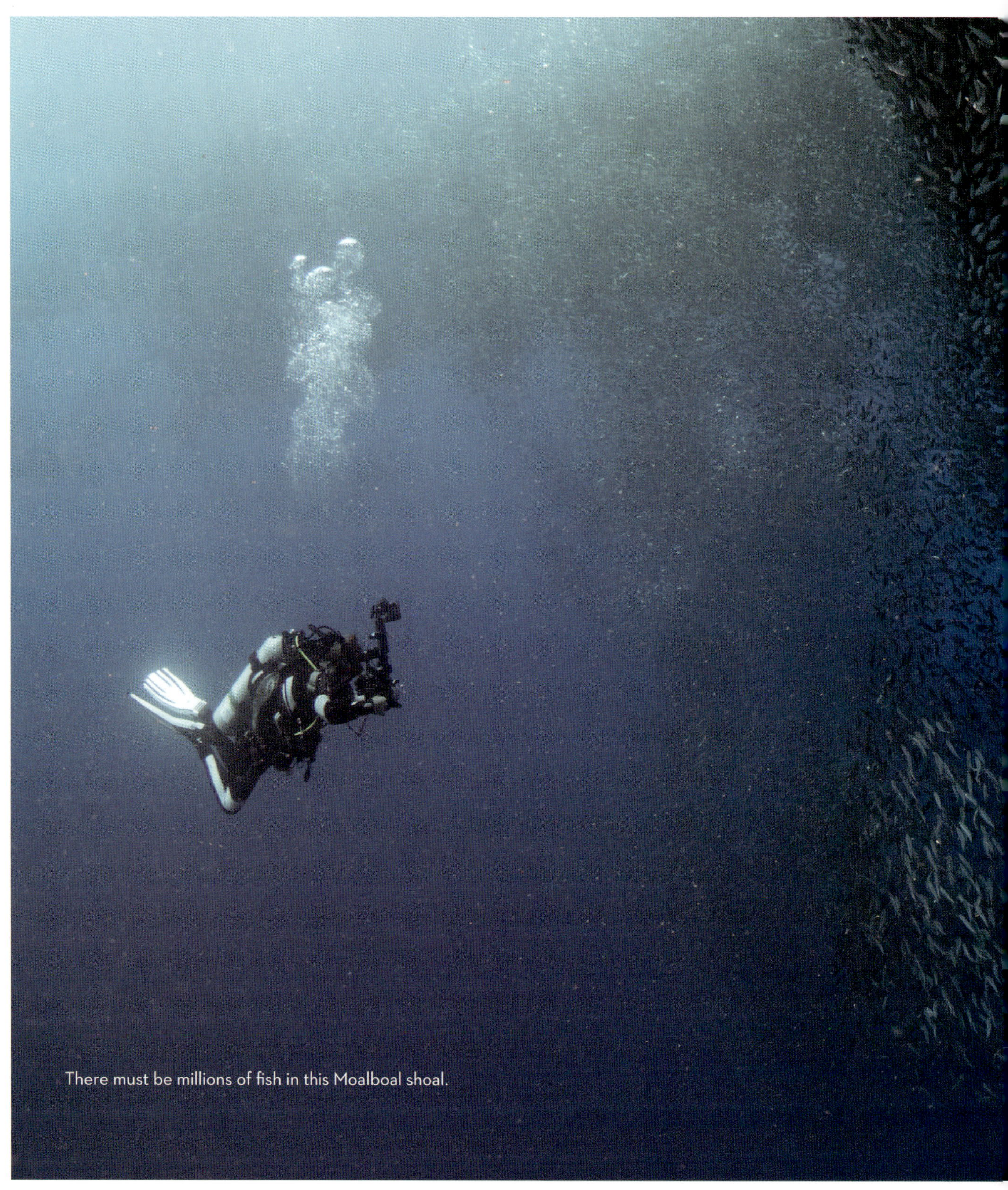
There must be millions of fish in this Moalboal shoal.

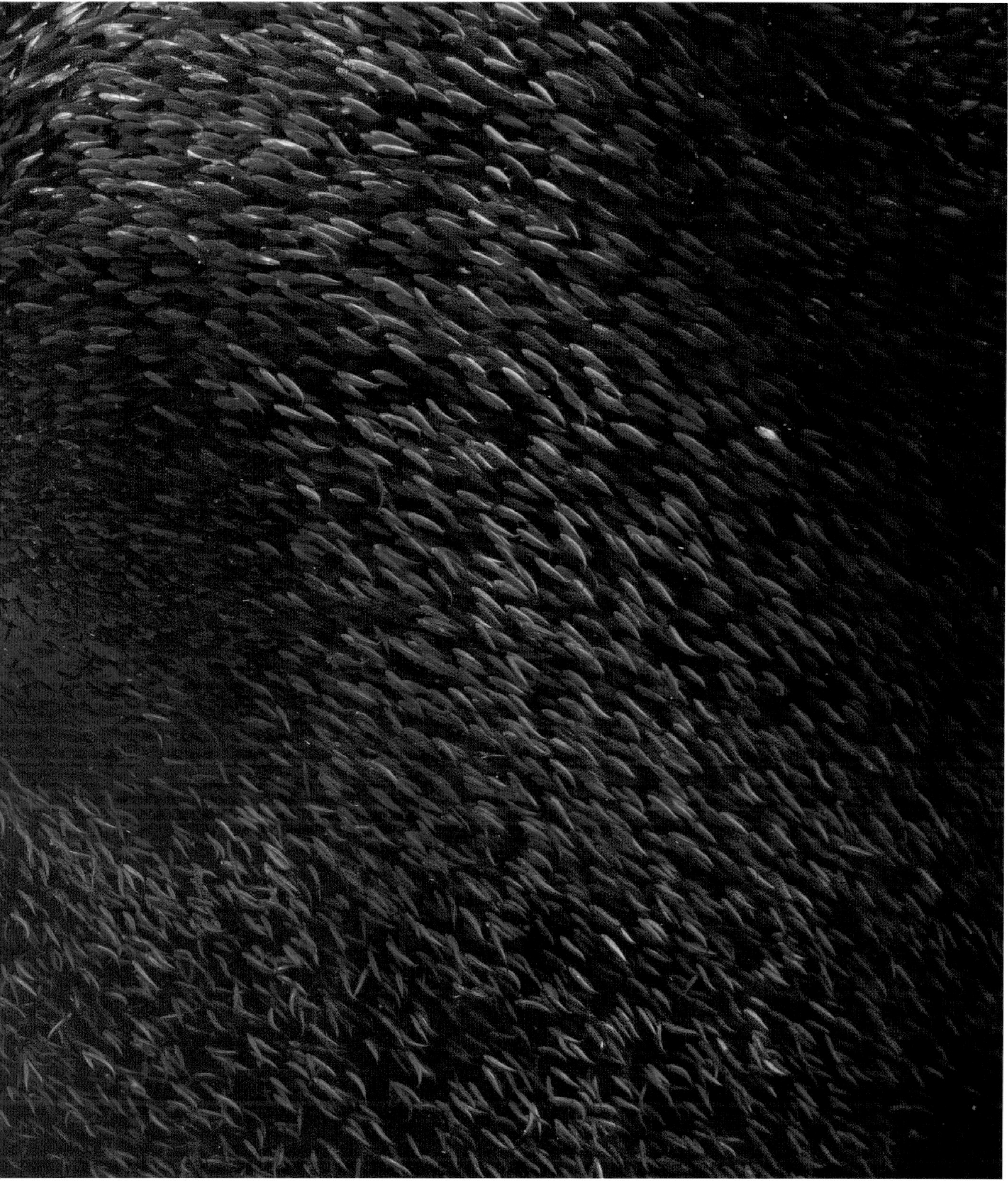

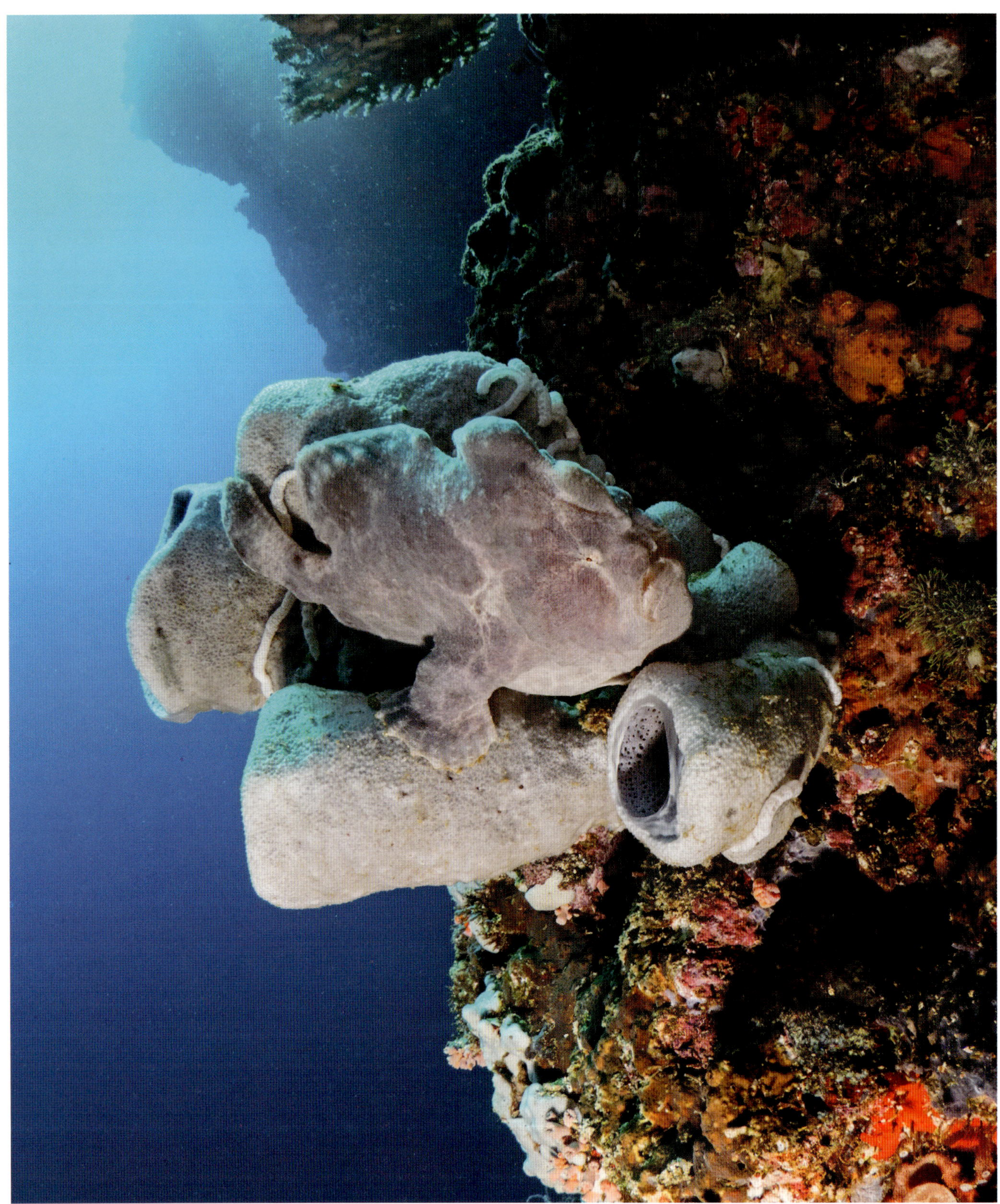

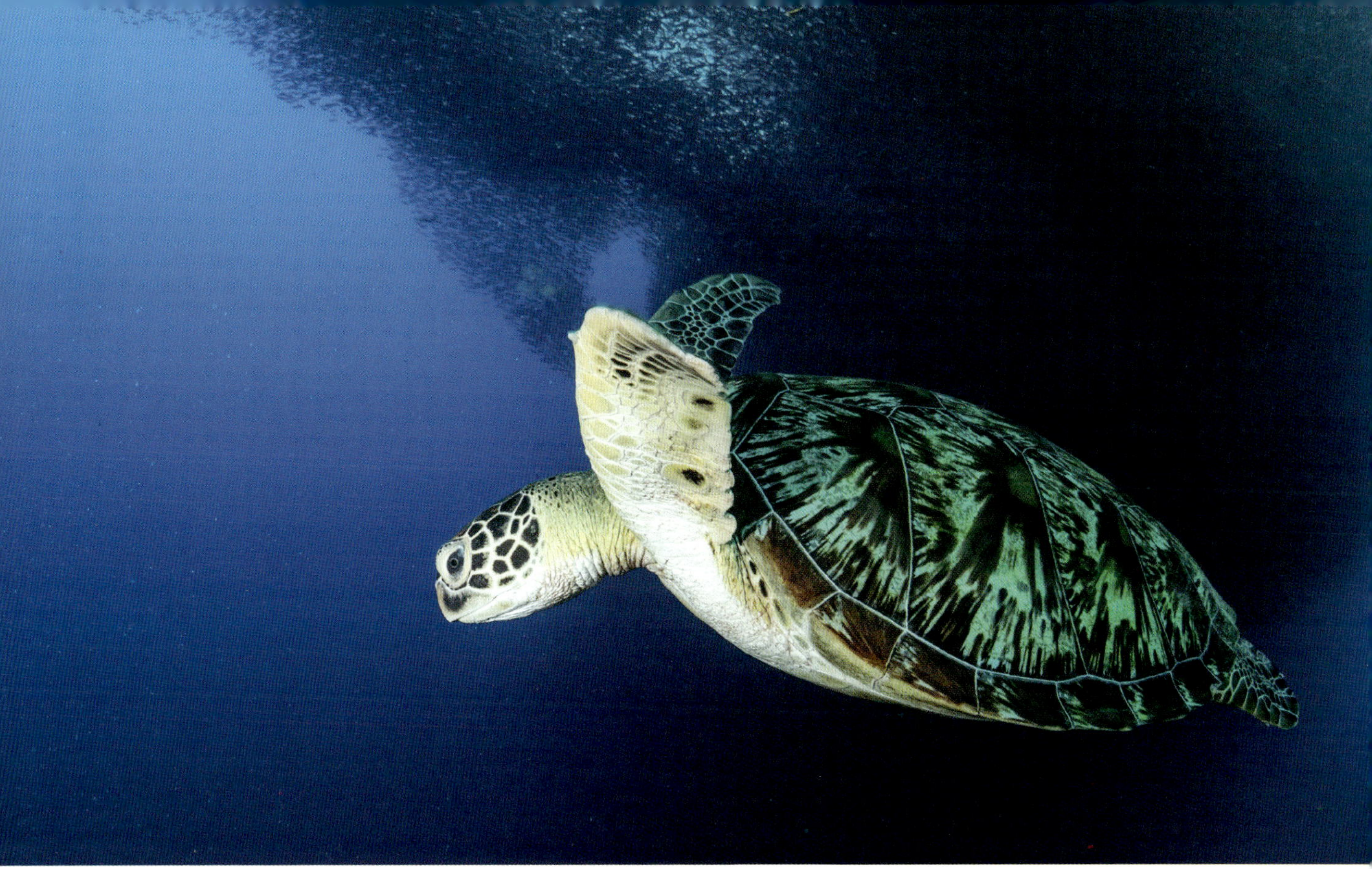

come to see – a super-shoal of sardines swimming as one, shifting and changing shape as predators or divers swam among them. There must have been more than a million individuals creating this cloud of sardines that stretched from the surface down to about 8m (26ft) deep and covered tens of metres in length. I moved below the school and looked up. It was at this point that I realised just how much fish poo a million fish can make! This is perhaps the worst aspect of diving the sardine shoal, as the current can be non-existent and hence the visibility can be quite badly impacted. However, if there is a small current running, keeping upstream can give some superb photographic opportunities.

At one point a turtle cruised into midst of the fish, causing them to part and make way as it headed to the reef to feed. As divers drifted into the school, the fish would swoosh and swirl around them. I had heard that thresher sharks sometimes hunt here, and recently scientists have discovered that they use their tail like a whip to stun their prey. Such an encounter would truly have made this one of the best dives I had ever done, and while I kept my eyes open looking out in the blue, I had no luck this time. That just means I will have to go back and try again.

OPPOSITE
A frogfish sits on a sponge, perfectly camouflaged.

ABOVE A turtle swims through the huge school of fish.

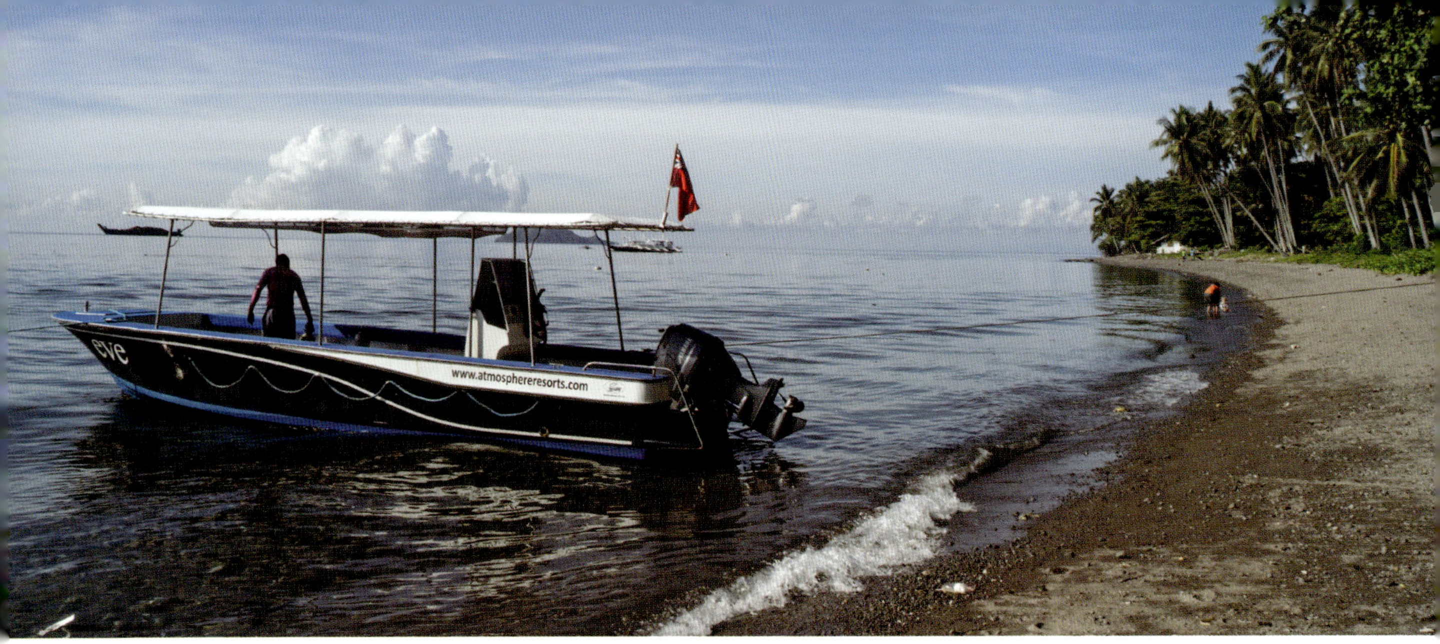

09 MATING MANDARINFISH IN THE MOONLIGHT

PHILIPPINES

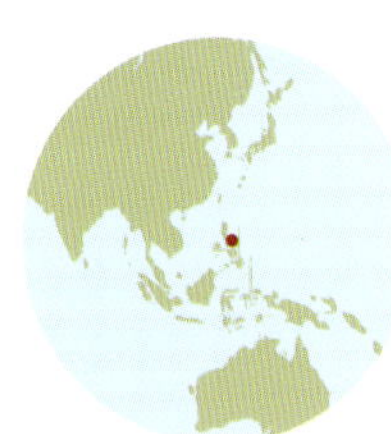

ABOVE A small speedboat waits to take us out to the dive sites.

OPPOSITE Mating Mandarinfish only brave open water for a matter of seconds.

Dauin is a small coastal town close to the city of Dumaguete on Negros Oriental island, not far from Cebu in the Philippines. The majority of dive sites in this area are based around Dauin, and it is essentially one huge site along the shore where muck-diving and critter-hunting are the norm. Whilst muck diving is what this region is known for, a short journey by road or sea can take you to some fabulous reefs where wide-angle subjects such as turtles and schools of jacks can be found.

The sandy beaches are lined with palm trees that reach out over the calm waters of the Bohol Sea. This is a volcanic island so white beaches are not a feature – instead it has darker volcanic sand. There is no shortage of relaxing resorts for non-divers to relax in, so it is a popular destination for divers with non-diving families. This region is ideal for a dual or multiple resort trip and the Whale Sharks at Oslob, while controversial to some, are a relatively short journey from here. The area is renowned for its outstanding beauty, but also offers the visitor an excellent chance to immerse themselves in local culture too. The city is known to be a safe place to wander and explore and violent crime is virtually unheard of.

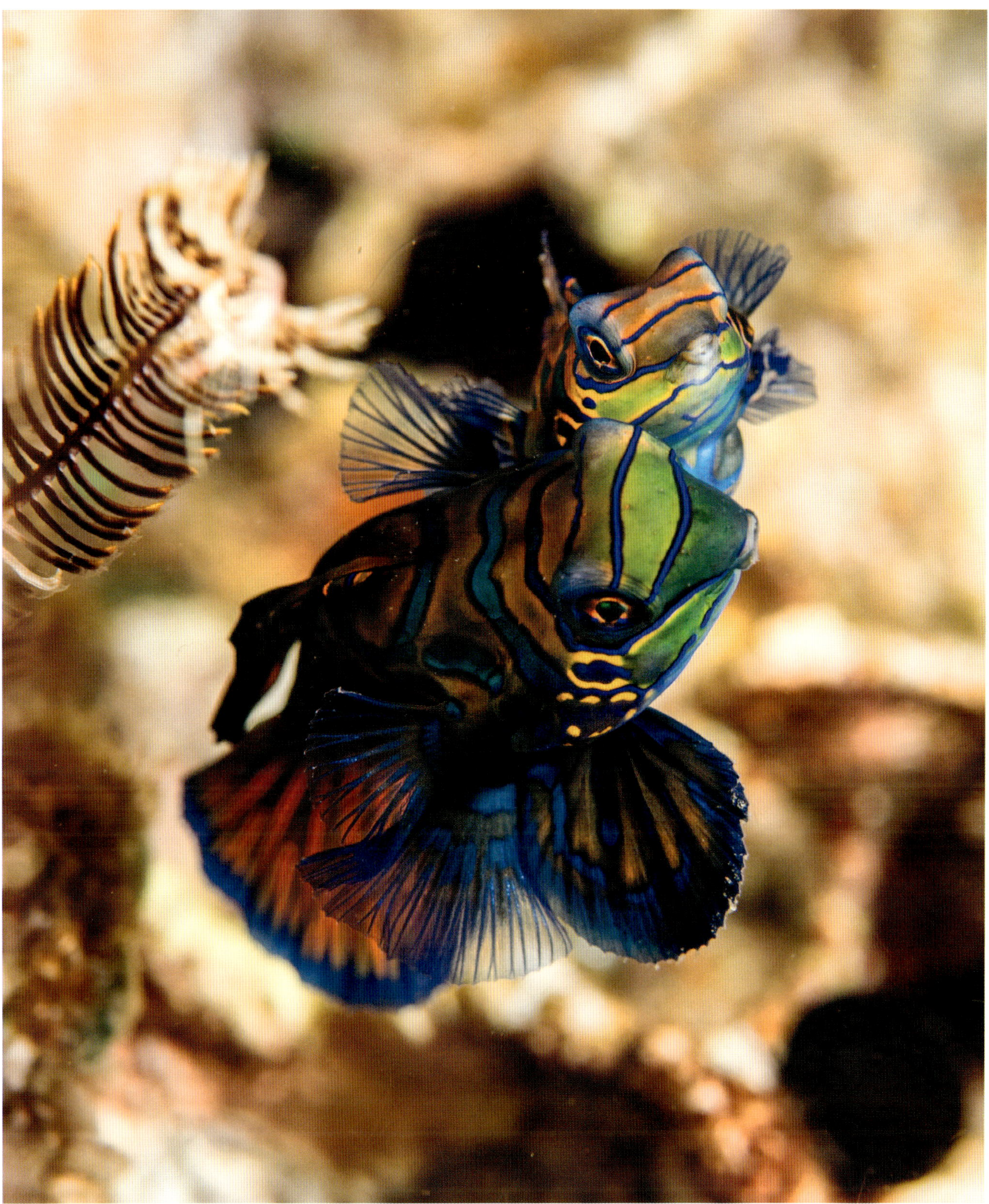

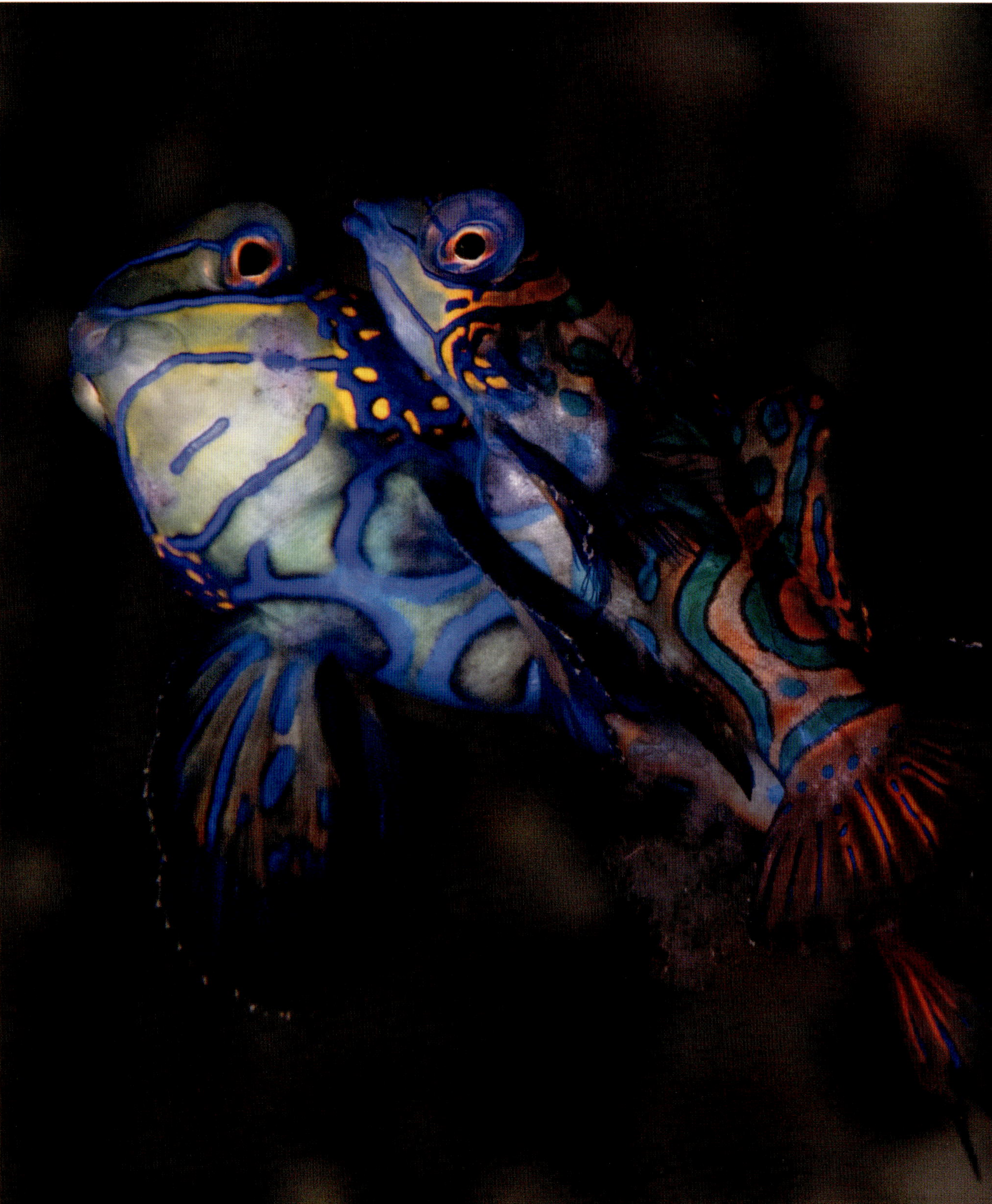

GENERAL DIVING

These waters are famous for rare macro critters and are act as a magnet for underwater photographers searching for marine life such as frogfish, nudibranchs, seahorses, crustaceans and cephalopods. The nearby Apo Island, with its stunning coral reef, steep walls, schooling fish and sleepy turtles, is just a short boat ride away. So this really is a destination that has something for everyone.

THE WILD DIVE

Mandarinfish are members of the dragonet family and are arguably the most beautiful fish in the sea. However, they can be particularly challenging to find and even more tricky to photograph. They hide amongst dense, usually dead, coral during the day, trying not to be spotted by divers and predators alike. They are extremely shy and are diminutive in size too. So many divers have tried to find and watch them, but rarely get the opportunity to see them.

When our guide offered us the chance to try and find mating Mandarinfish we were a little wary, having tried so many times in the past and frequently being disappointed at the lack of sightings or the frustrating time we have had trying to get a focus on these fast-moving, elusive creatures and to capture their images. However he seemed very positive, with the moon being in the correct phase and other divers having had success the previous evening, so we decided to do another dive just before the sun was beginning to sink in the sky. It is at dusk that these tiny but beautifully patterned fish come to life, and as they tend to hide from torchlight it requires patience and a keen eye to locate them. It takes a while to 'get your eye in' so it is advisable to arrive at the site while there is still some light so that you can acclimatise yourself and watch for the first signs of these elusive creatures flitting in and around the dead coral.

Trying to see and photograph Mandarinfish requires you to remain very still and simply wait for the action to happen, so as we approached the right area I found myself a spot away from the other divers and let my eyes adjust to the light as it started to fade. It was not long before I saw a larger female darting between the corals and then, within a few moments, I spotted another female lurking close by. They move their fins at incredible speed and seem to dart back and forth almost like they are imitating the way in which a feeding hummingbird would move between flower heads. It makes keeping track of their movements among the coral in gloomy lighting very difficult. I switched on a red light so that I was able to see the fish and follow them without disturbing them as they scurried around.

OPPOSITE The peak of the action as the female releases her eggs.

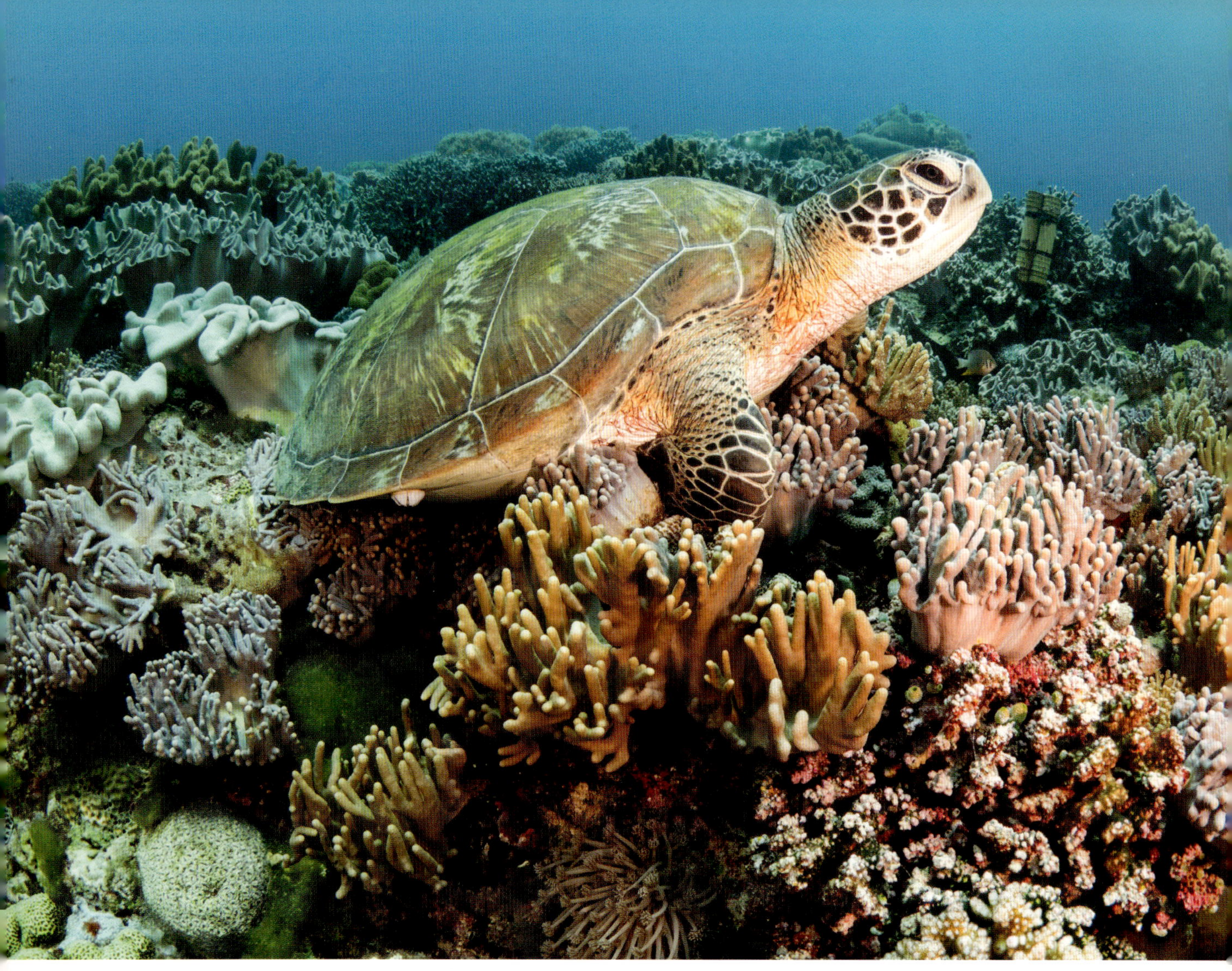

ABOVE Apo Island is famous for its turtles.

OPPOSITE Magnificent anemones are home to feisty anemonefish.

The bright white light I would normally use on a night dive would very likely have scared them away for the rest of the dive. As it grew ever darker it became obvious that the activity seemed to be increasing, and then I saw a male, much smaller than the females, weaving its way through the coral. The Mandarinfish only leave the protection of the coral for a split second for the actual moment of mating. First they meet in the shelter of the reef, and then if the female decides that the male is a suitable and fit subject to fertilise her eggs she darts vertically upwards, just a few inches out into open water. The male follows as if glued to her side and then, in an instant, they simultaneously release their sperm and eggs into the water before swiftly retreating back to the relative safety of the hard coral below. For a photographer this moment is known as 'peak of the action' and is one of those

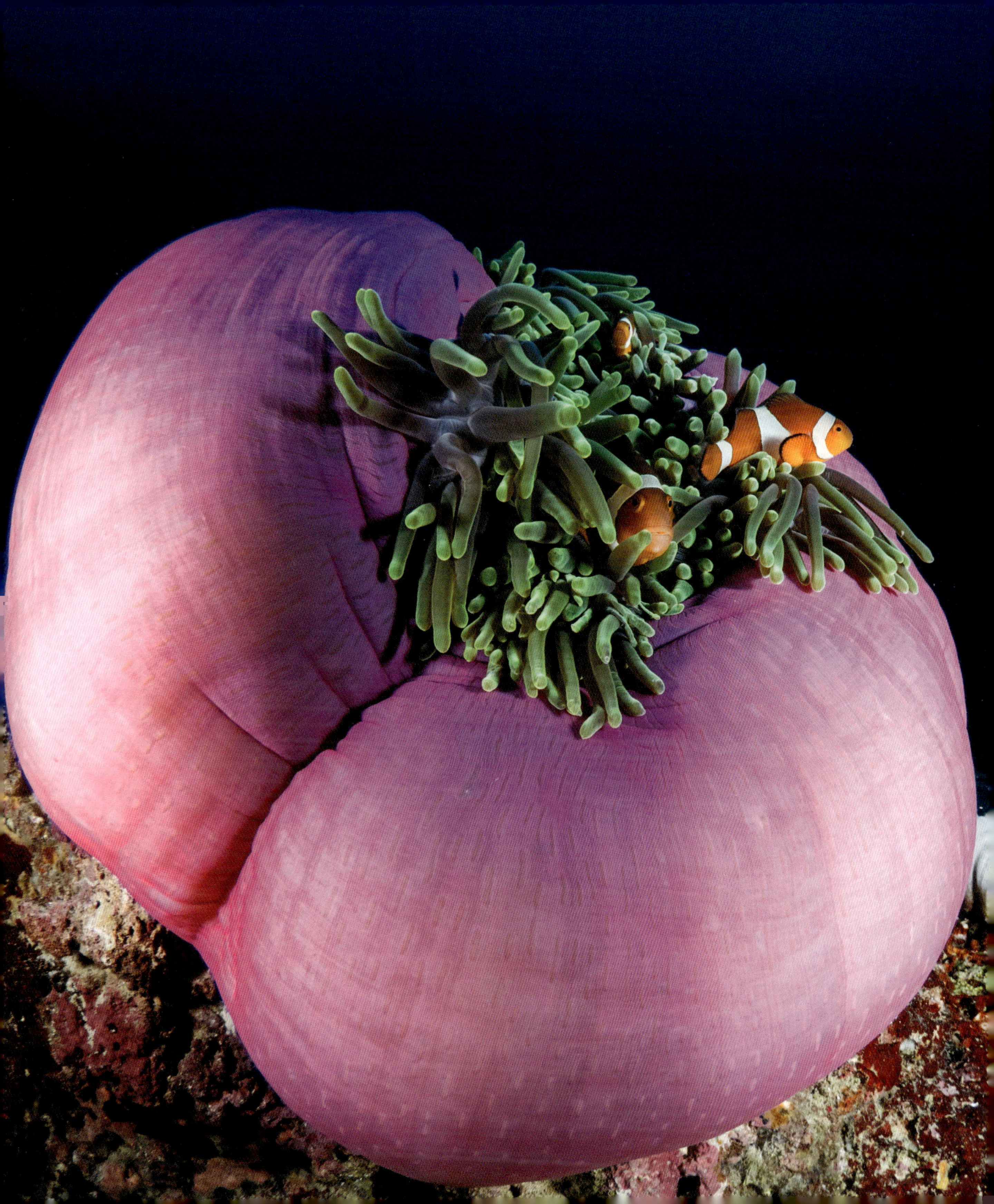

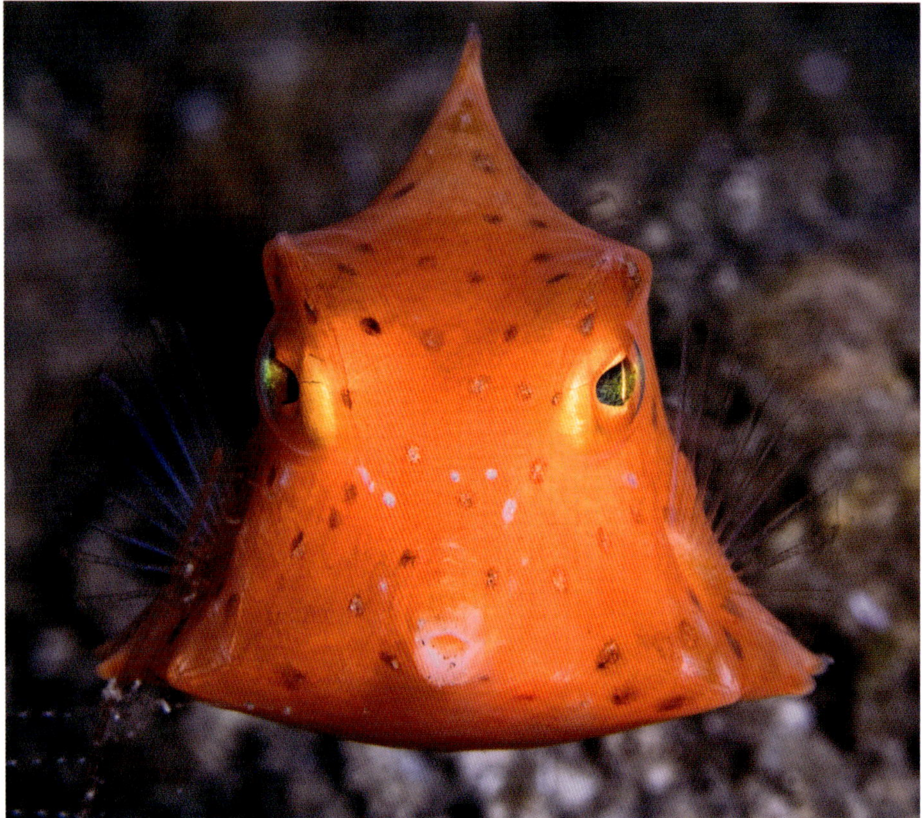

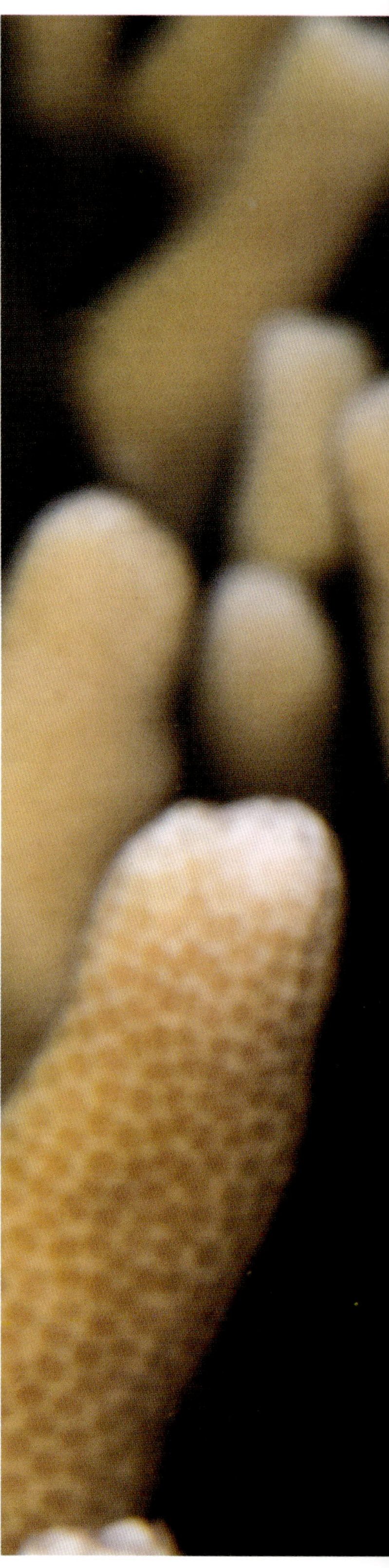

opportunities that all serious photographers are hoping for. This process is usually repeated several times and the male may mate with a number of females in the same evening. This at least gives the photographer a second chance at capturing that fleeting moment.

However, trying to follow the action in near darkness with fast moving fish that only stay in the open for a fraction of a second can make underwater photography frustrating. This time, however, I was in the right place at the right time, with an excellent position in the water for photography. Finally, when I looked down at my screen, I was delighted to see that I had managed to capture the very moment of release. It was an incredible bit of behaviour to watch and to capture on camera, and made up for all the frustrating dives I had done in the past in search of this species.

ABOVE Some fish make super cute models.

RIGHT Mandarinfish have to be the most beautiful fish in the sea.

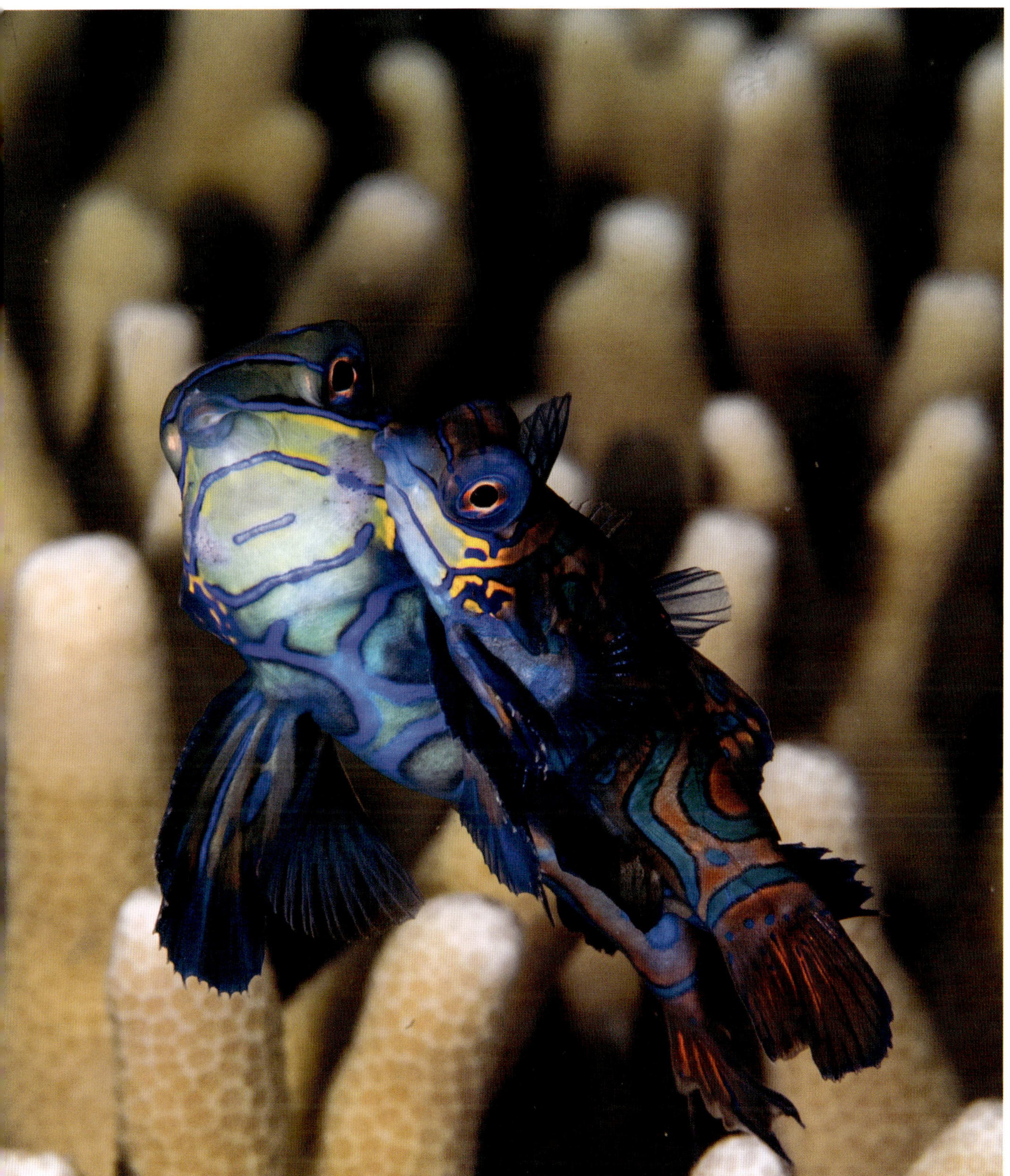

10 A SEA SLUG CALLED SHAUN THE SHEEP

PHILIPPINES

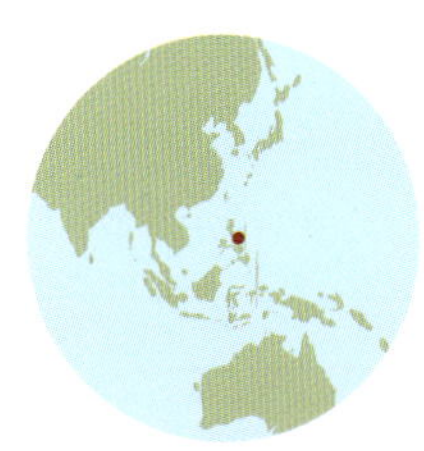

BELOW A headshot of Shaun the Sheep.

ANDA IS A QUIET DESTINATION on the leafy island of Bohol in the Philippines, located on a small peninsula on the eastern tip of the island. This is a place to relax, taking in the sights and smells in slow time, adapting to the more leisurely pace of life you encounter here. The forests nearby are one of the few remaining habitats for the Tarsier, a charismatic, diminutive primate with huge appealing eyes, and when here a visit to see them is well worth the effort. The area is also famous for its Chocolate Hills – more than a thousand hills that turn brown in the dry season, hence the name. It is usually the coastline and water that bring people here, though, and for good reason.

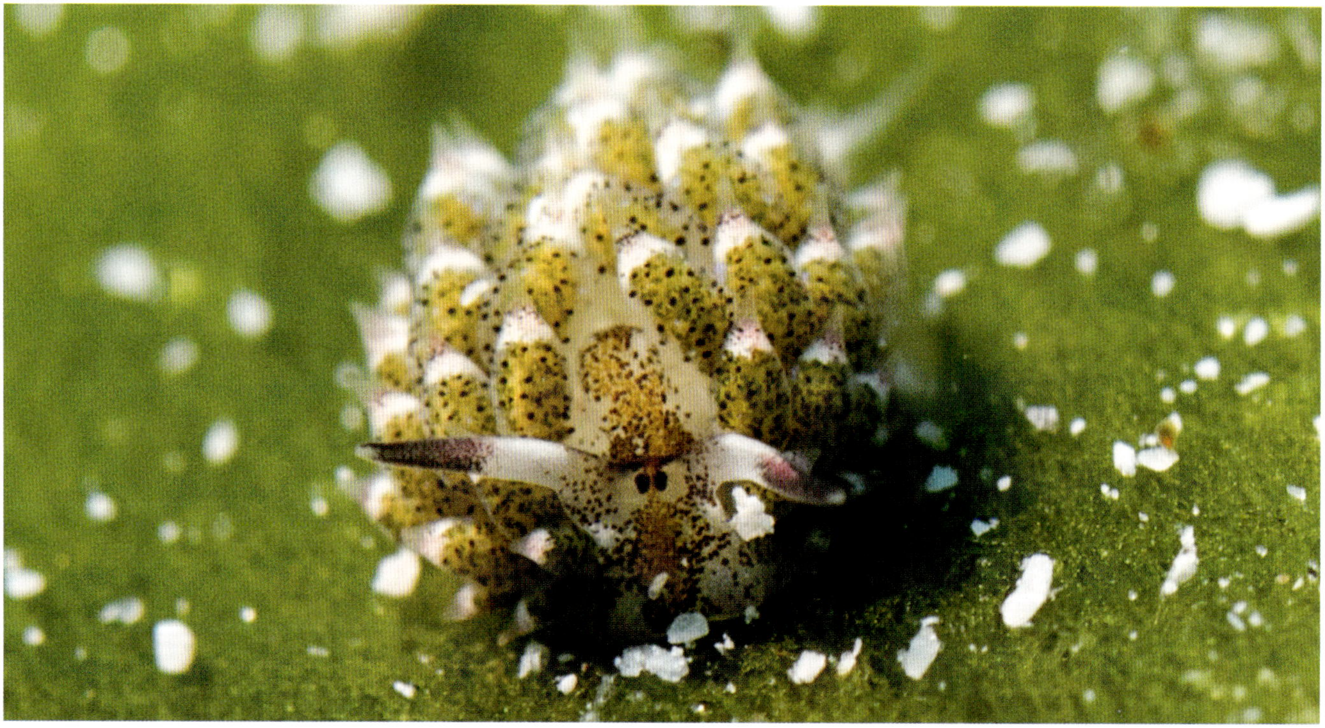

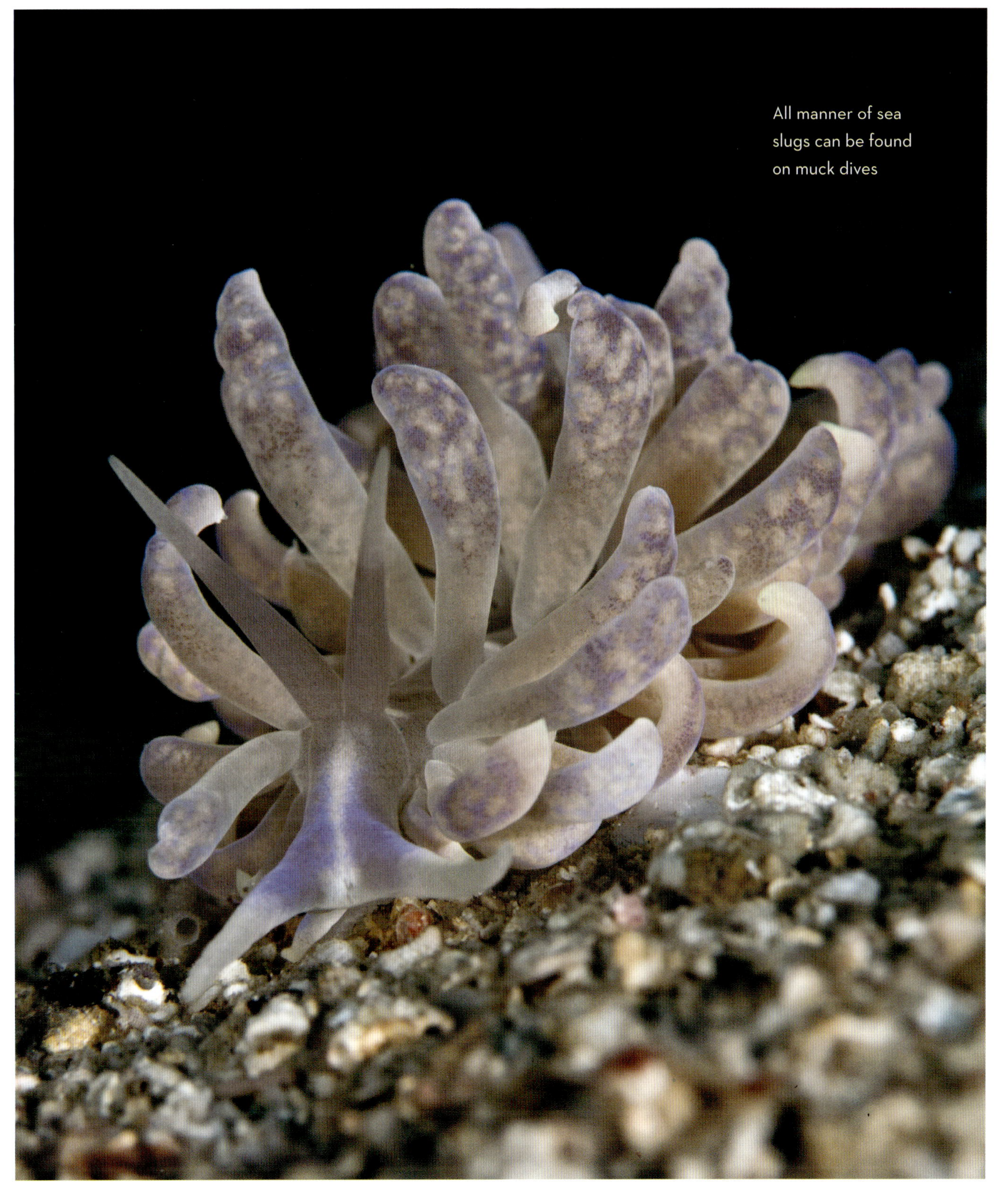

All manner of sea slugs can be found on muck dives

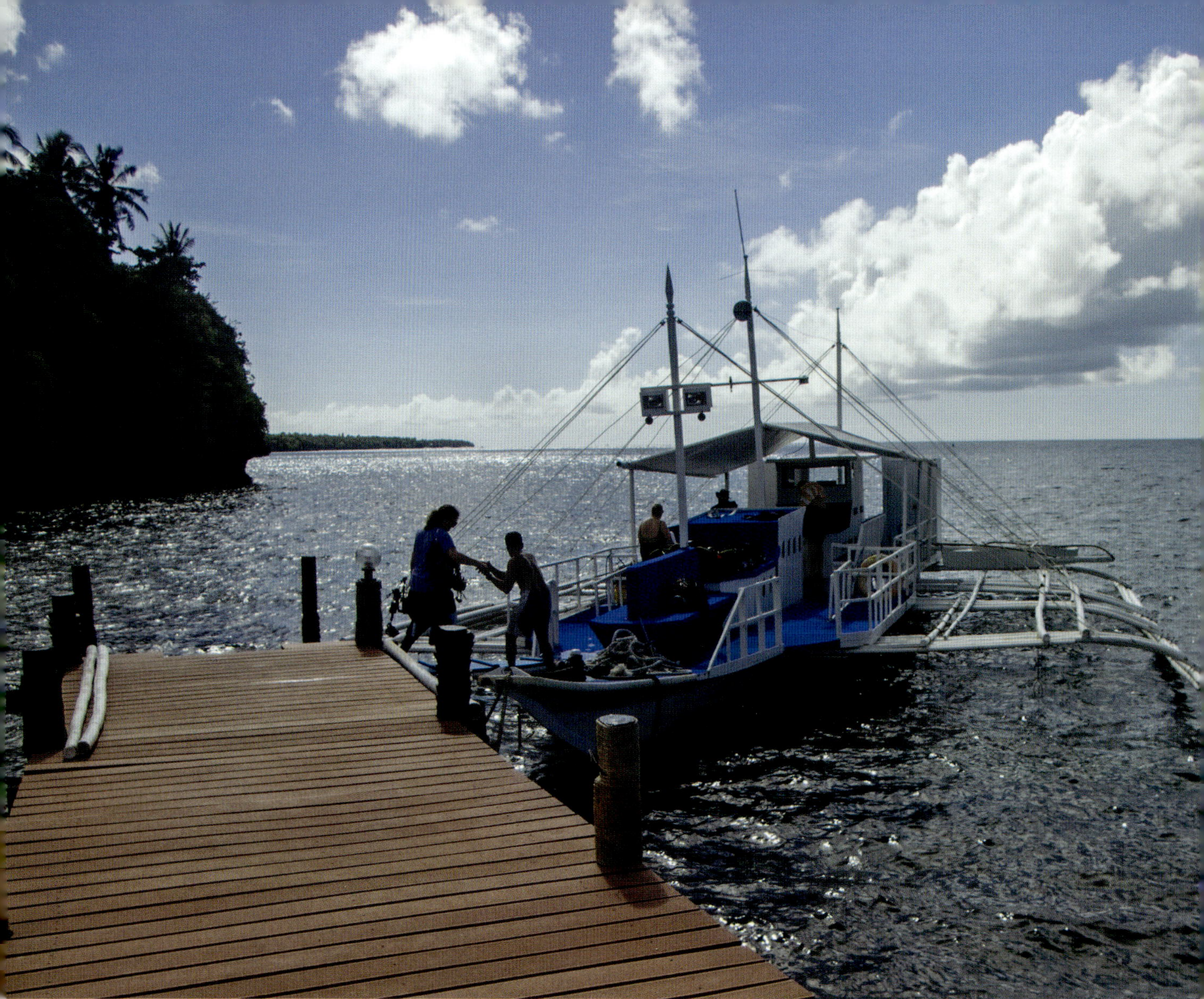

GENERAL DIVING

The diving here offers incredible opportunities for finding an array of tiny marine critters. The reefs are healthy and teeming with life, and the key to having a great dive in this region is to take it slowly. The more time you take to look at the reef in front of you, the more you will find. Turtles lounge on the top of the reef, making soft corals their bed as they snooze, only moving to head to the surface on occasion in order to breathe. As well as reef, there is great 'muck diving' here, where you find some of the most treasured macro subjects superbly camouflaged in the grey volcanic sand. Tiny frogfish, barely larger than a grain of sand, sit on

ABOVE A traditional dive boat waits to take us out.

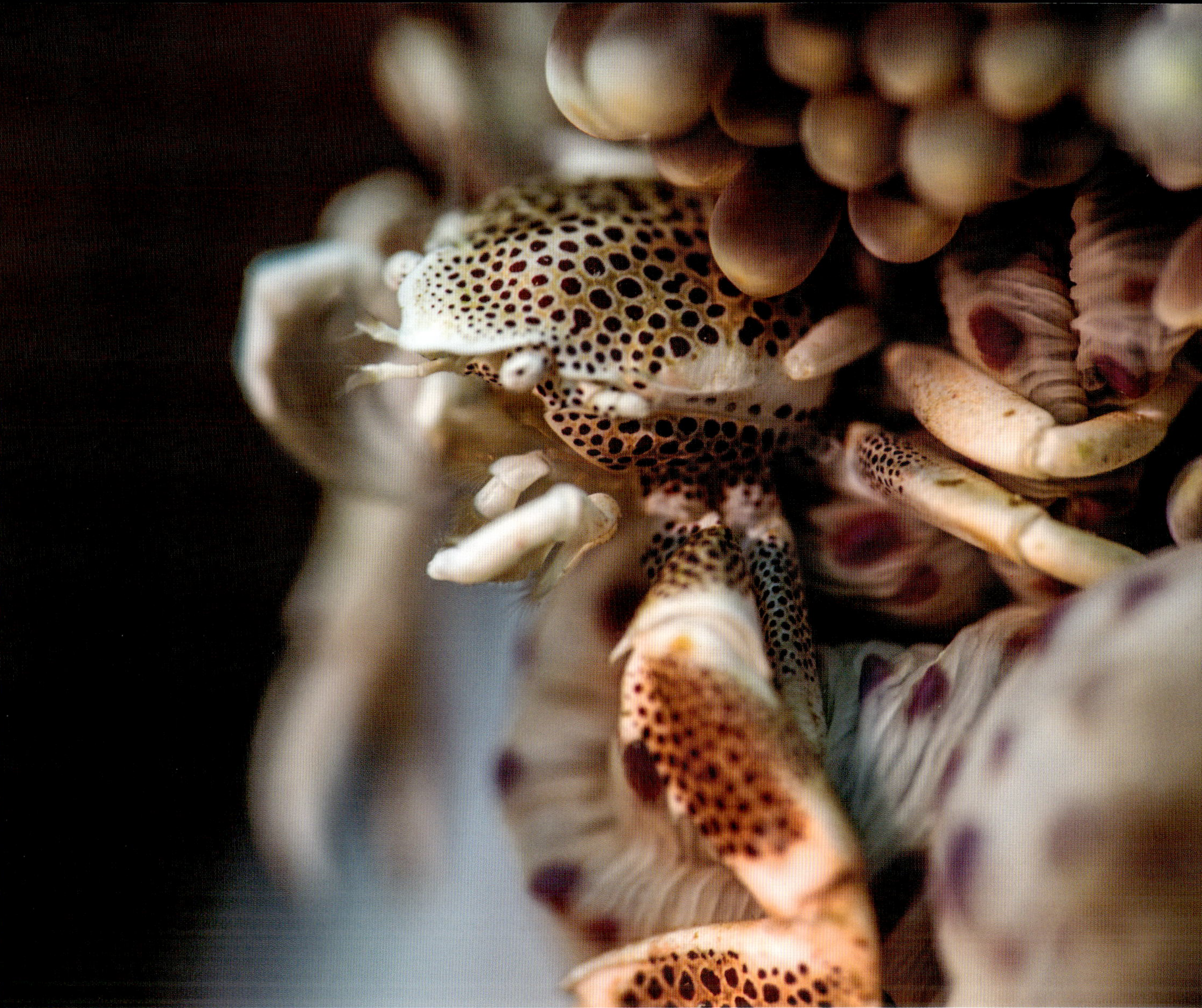

the seabed hoping for passing food to stray close enough to be within range of their lightning-quick strike. Their larger hairy cousins can also be found here, camouflage wafting with the gentle currents. Nudibranchs of all shapes, shades and sizes roam over the seabed and reefs, looking for their next meal. Pygmy seahorses hide in the beautiful pink sea fans. At night a whole new set of creatures take the stage, including elusive octopus and bright crustaceans. Hidden in the day, these weird and wonderful creatures come out as the sun goes down, and a whole new cast of characters take over the reef.

ABOVE A porcelain crab filters out food from water flowing past

THE WILD DIVE

Our guide asked what special marine creatures we might like to see as we set up our diving equipment for our first dive here. It is a tough question to answer as there are always so many things we would like to encounter. This time it was easy – I had been wanting to see a 'Shaun the Sheep' nudibranch for quite a while and I had heard that this was a good place to find them. As I asked the guide, he laughed and told me that was too easy as they were very easy to find on the dive site we were heading to. He agreed that it would be the first thing he pointed out as we headed under water. I was delighted and excited to finally be able to see and photograph this now iconic sea slug.

The Shaun the Sheep nudibranch is named after the animated character created by Aardman Animations. They are usually found in shallow water on seagrass and harness the plant material they eat to be able to photosynthesise. They are also known as the 'solar powered nudibranch'!

The dive was to begin in a shallow sandy seabed with patches of seagrass. Our guide was keen to get started and agreed to meet us underwater in a few minutes, so that he could start looking for Shaun the Sheep while we got ourselves ready for the dive. We jumped in and soon heard the familiar sound of the guide banging on his tank with his metal pointing stick. We finned over to him and he pointed carefully down to a single blade of seagrass. I peered down to where he was pointing, but I could not see what he was getting so excited about. I signalled

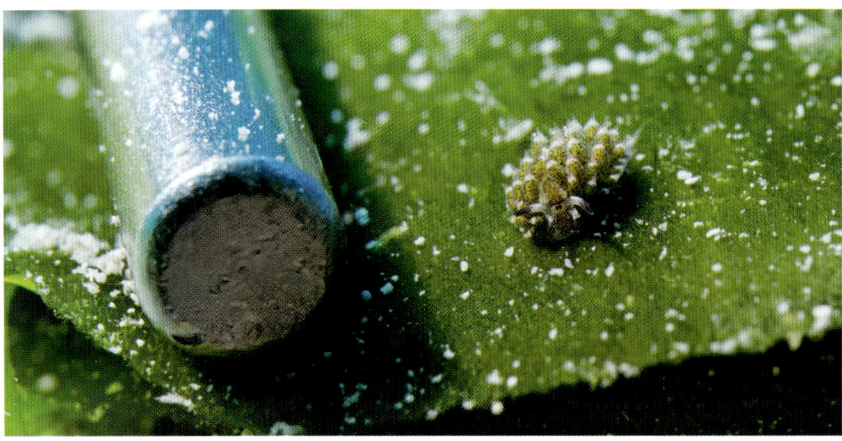

ABOVE Our guide lowers his pointing stick to show us Shaun the Sheep for the first time.

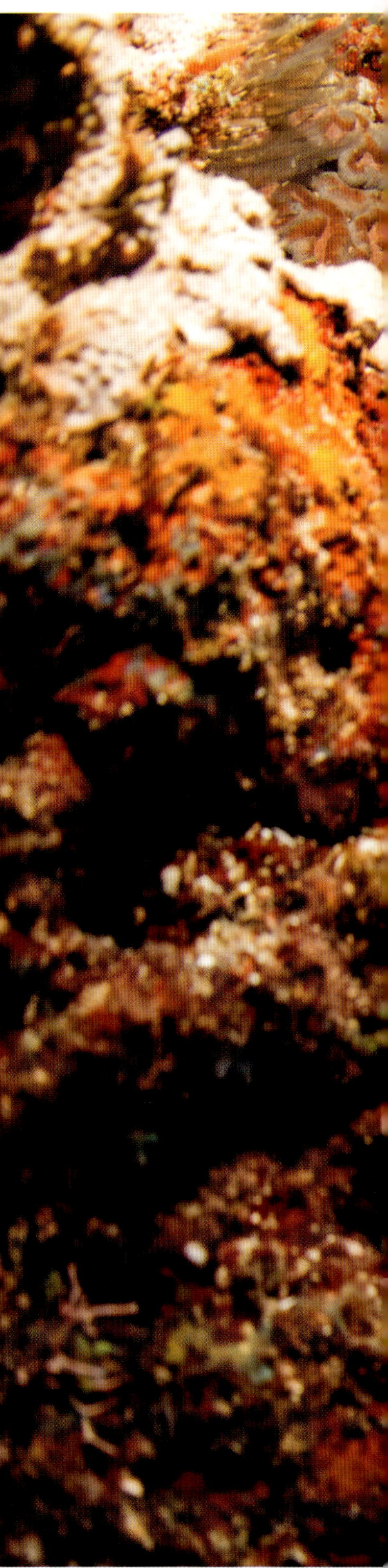

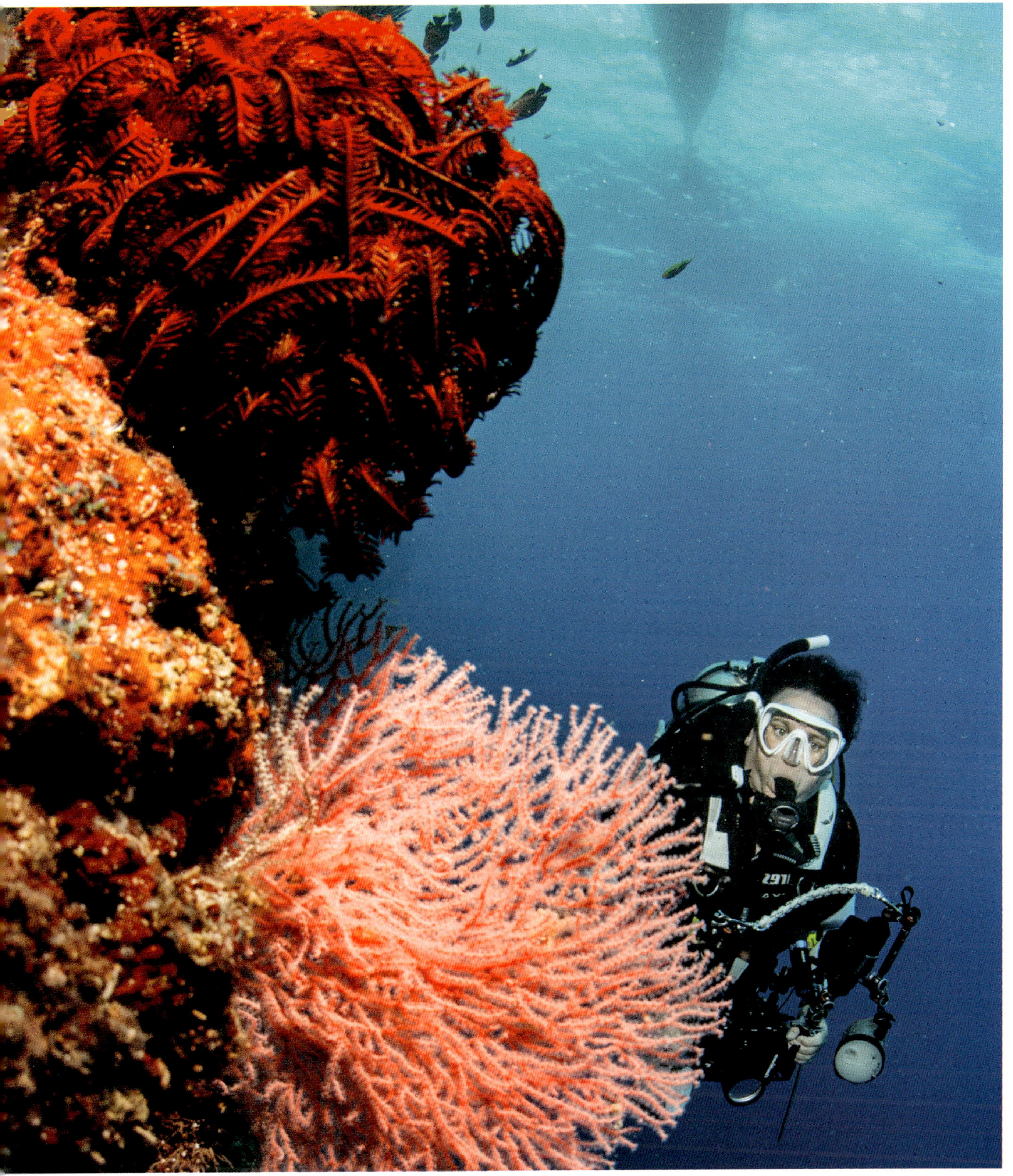

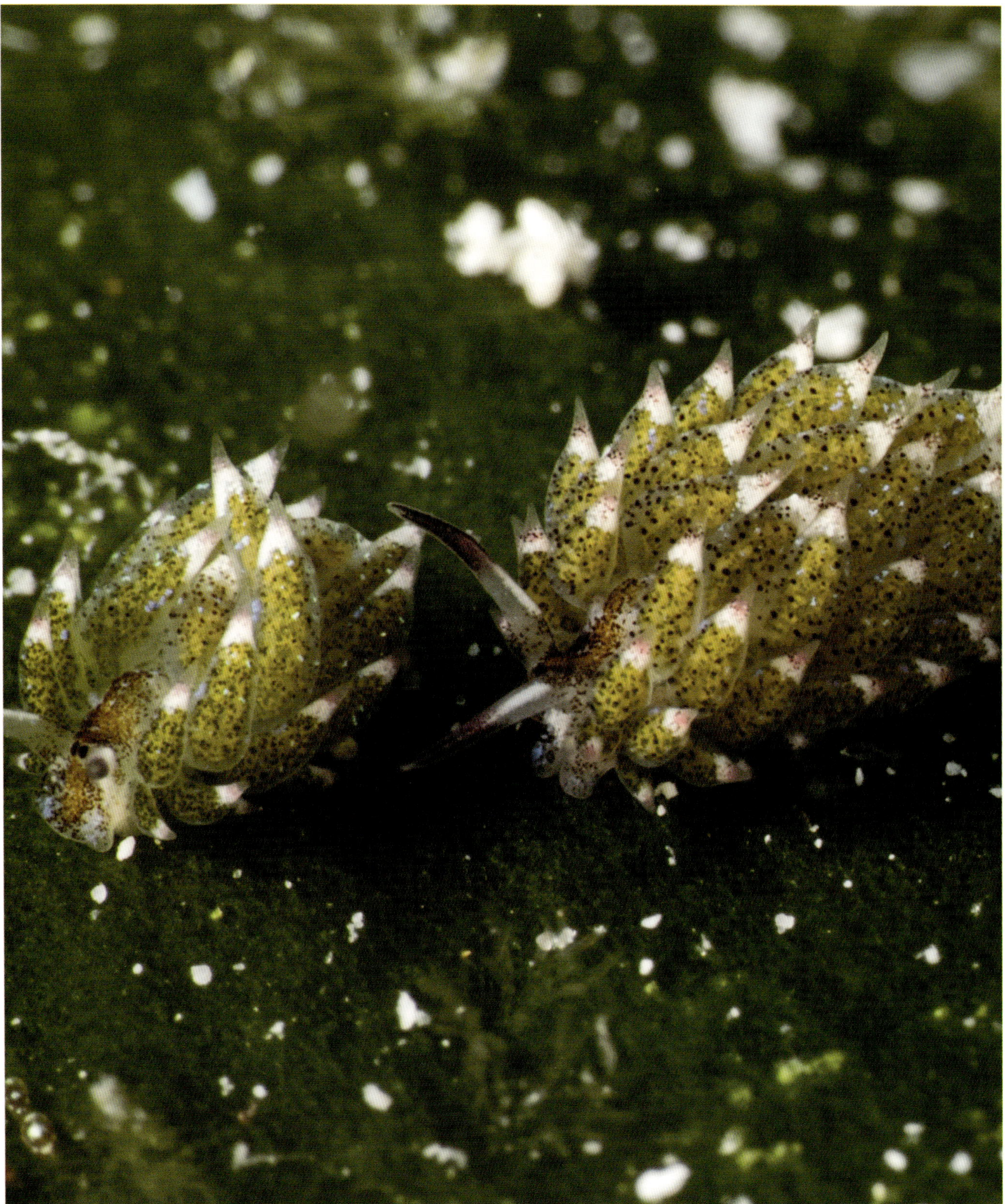

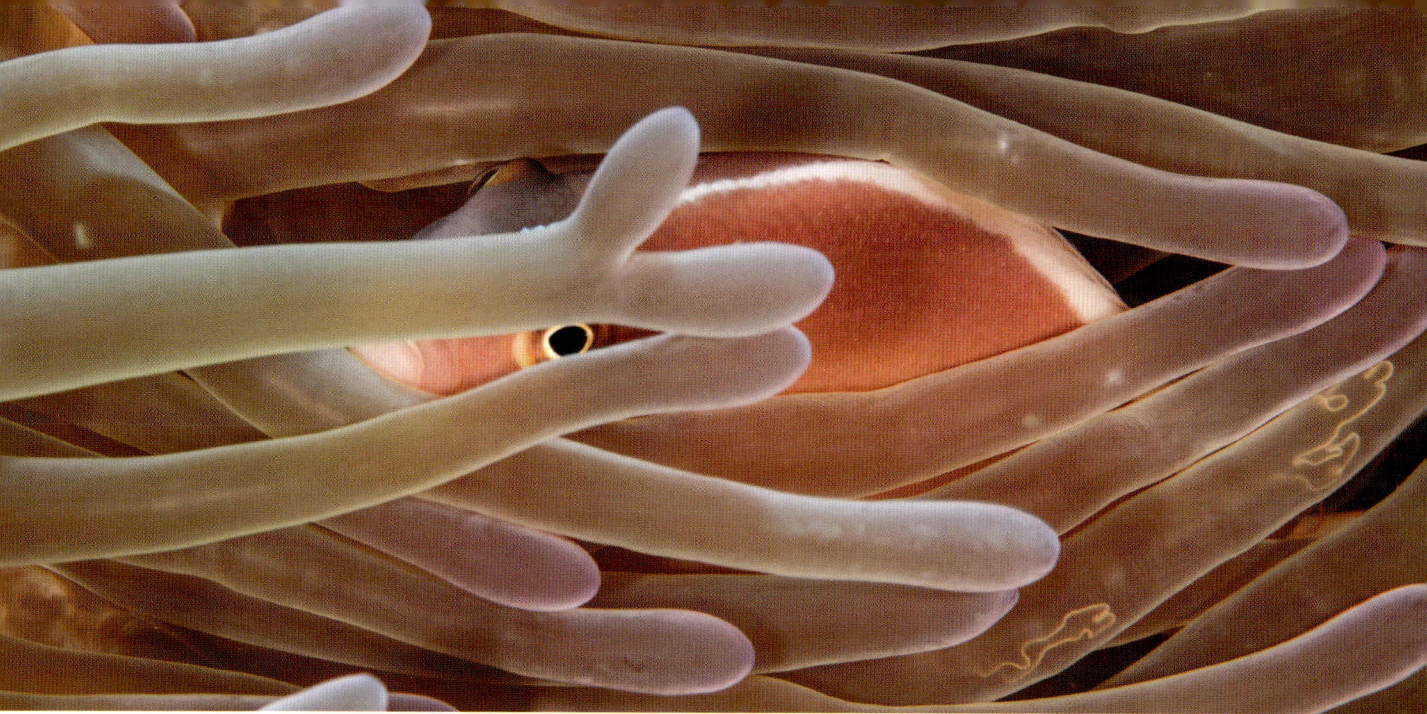

this with a shrug of my shoulders and he pointed again, this time more vigorously. I carefully got closer, trying to see it without disturbing the sand below. If I really squinted I could see a tiny white blob. I had not appreciated just how small these nudibranchs were. Much smaller than a grain of rice, I finally started to make out the outline of the nudibranch. I raised my camera to try a take a photo, but the guide started waving frantically at me and shaking his finger as if I was doing something wrong. I paused and waited for him to give me another signal. He pointed at his bottom – not a normal underwater signal, but I quickly caught on that I had been about to photograph the backside of the nudibranch. Slightly embarrassed, I gently repositioned myself at the other end of the small blurry blob and looked at the guide for approval. He gave me a nod and I positioned my camera, very carefully lined up the focussing cursor and fired. Then, as things started to come more into focus, I saw that there were actually two of these tiny but delightful sea slugs on this single frond of sea grass. The guide swam off and found several more Shaun the Sheep nudibranchs so that soon each diver had their own subject to marvel at and attempt to photograph. As a fan of the animated films, I was really chuffed to have finally seen this perfectly named sea slug.

A good guide can be invaluable, especially if you are of a certain age with eyesight that could be better. They know where to find the critters you want to see and have razor sharp, well trained eyes for spotting the amazing tiny critters that have divers flocking to this region. I was certainly very glad that mine was so understanding.

OPPOSITE Two Shaun the Sheep Nudibranchs.

ABOVE An anemonefish peeks out from behind the protective tentacles.

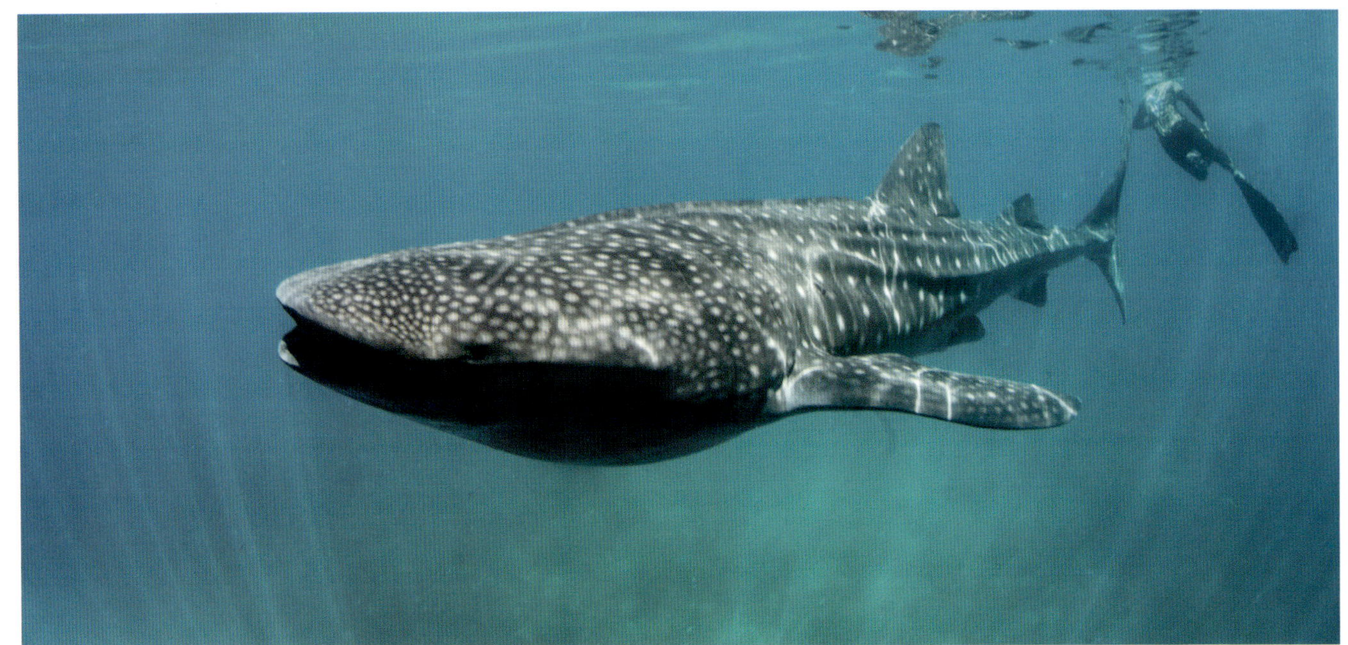

11 UP CLOSE WITH THE WHALE SHARKS OF OSLOB

PHILIPPINES

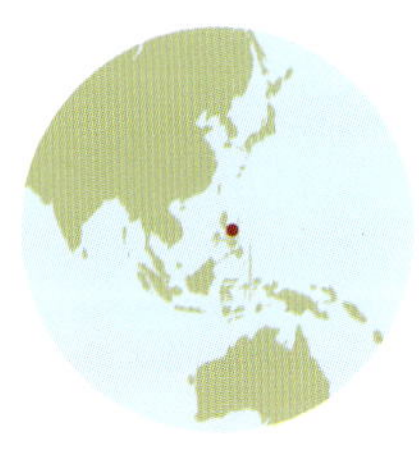

ABOVE A free-diver from the research team gets up close to one of the sharks.

OSLOB IS A MUNICIPALITY of the island of Cebu in the Philippines. It is surrounded by rolling hills covered in lush vegetation. Now a small village on the coast has become a buzzing hive of activity as people flock to see the biggest fish in the sea – the Whale Shark. Tourists tend to come from Bohol and Dumaguete, as well as from Cebu, so arrive by both car and boat to get a chance to see the star attraction.

GENERAL DIVING

The Whale Shark attraction has caused controversy since its inception in the early 2010s. In the 1980s (and for many years before) these majestic sharks were hunted mostly for their fins. Records show that some 100 Whale Sharks per year would be

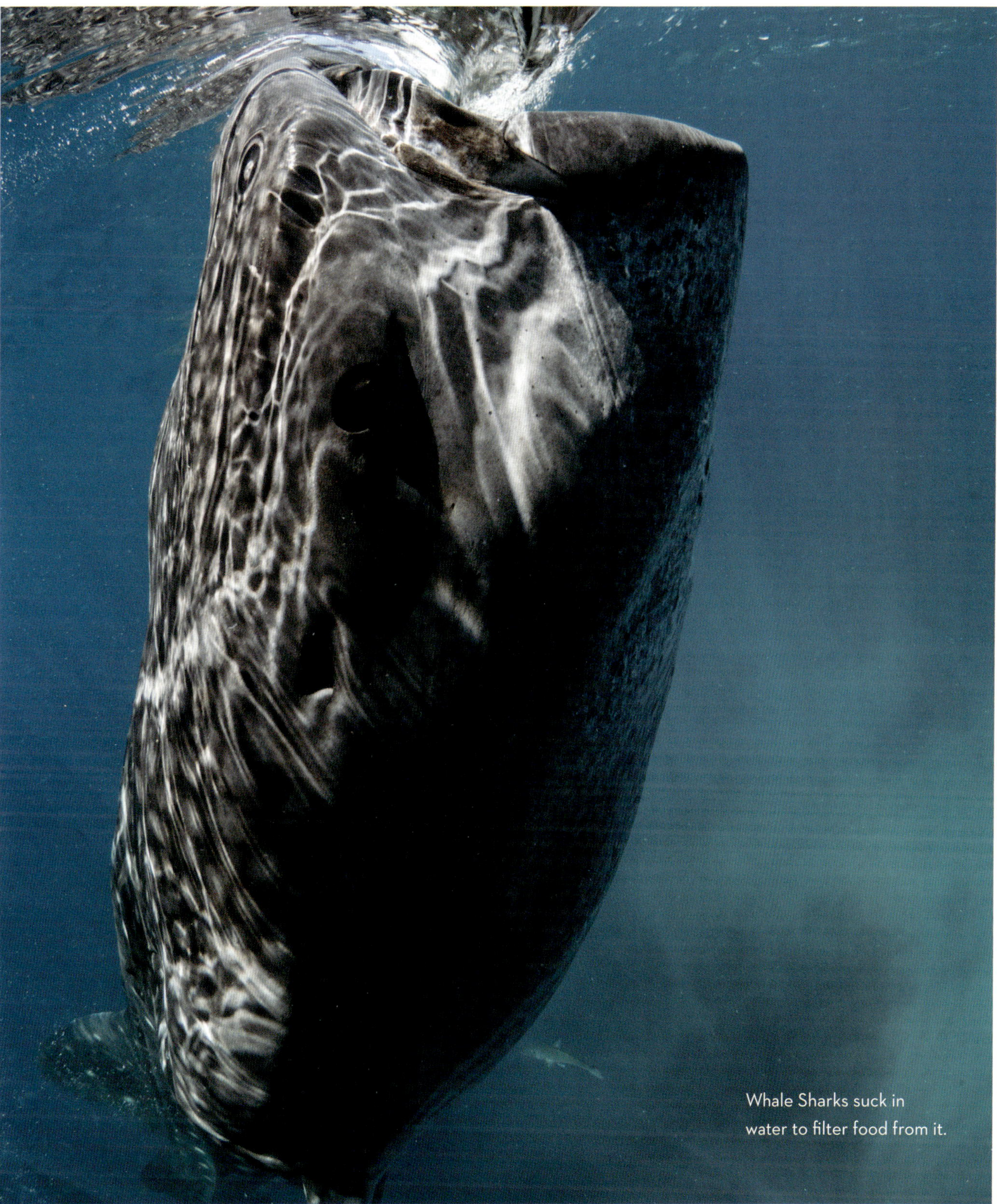

Whale Sharks suck in water to filter food from it.

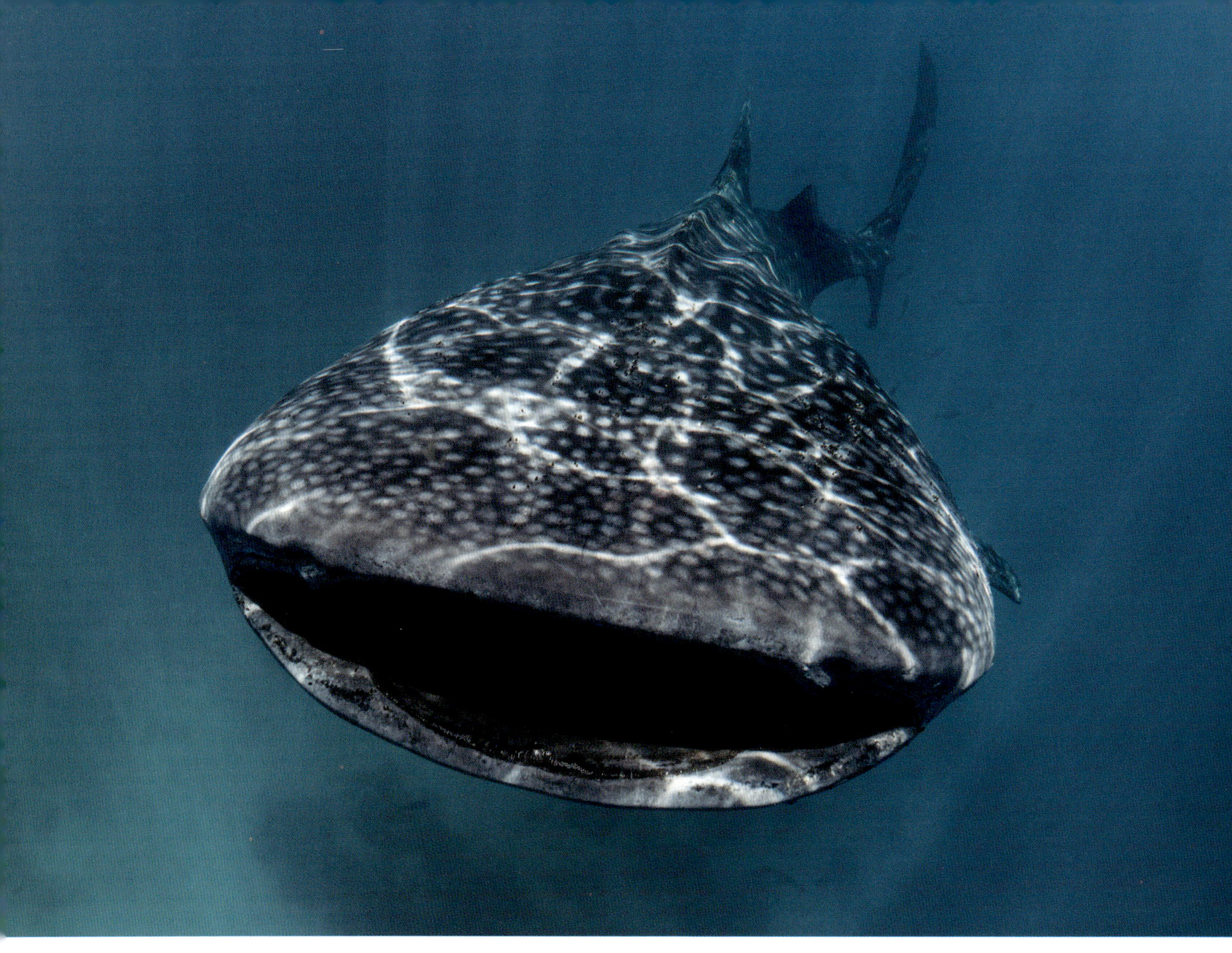

ABOVE They have huge mouths to filter as much food as possible.

killed in the region and so it was widely welcomed when the Philippines became one of the first countries in the world to protect them in 1998. The local fishermen here use small shrimp as bait and the Whale Sharks would 'suck' at their fishing gear and nets to try to steal this food. As it was now illegal to hurt or kill them, instead they discovered that a single boat could lead them away by throwing small handfuls of the shrimp into the water, leaving the other fishermen to work without these giants getting entangled in their nets. Soon the story was out, and divers, tourists and underwater photographers flocked to the area to see the Whale Sharks. An industry was born.

Initially there were no regulations and stories of people riding and grabbing the Whale Sharks were common. This is how this experience acquired its poor

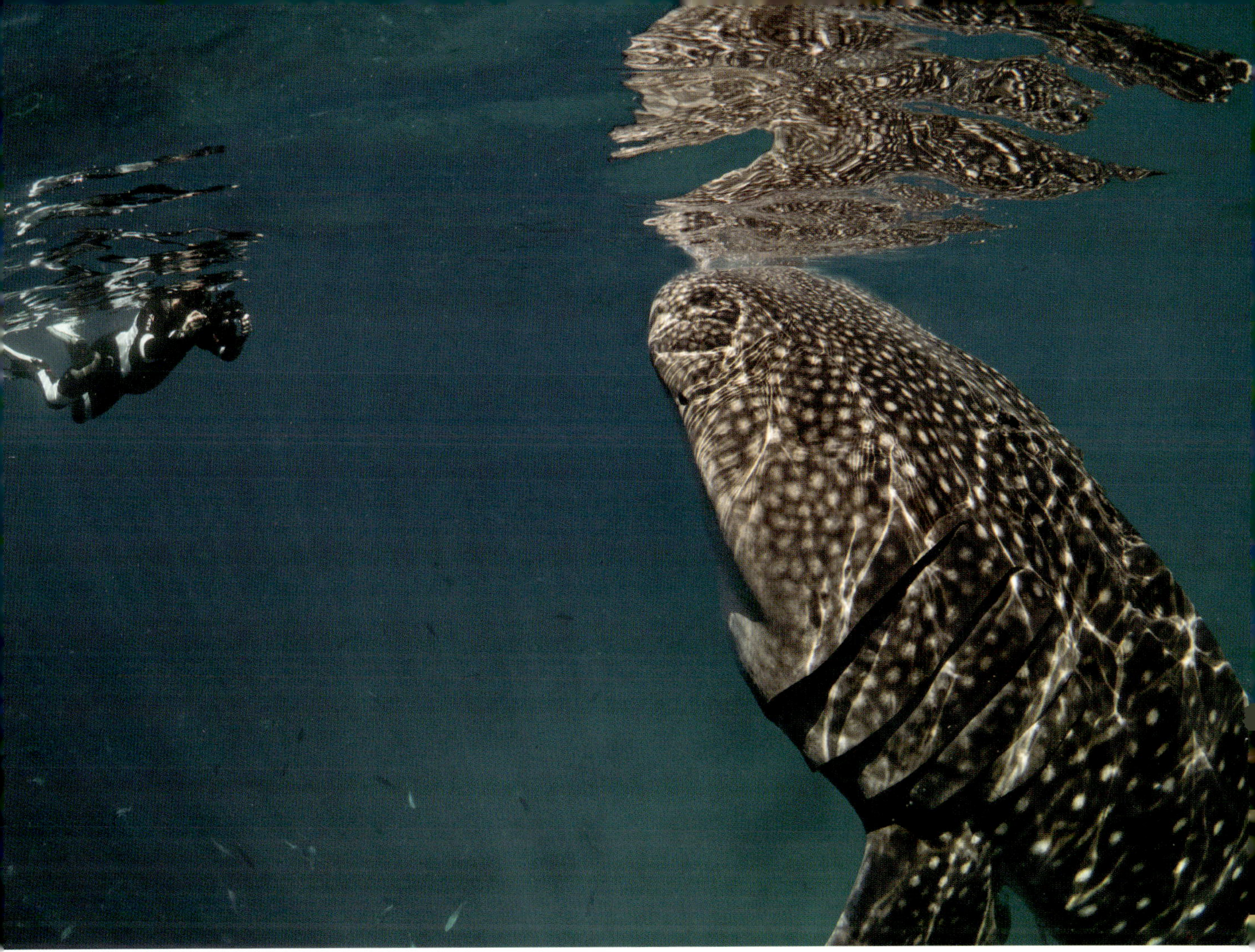

reputation. Nowadays there are strict rules laid out to try to protect the Whale Sharks more effectively. A specialist charity is on site to document interactions and to add each individual shark to a database of visitors to the area.

It is certainly much better organised now than in the past, with more protection for the sharks, but there are still concerns for the Whale Sharks including behavioural changes, dietary concerns and injuries. However, this phenomenon has brought much needed money to an impoverished area and allowed many people who would never normally get the chance to see a marine giant up close – an experience of a lifetime that may encourage greater conservation efforts in the future. One thing is for sure – it is not for everyone.

ABOVE A snorkeler is dwarfed by the biggest fish in the sea.

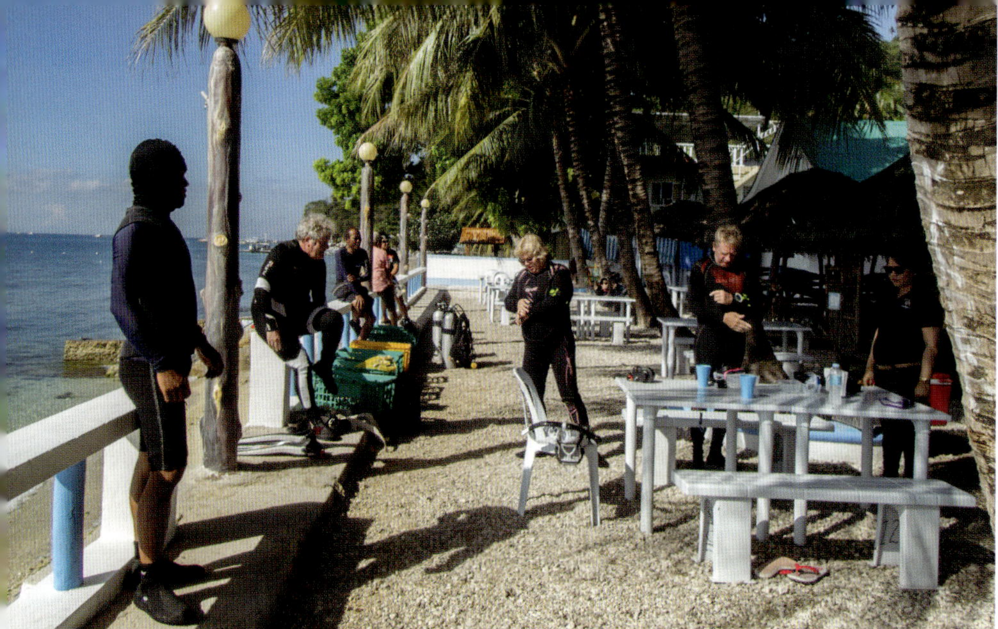

THE WILD DIVE

The dive operator we went with was situated in Moalboal, which meant that we could reach the village via minibus and they had us set up our diving equipment away from the main tourist area, in a calmer more relaxed beach area a short distance away. Our day was going to include scuba diving with the Whale Sharks, doing a shore dive right from the beach we were based at, and then also getting to see what the majority of tourists experience – going out on a local boat to snorkel with the sharks. The Whale Sharks are only fed in the mornings, so it was an early start for us in order to be able to fit all this in.

As I walked into the water, geared up for our first dive, I could see all the small boats, each with around 10 people on board. Each boat usually has a Whale Shark with it, giving you a rough indication of how many there might be at the site. There were 12 boats out on the water, so I hoped that there were plenty of sharks around. The rules mean that we only had one hour for the dive, and so as soon as we submerged we were on the clock. I finned hard to keep up with our guide who wanted us to get the long swim out of the way as swiftly as possible. In just a few minutes I looked up to see we were nearing the boats and there in the distance was a Whale Shark. We stayed relatively shallow, as these huge sharks spend a large amount of their time feeding next to the boats but would occasionally drop deeper to check out what we were up to. While individuals of this species can grow to more than 10m (33ft) in length, these sharks seemed to be fairly young and nowhere near as big as some of the adults I had seen in other locations.

ABOVE Gearing up away from the crowds to go diving with the sharks

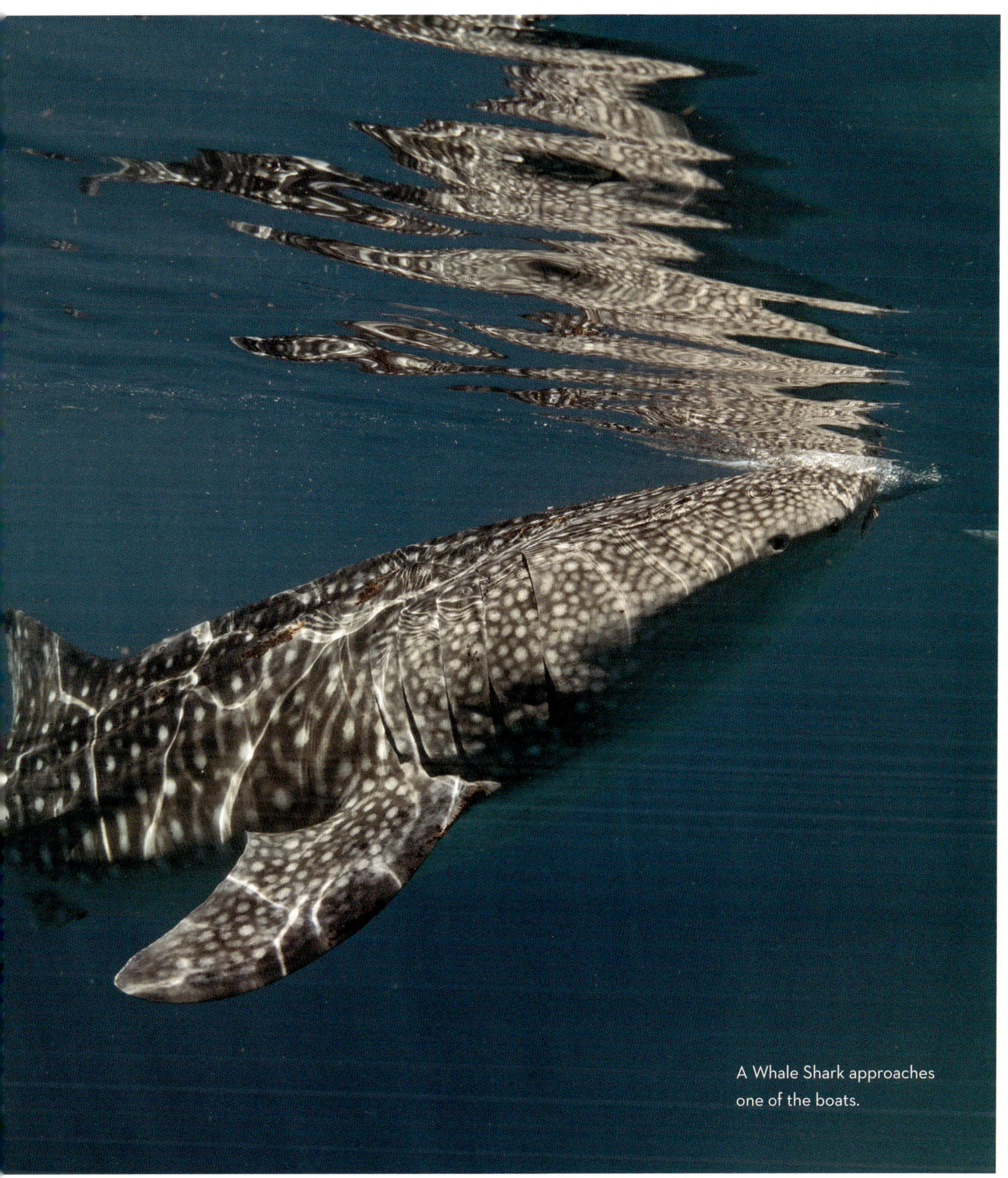

A Whale Shark approaches one of the boats.

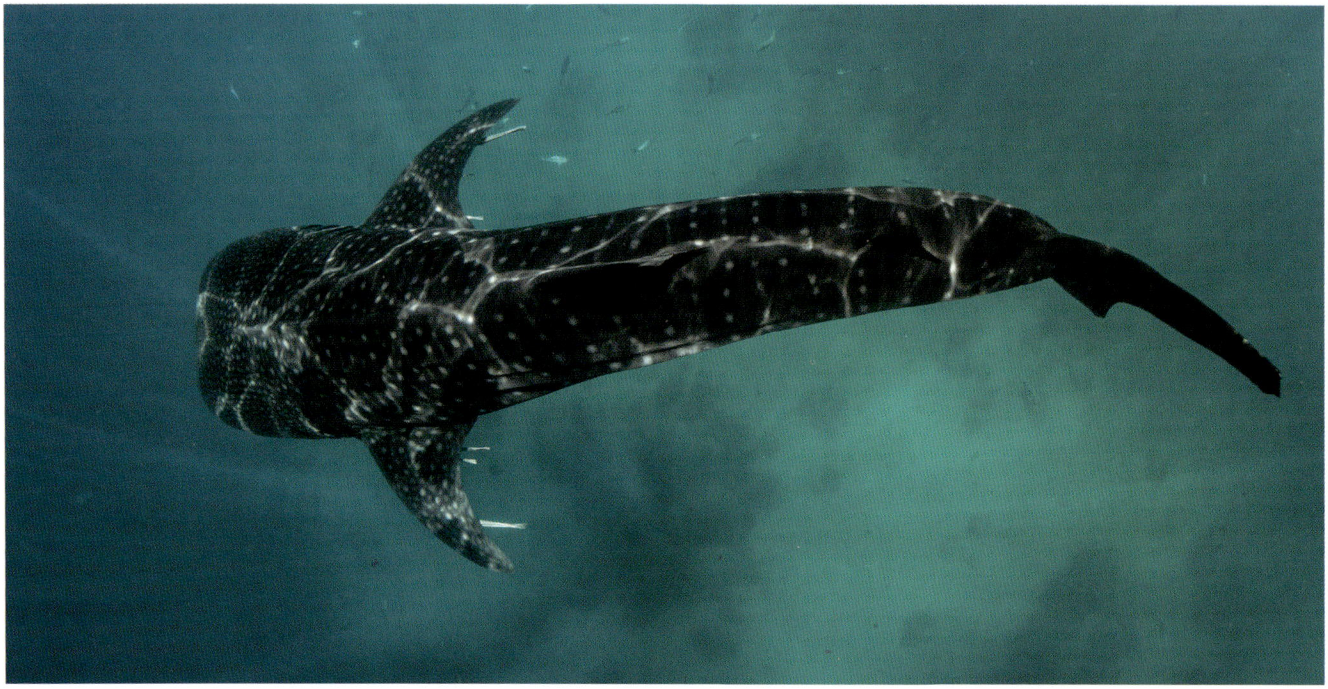

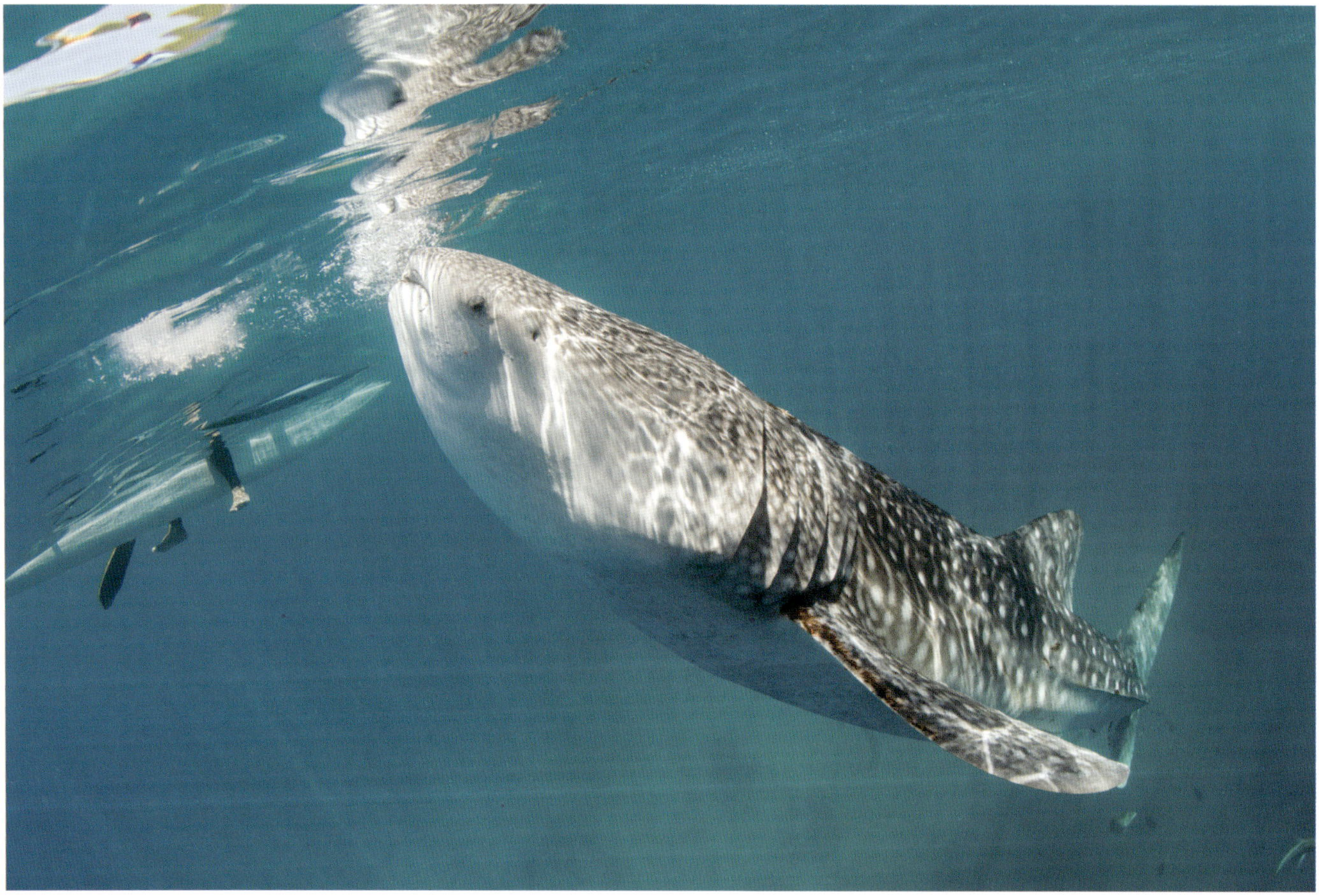

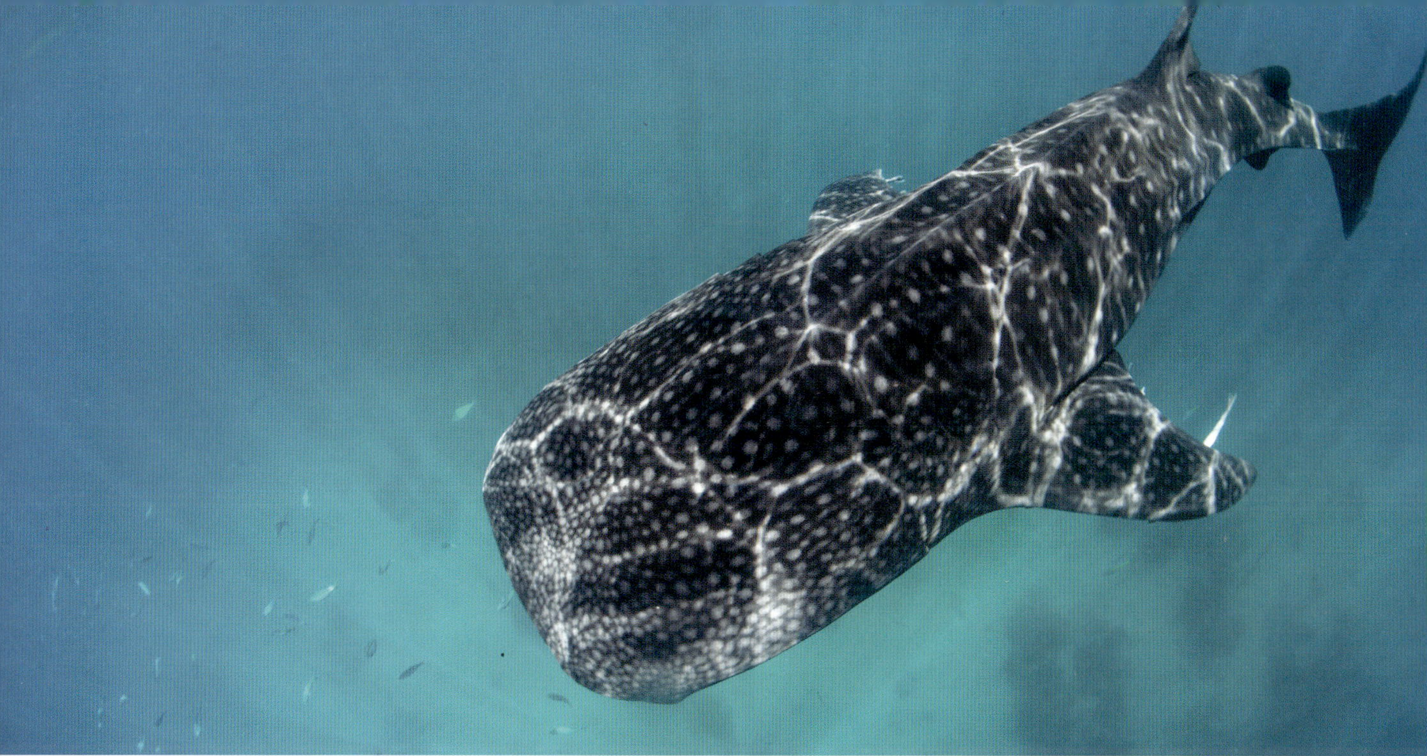

Our second shot at getting up close with the sharks saw us get into a small wooden boat that was rowed out by our guide. We were told the rules – no flash photography, no touching the sharks or getting within 3m (10ft) of them. We took up our place among the other boats and the most striking thing was how few people actually got into the water, with most watching from the safety of the boats. You get a completely different view up close from the boat, seeing the gorgeous spotted patterns on the sharks' backs. In the water the sharks were feeding, sometimes going completely vertical sucking in huge volumes of water with tiny handfuls of shrimps that are thrown in to attract them. I saw a tourist near me reach out to grab a fin, but instantly he was hauled out of the water and banned from going in again. While there were lots of tourists here, it never got crowded in the water with the sharks. I checked with our guide where was okay to stay in the water and just waited for the sharks to come to me, in the process getting some of the closest encounters you could ever wish for.

There is a fine balance between trying to make sharks worth more alive swimming in our seas and oceans than dead and their fins sold for soup. In an ideal world, all these sharks would be protected across the world, but this is currently far from being the case. If projects like this one help us to learn more about the sharks and bring an income to people who might be tempted to make money from them in alternate ways without tourism, then in my opinion this must be an improvement from the sometimes horrible alternatives.

OPPOSITE TOP These majestic sharks have beautiful and unique patterns on their backs.

OPPOSITE BOTTOM Oslob, whilst controversial, does offer incredible close encounters with Whale Sharks.

ABOVE In shallow water the sun creates further patterns on their backs.

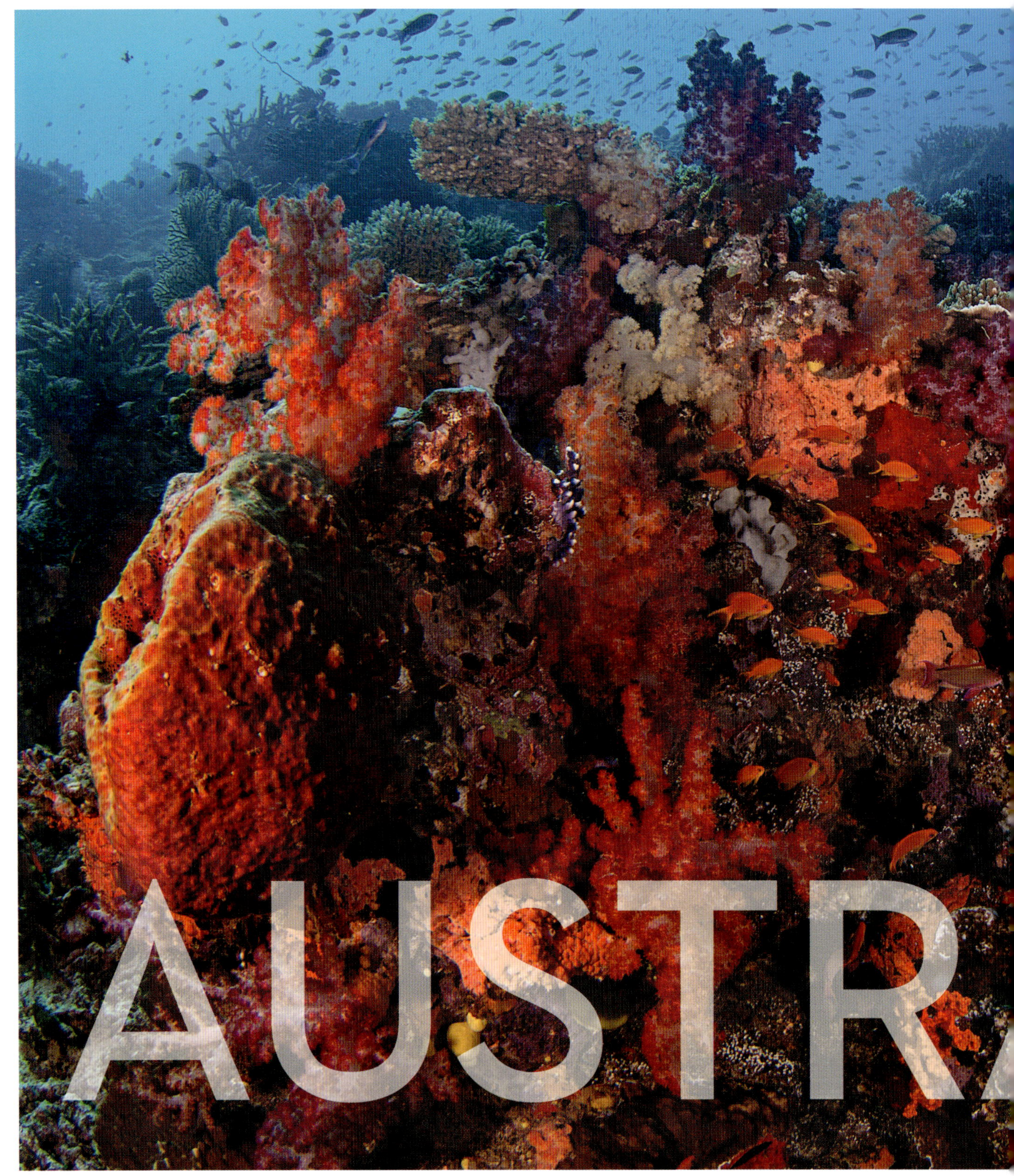
AUSTRA

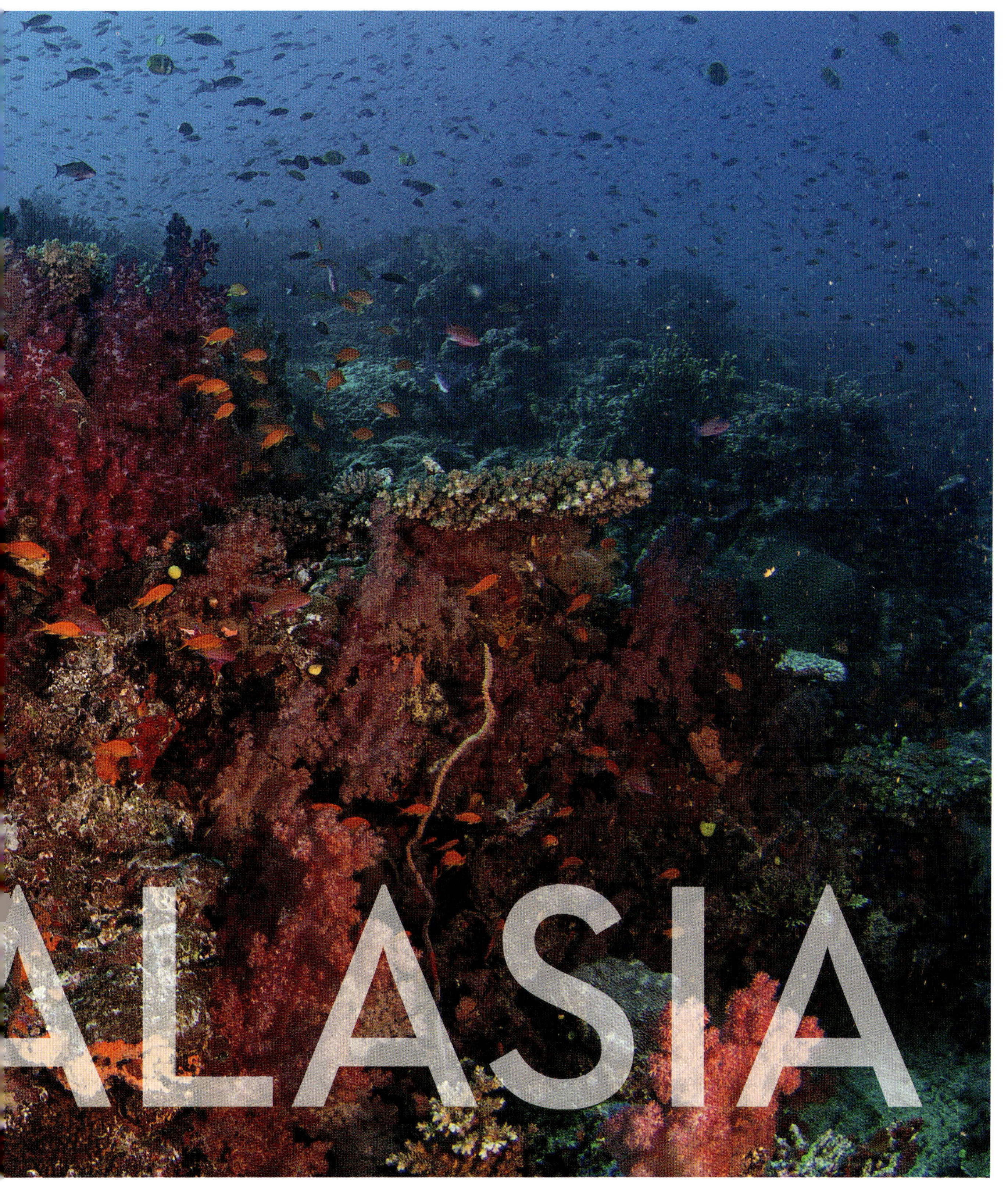
ALASIA

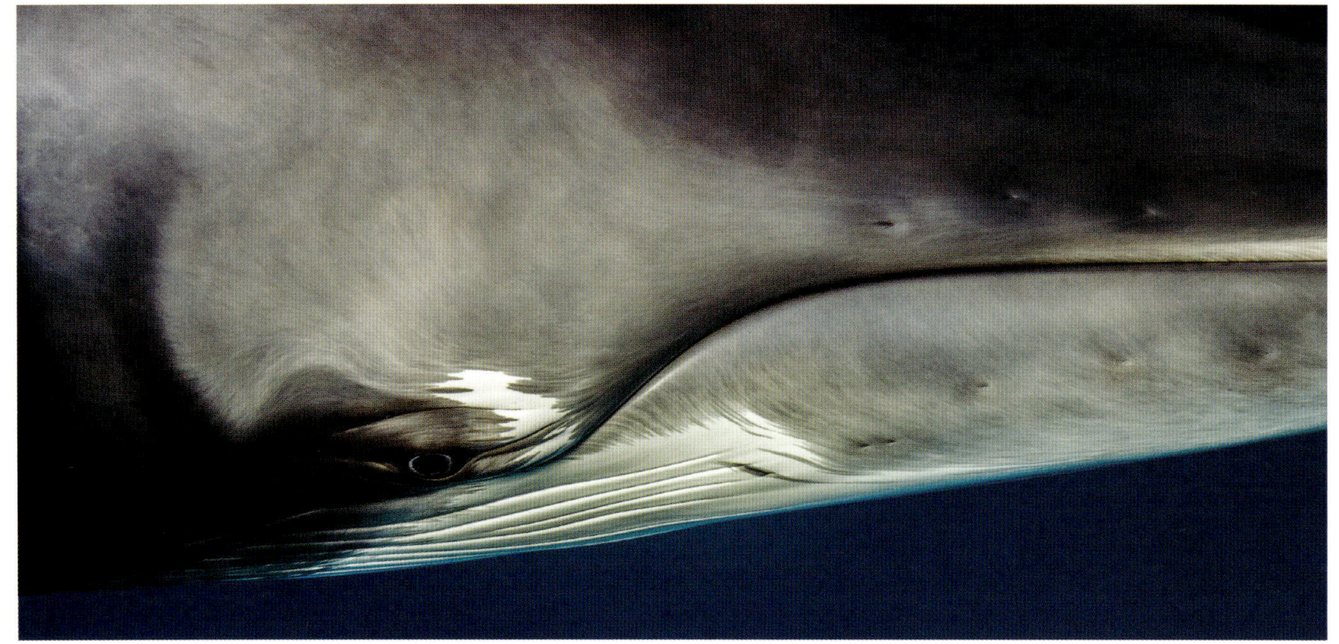

12 EYEBALL TO EYEBALL WITH DWARF MINKE WHALES

AUSTRALIA

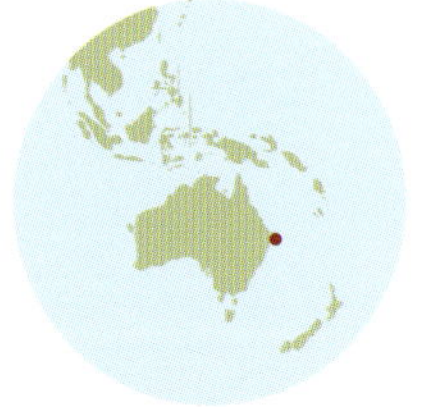

ABOVE Close up of a Dwarf Minke Whale.

OPPOSITE Snorkellers watch a whale swim below them.

THE GREAT BARRIER REEF OF AUSTRALIA has to be one of the world's most iconic diving destinations. It is the largest barrier reef in the world, stretching for more than 2,300km (1,430 miles) and it is a national as well as natural treasure. The most common departure point for divers and snorkelers to visit the Great Barrier Reef is Cairns, which is situated in Queensland in the north-east corner of Australia. The incredible reef stretches for hundreds of miles and divers from all parts of the world flock here to see it and the marine life it supports. The reef is like the lungs of our oceans, vitally important and people are justifiably passionate about protecting it.

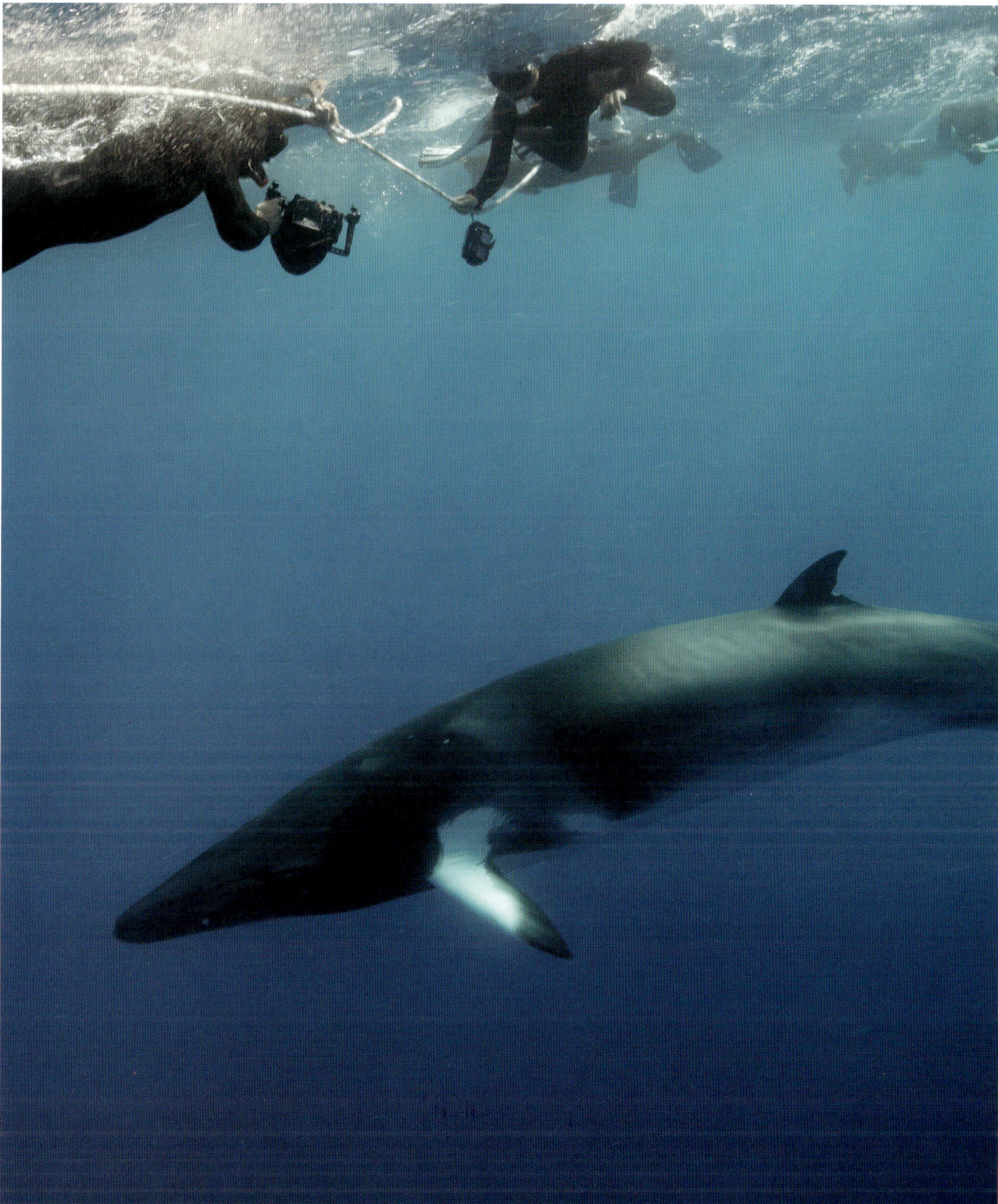

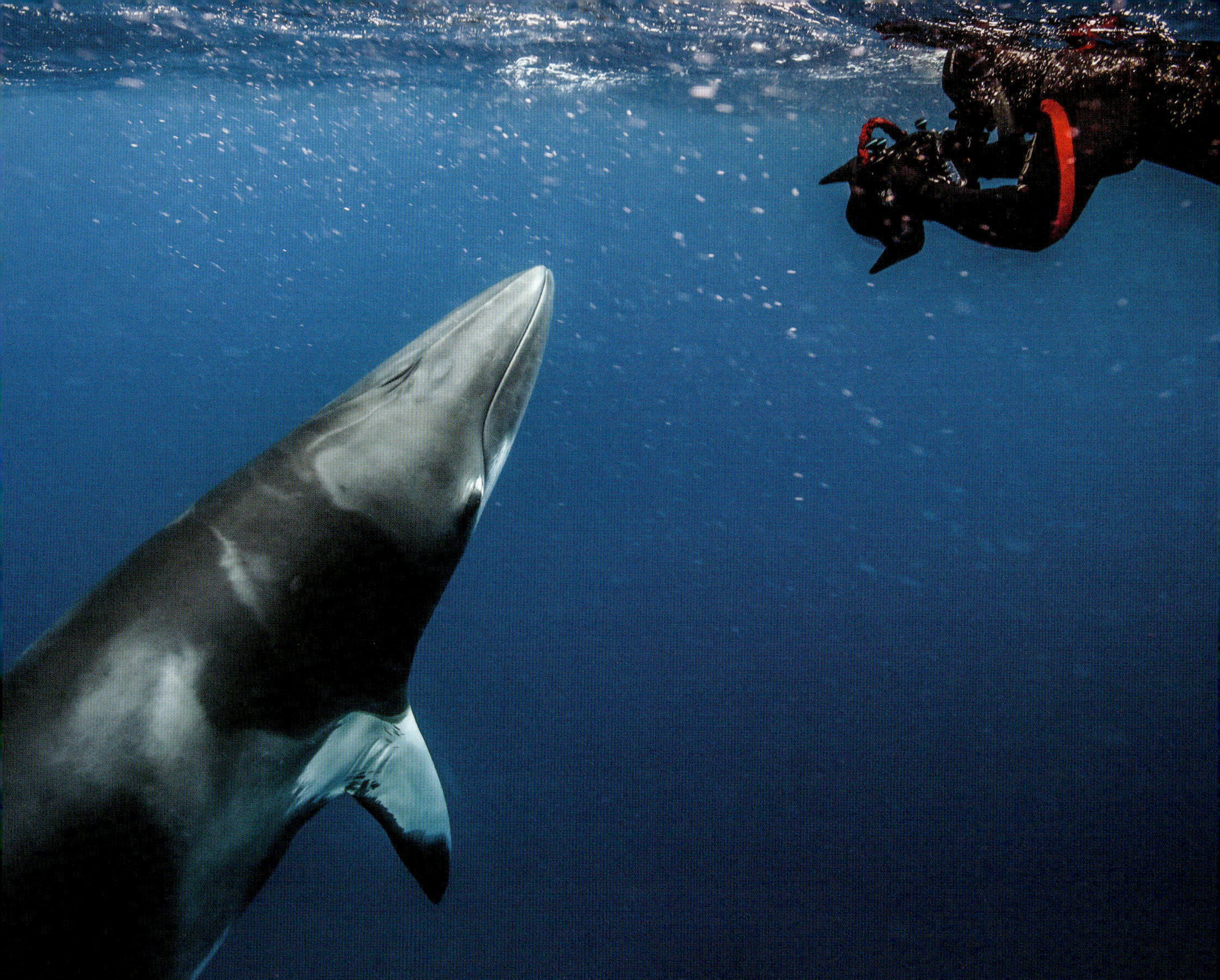

ABOVE An underwater photographer gets an incredible close up.

GENERAL DIVING

The Great Barrier Reef mirrors the effects of human activity all over the world and, perhaps because it is so loved, always makes the headlines as it falls into decline due to climate change, sea temperatures rising and pollution choking the very life out of it. And yet, there are still huge areas of vibrant and healthy reef with a huge array of creatures. Some of the more spectacular diving can be experienced a little further off the coast and so the best way to see some of it is by going on a mini expedition on a liveaboard. Cod Hole provides divers with the chance to get face to face with a fish as large as any diver. Pinnacles reaching up from the seabed attract

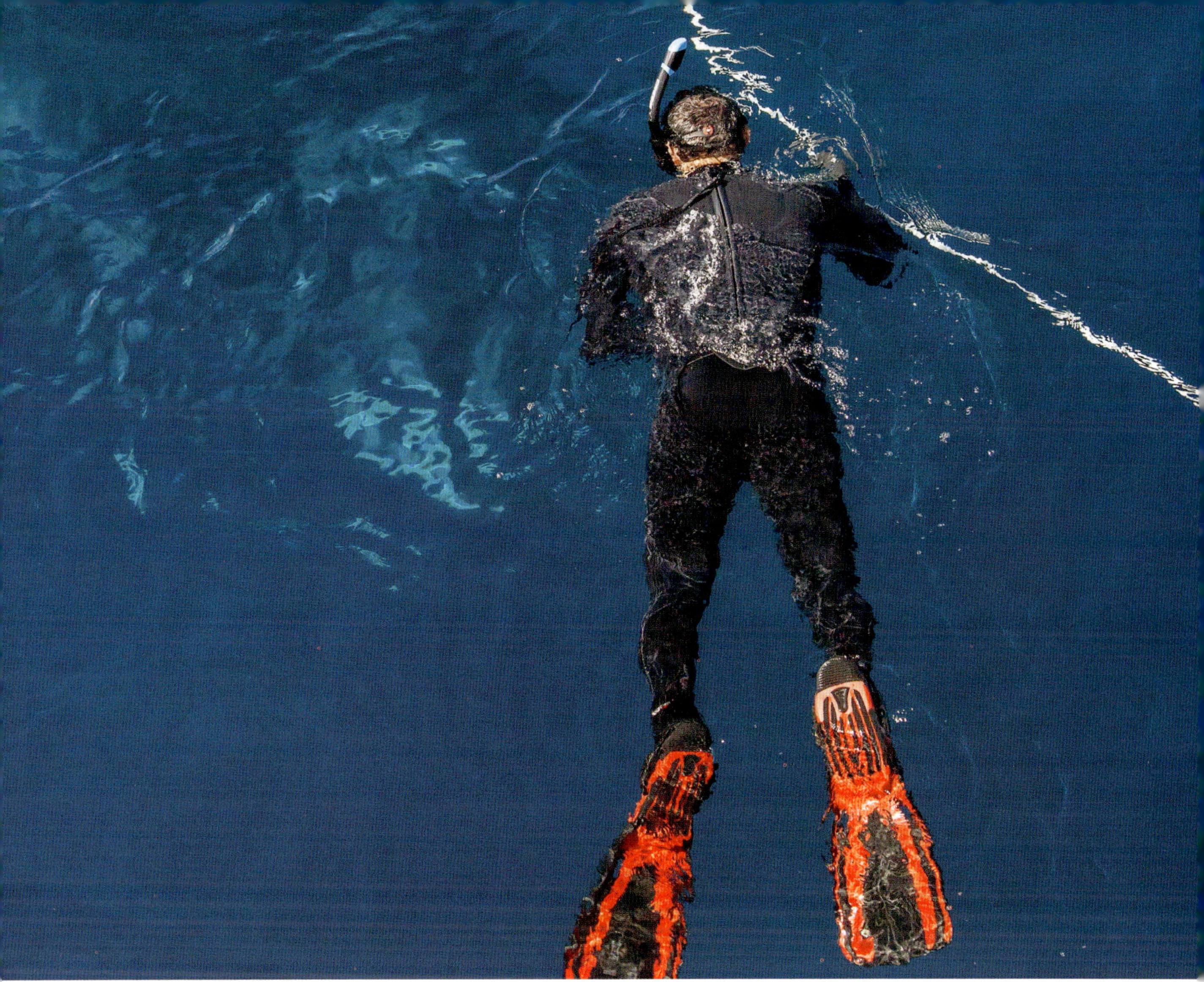

sharks, turtles and rays, and there is an abundance of great and diverse diving to be explored.

THE WILD DIVE

This trip was like no other we had been on as it started out with an incredible plane ride, giving us a birds-eye view of the reef as we headed to Lizard Island. Here we were met by the crew and set out for Ribbon Reef, a site famous for having some of the best diving in the region. Nutrient-rich waters from the Continental Shelf

ABOVE A snorkeller looking down at the whales.

attract abundant fish and coral species. However in winter, particularly in June and July, this area is especially famous for attracting Dwarf Minke Whales and this is one of the few places in the world where you can get into the water and snorkel with them. Anyone can take part in this experience, but there are strict rules to ensure that the whales are not disturbed and that any interaction is on their terms.

As our boat approached Ribbon Reef the excitement grew on board. The crew and passengers peered out onto the flat water to look for tell-tale signs of whale activity. Everyone knows that if no whales are spotted then diving the reef is always an option, but we could tell that everyone was hoping for a sighting and no one was disappointed because soon the word was out – the whales had been spotted in one of their usual haunts. We slid into our wetsuits and grabbed our masks, snorkels, fins and cameras and headed towards the water. I knew that these somewhat diminutive whales had a reputation for curiosity, but I still had everything crossed as I entered the water, in the hope that we would get a good look at a whale I had never seen before. Dwarf Minke Whales grow to about 8m (26ft) in length and travel in social pods, so chances were that if one was seen there were more nearby. As I got into position I heard my buddy squeak with delight – she had clearly seen something of note and so I scanned the clear blue water around us. Then I saw what her squeaking was all about, a whale was swimming directly towards us. A shadow in the distance had to be another, and soon we had three in the water around us.

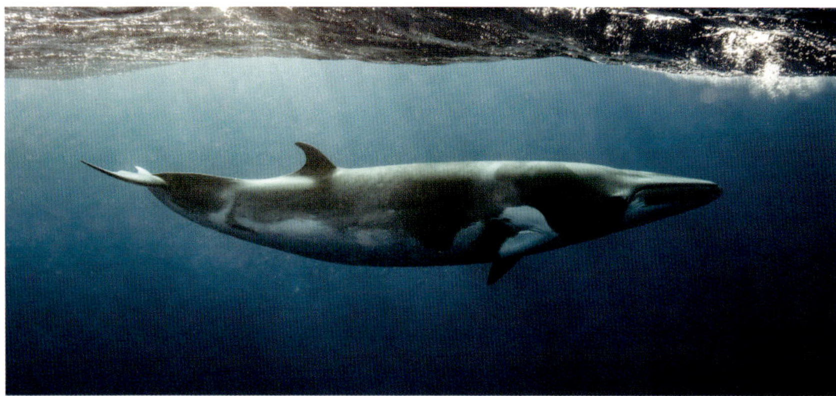

ABOVE A Dwarf Minke Whale swims past.

OPPOSITE A whale dives below the dive boat.

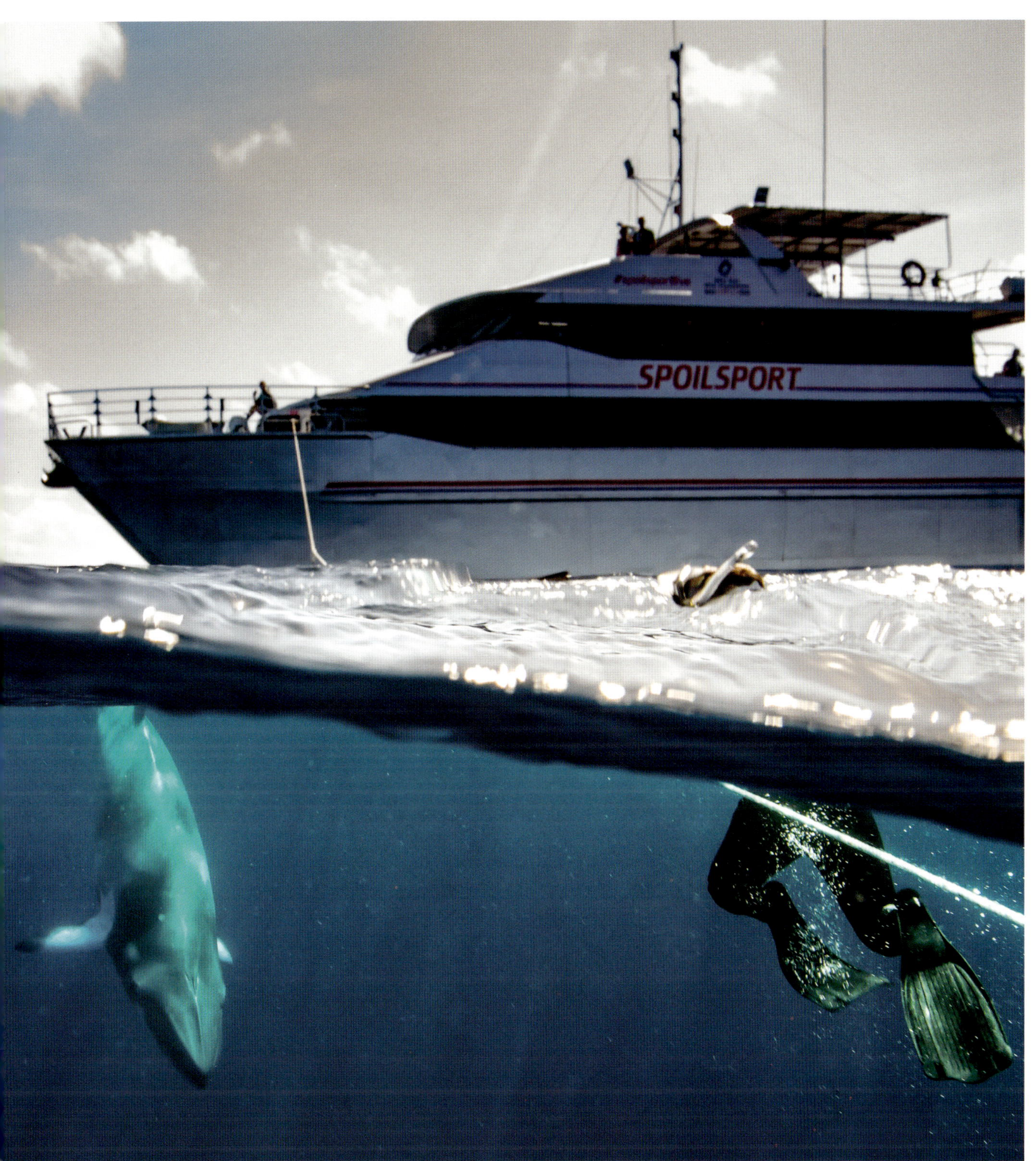
SPOILSPORT

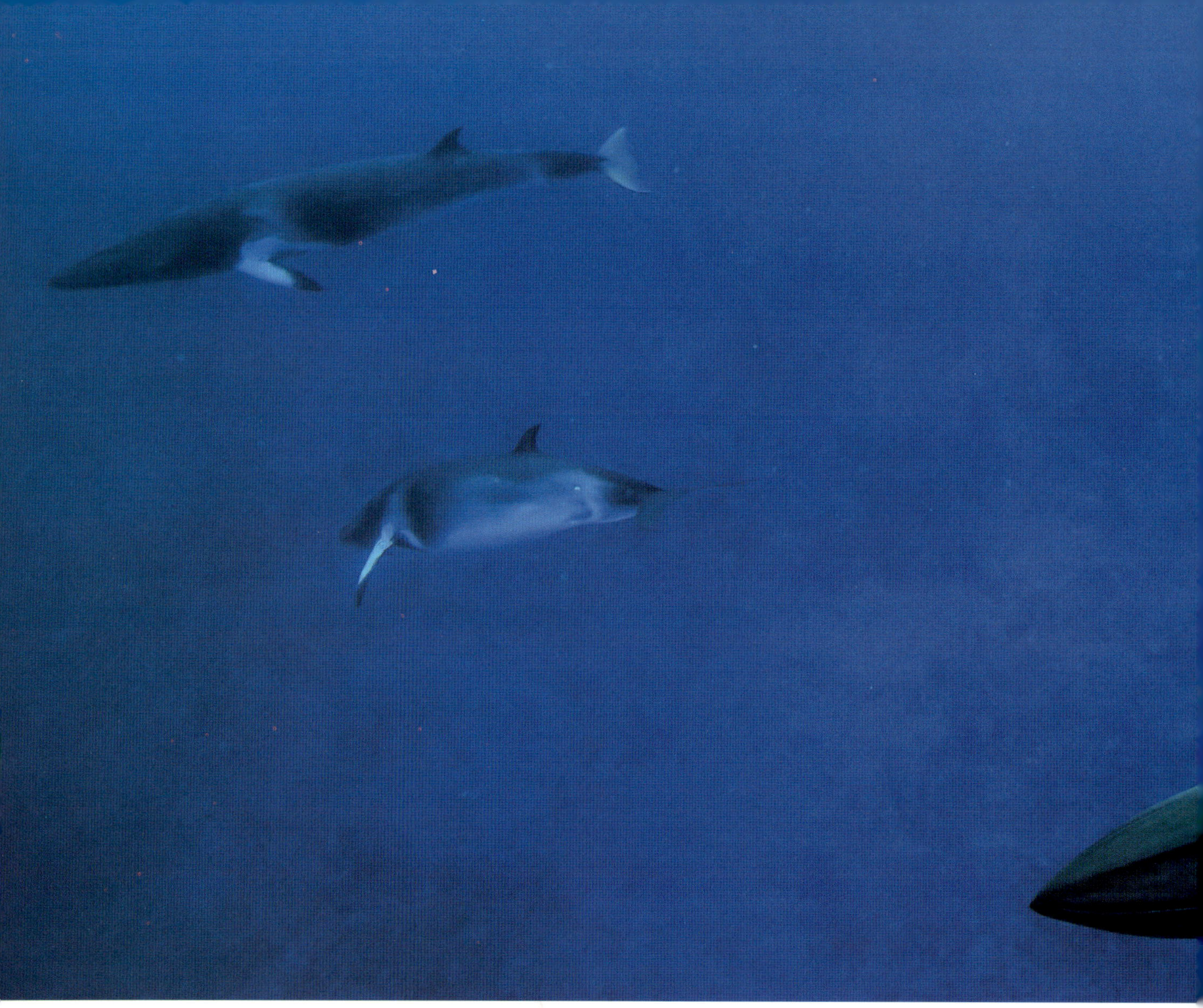

The most curious of the whales approached even closer and now I could get a genuine feel for their size, even getting the chance to look this clearly intelligent cetacean in the eye. With one hand holding the line attached to the boat to keep all the snorkelers within reach and in check, I pulled my camera round in front of me and fired off a few shots. Every now and then I could hear a whoop from the boat, as those watching from the surface experienced other whales spy-hopping or even breaching in the distance. Our three whales seemed content to watch us, swimming around and underneath us, allowing us to just revel in their presence. I wanted to stay in the water with the whales for as long as possible and they

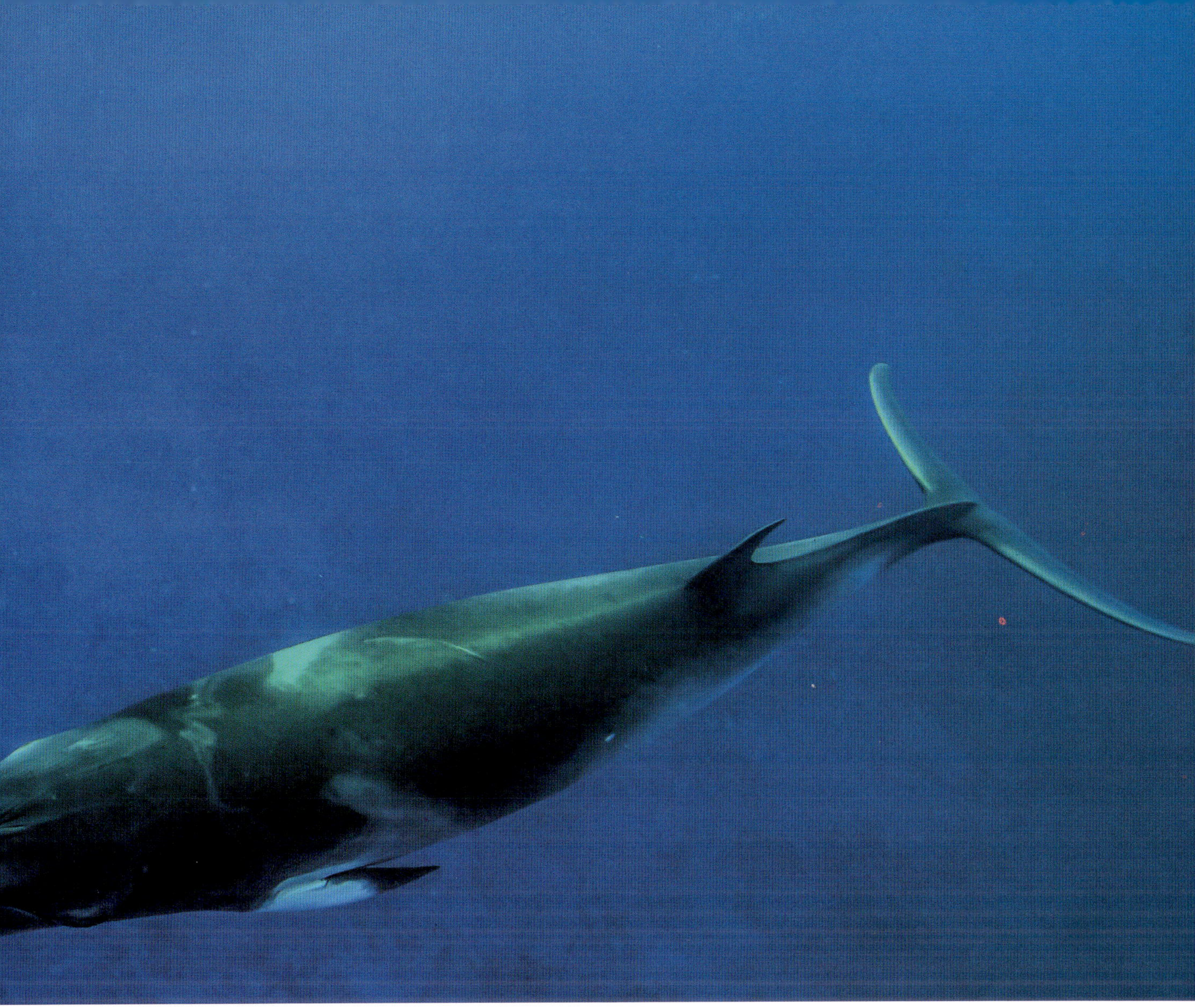

seemed happy to hang around too. Then one of the pod got so close that it took my breath away, slowly cruising past me less than 1m (3.3ft) from my mask. An hour or so later, as I headed back towards the boat with my buddy, this particular whale seemed to follow us, drifting along beneath us as we pulled ourselves along the rope to the back of the boat. It was hard to leave it, and even once we were out of the water we hung around on the back of the deck and watched as occasionally our whale lifted its head clear of the water, still looking at us with as much curiosity as we had for it. It was a magical experience – one that still makes my hair stand on end when I think about it.

ABOVE Three whales remain below us for the entire time we are in the water.

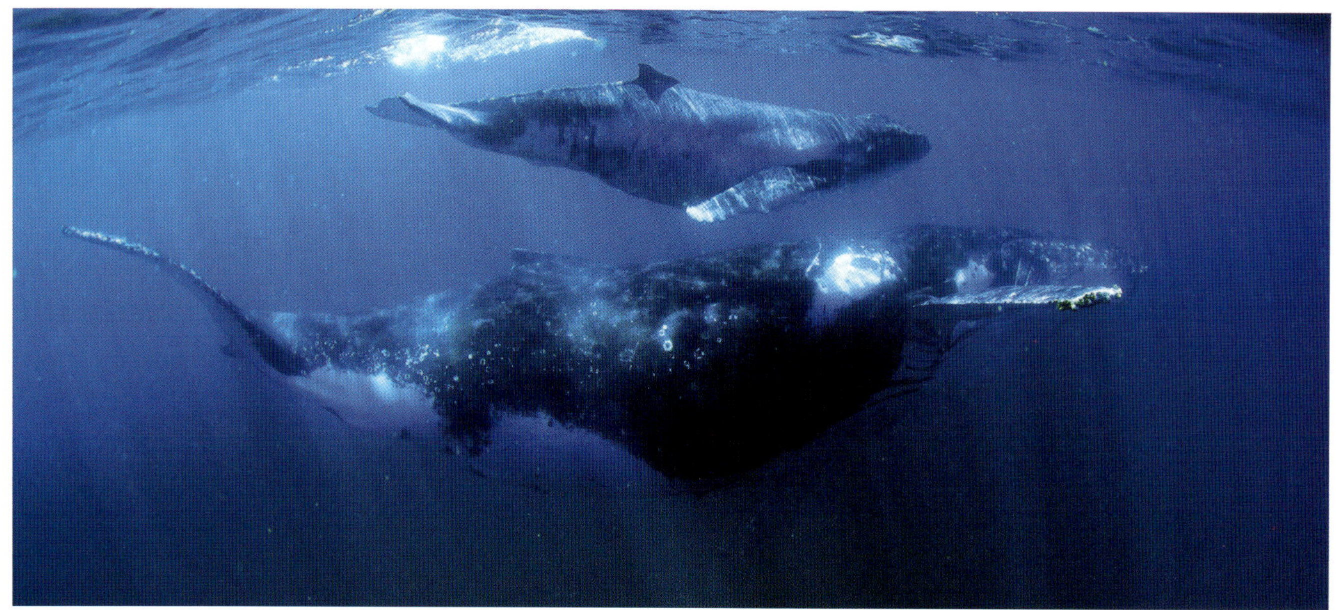

13 THE GIANTS OF NINGALOO REEF

AUSTRALIA

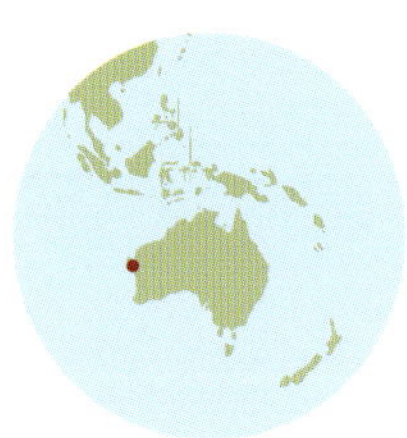

ABOVE It is possible to see Humpback Whales and their calves at Ningaloo Reef.

NINGALOO REEF IS SITUATED on the north-west coast of Australia in the state of Western Australia. The Ningaloo Coast is a World Heritage Site that includes Ningaloo Reef in the waters of the East Indian Ocean. It is Australia's largest fringing coral reef – situated close to the mainland – and was declared a marine park in 1987 in an effort to protect the reef and its inhabitants. The reef stretches for 260km (160 miles) and is one of the world's most biologically diverse marine habitats. More than 250 species of coral can be found here while more than 500 species of fish populate the region. Exmouth is the town known as the gateway to Ningaloo, and is where most people travel to in order to begin their Ningaloo experience.

GENERAL DIVING

Ningaloo Reef is famous for the annual visits, between March and June, of the biggest fish in the sea; the Whale Shark. These majestic, gentle giants arrive here

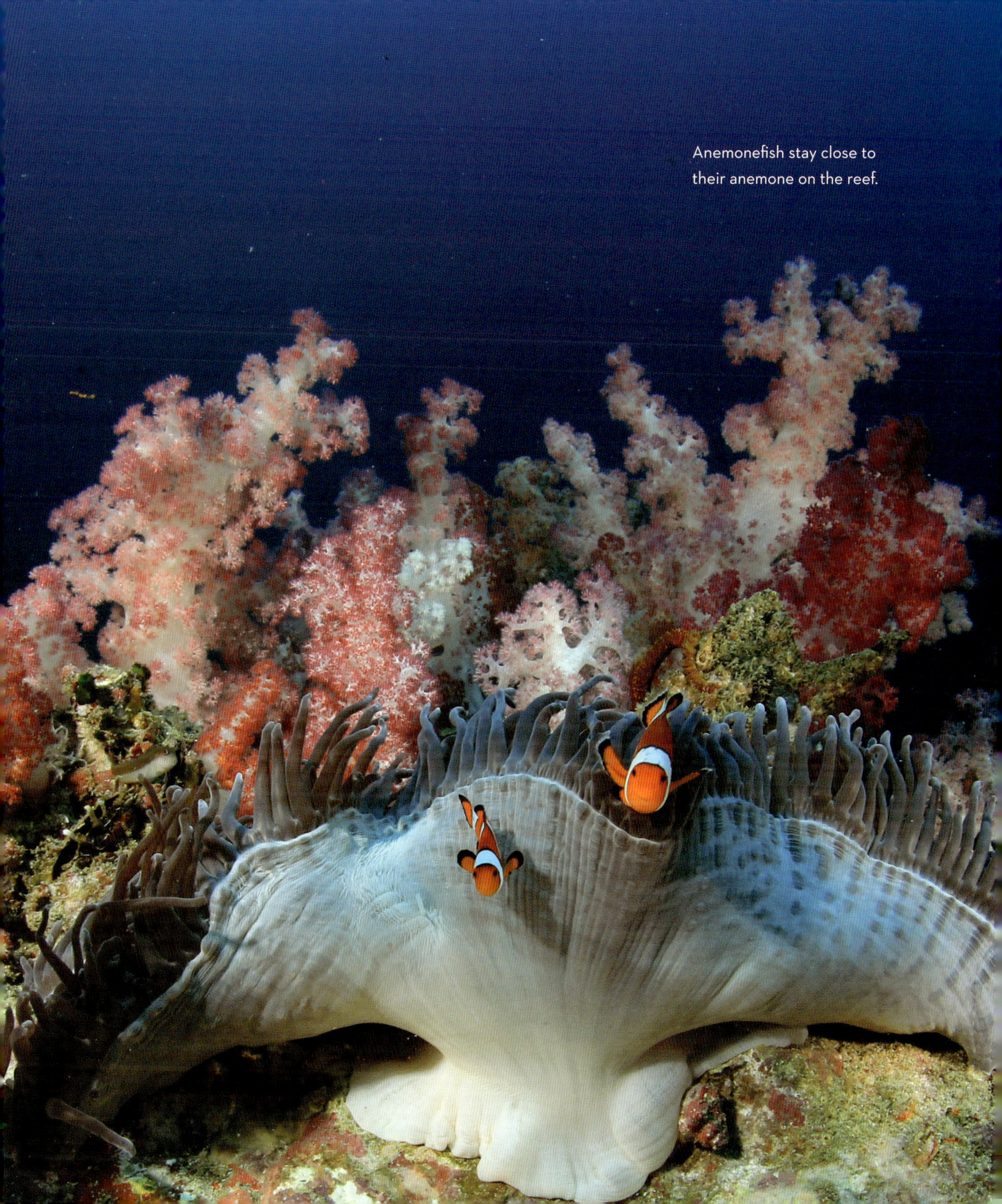

Anemonefish stay close to their anemone on the reef.

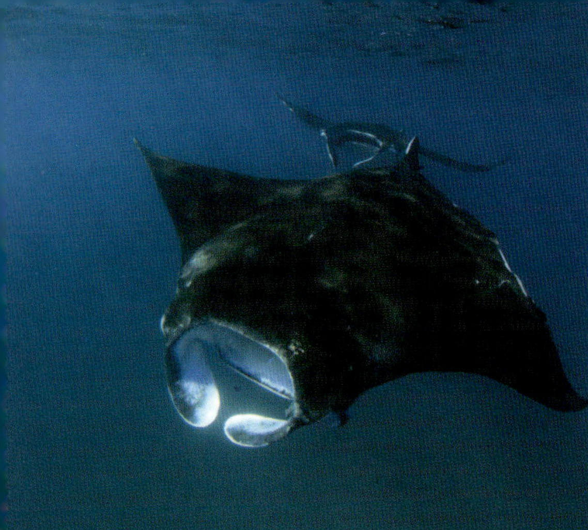

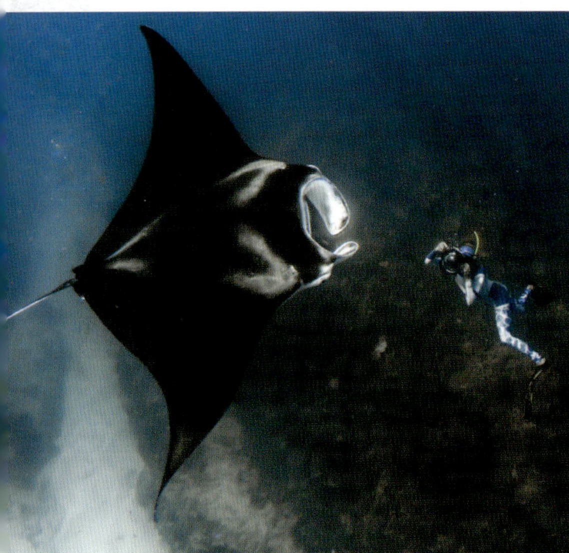

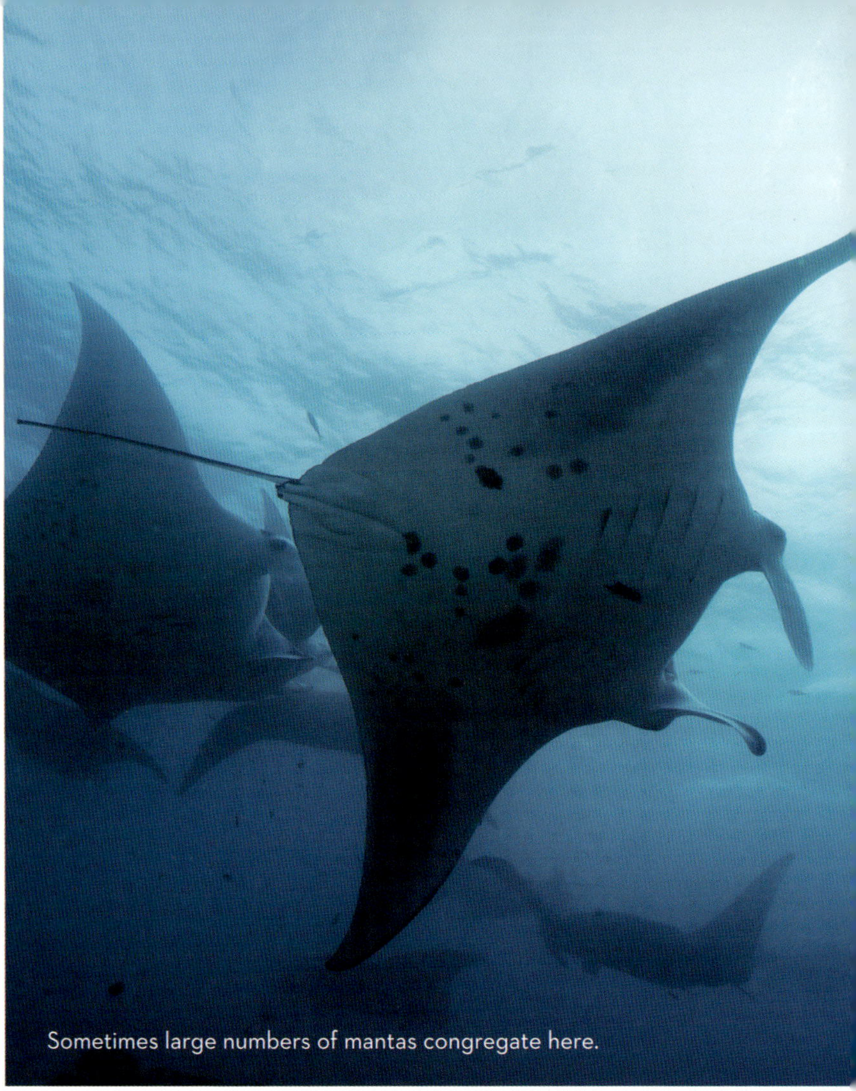

Sometimes large numbers of mantas congregate here.

ABOVE TOP A manta follows the lead manta as a chain starts to form.

ABOVE A freediving underwater photographer gets in close for a manta shot.

in time to take advantage of the coral spawning. However, Ningaloo is also rich in a variety of marine life, with hundreds of fish species and types of coral and hosts other migratory marine animals such as manta rays, dugongs, dolphins and Humpback Whales. The area is also an important nesting site for a number of sea turtle species. The clear water and abundance of marine life means that diving here is popular all year round.

THE WILD DIVE

Ningaloo Reef is one of those dive site names that stick with you. We had heard stories of this being the place to go to see Whale Sharks in clear water and without having to travel far from shore. The Whale Sharks travel here in groups and the

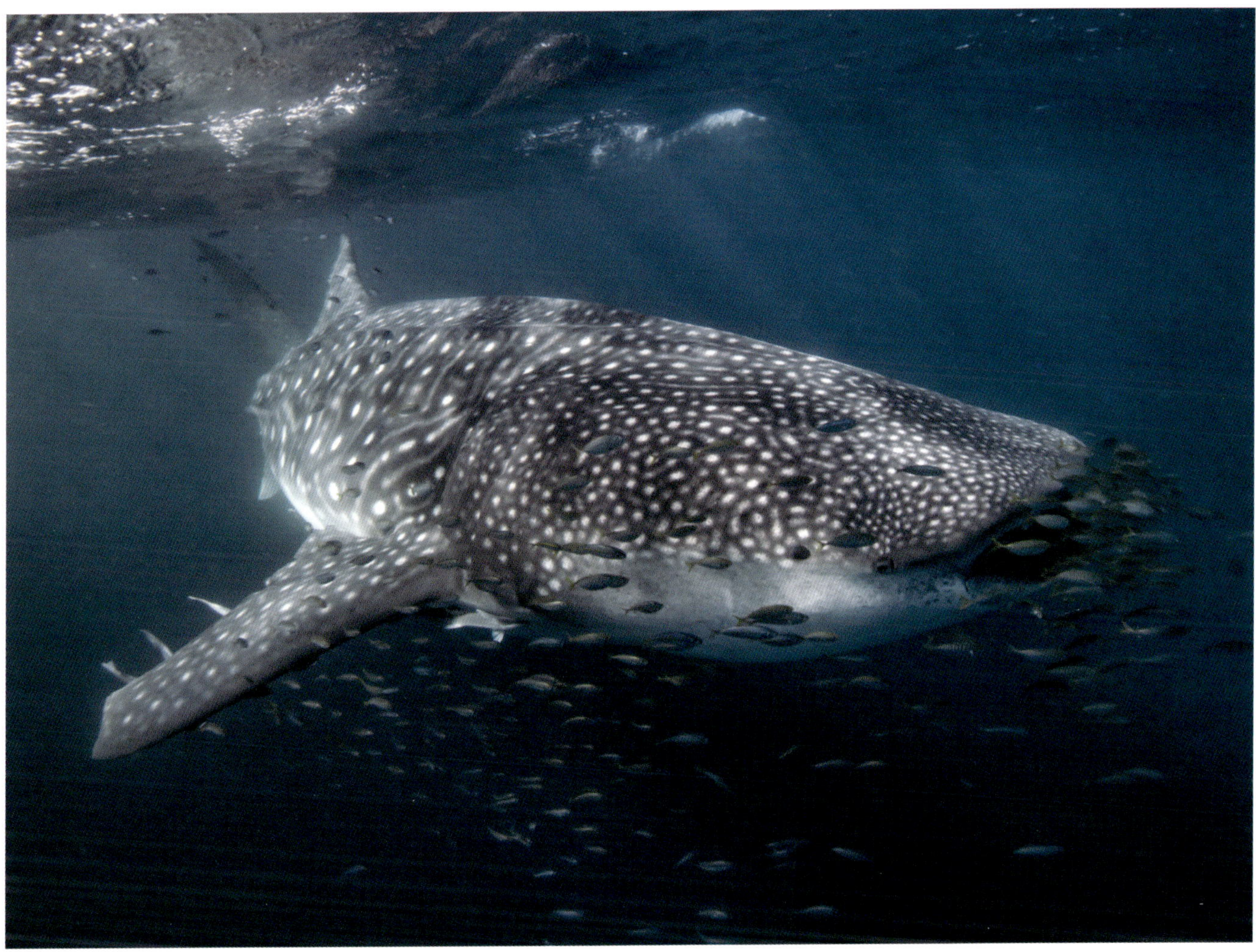

area is proud to claim that it is the most reliable place to see these gentle giants in their natural environment. We wanted to swim with Whale Sharks and also take in some of the diving, and so booked a combined tour that would allow us to do both. Our tour saw us spending the morning slowly exploring the flat clam waters looking out for Whale Sharks. There are strict laws in force here about shark encounters, so you have to keep your fingers crossed that the sharks are feeling playful and curious, as the only way that you can get a close-up view is if they feel comfortable enough to approach you and not the other way around. It was not long that morning before the first cry of 'Shark' rang out and I slipped into the water. The bright sunshine overhead made the light dance on the back of the Whale Shark swimming below us and soon it was joined by a second, with both swimming quite close and staying around us for 20 minutes or so. I was mesmerised by the beautiful

ABOVE A huge Whale Shark with a host of fish surrounding it.

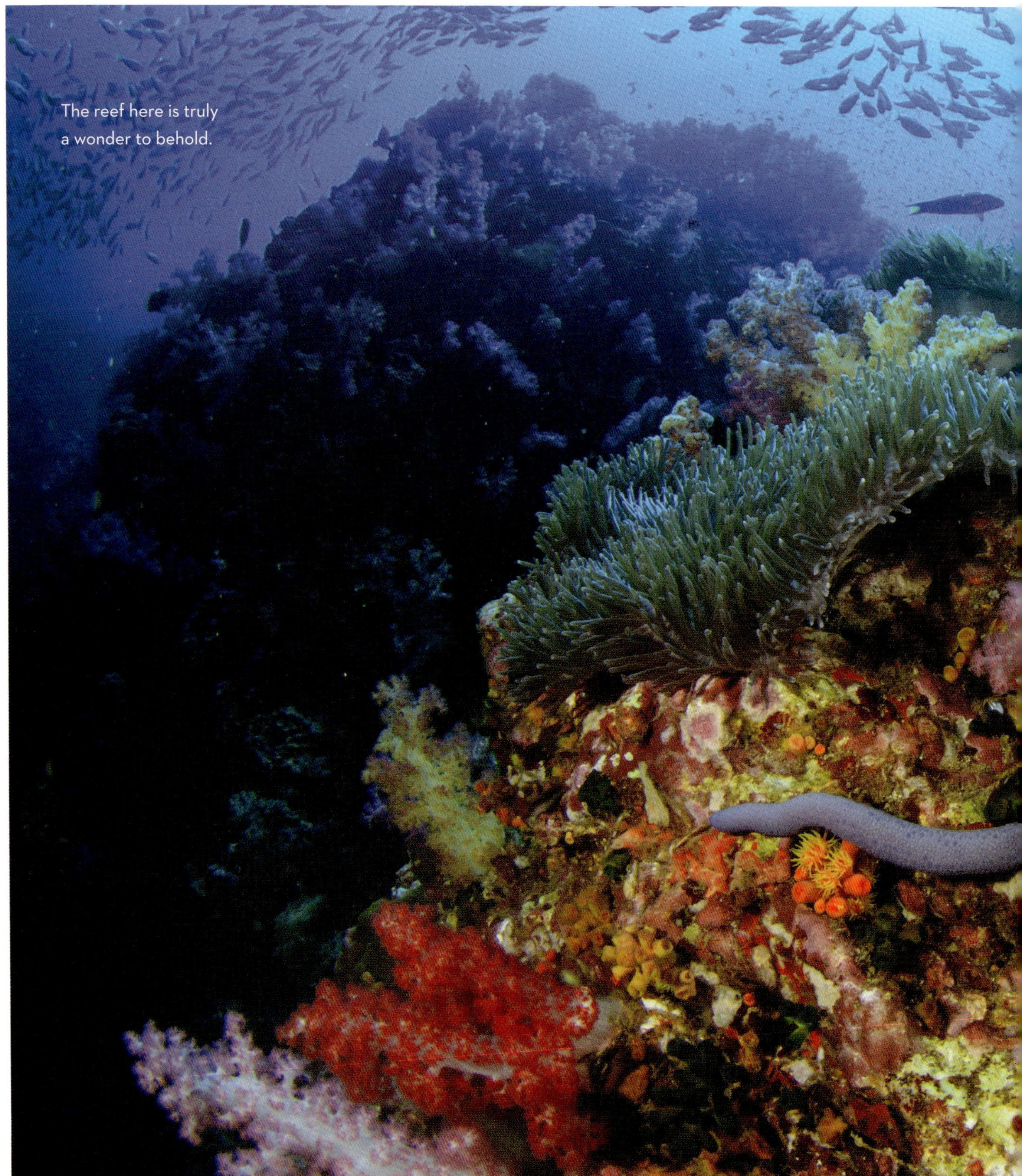
The reef here is truly a wonder to behold.

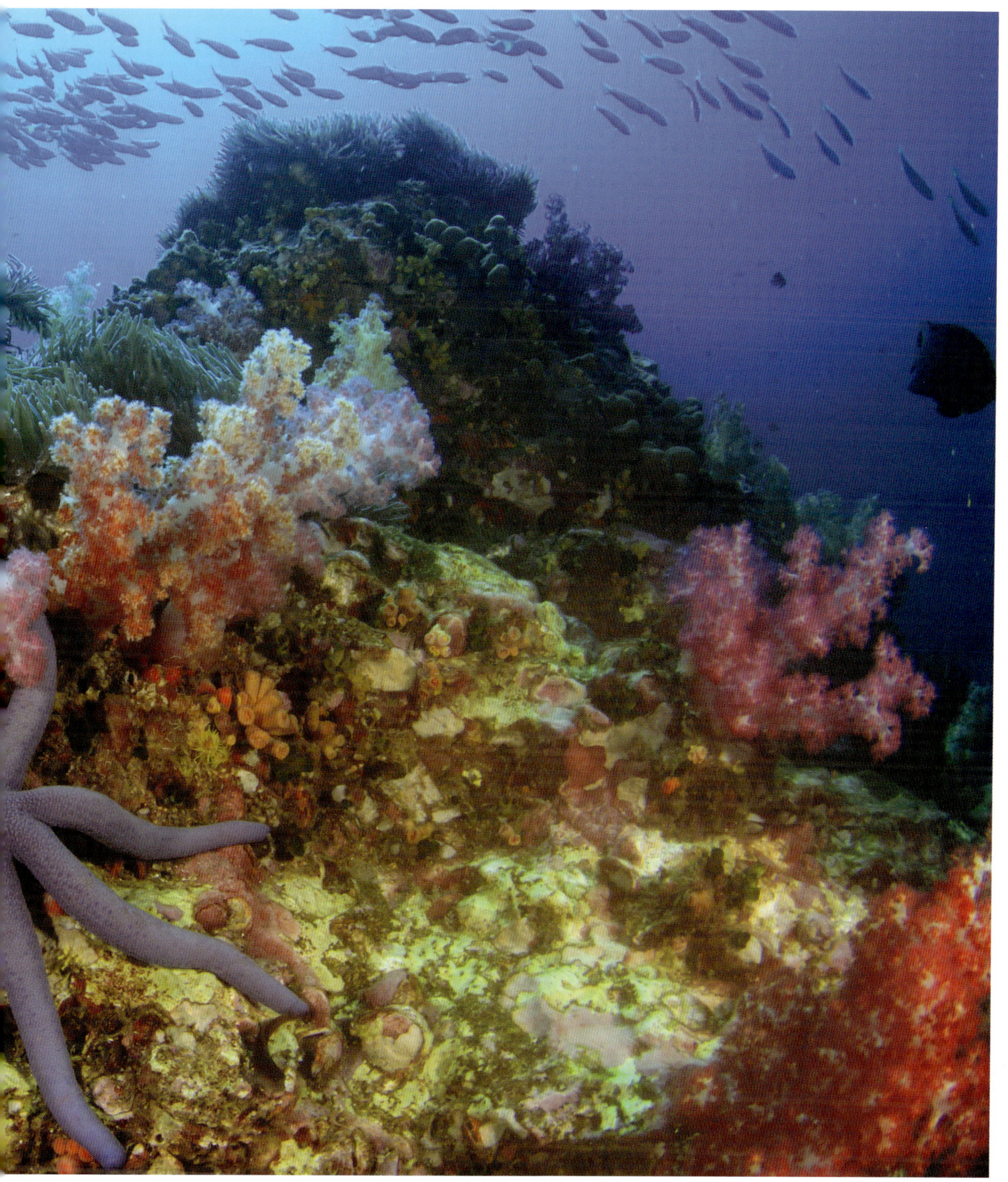

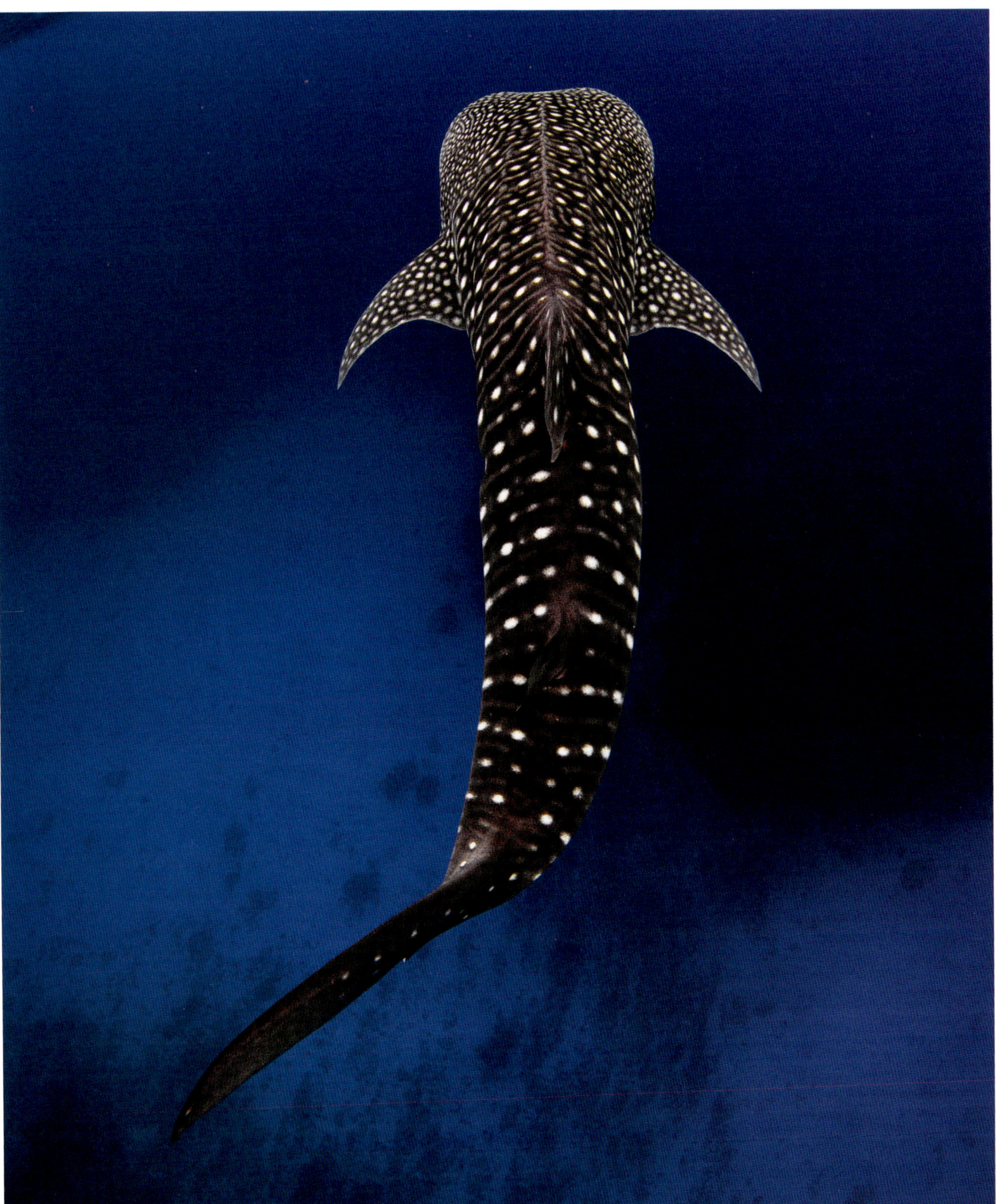

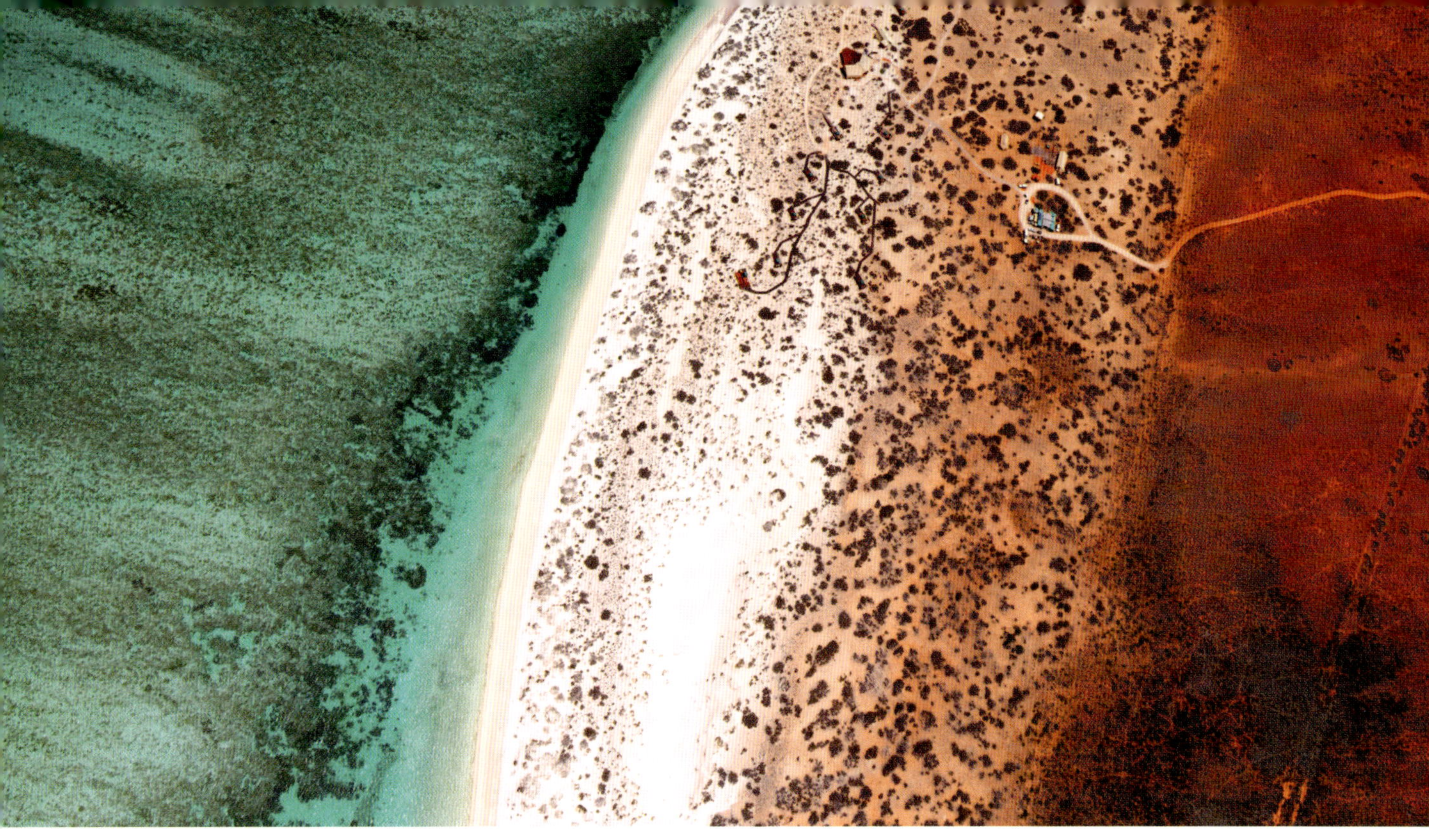

spotted patterns on their backs and their huge size as they cruised along beneath me. It was a fantastic morning which slipped away quickly with multiple Whale Sharks being spotted by the crew. Some of the sharks were accompanied by an entourage of other fish, while others swam alone.

The afternoon saw us don our diving gear to explore the reef. Ningaloo is famous for its healthy reef and abundant marine life, and so while the highlight of the trip was the time in the water with the Whale Sharks, I was also looking forward to this experience immensely, due to the fact that I had heard such great things about the diving here. I was not at all disappointed. As soon as I descended, the vibrancy of the reef screams out at you. Schools of fish appeared to fill the water just above the reef structure and the coral itself was like a multi-hued blanket covering the reef for as far as the eye could see. Then I heard my buddy call me and I turned in time to see a huge manta ray cruise along beneath us. As we signalled that it was time to come to the surface I thought to myself that diving days do not get much better than this. Just as we were about to head back to the boat I heard my buddy once again calling me, this time from the surface, and I ducked my head back under water to see two feeding mantas heading right towards us. We hung motionless as they glided past us, mouths gaping open to scoop up the food in the water. It was the perfect end to our day.

OPPOSITE A Whale Shark swims below showing off its beautiful and unique pattern.

ABOVE The coastline in this part of Australia is stunning.

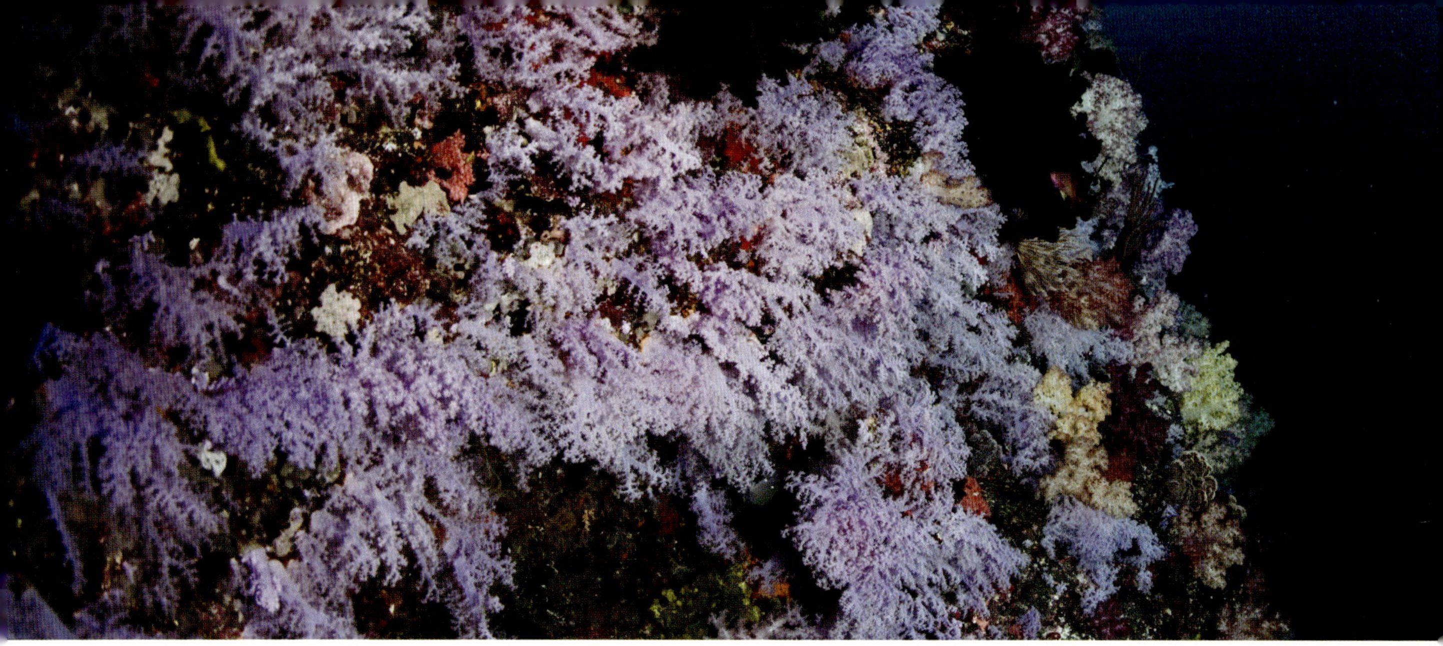

14

MIND-BLOWING CORAL

FIJI

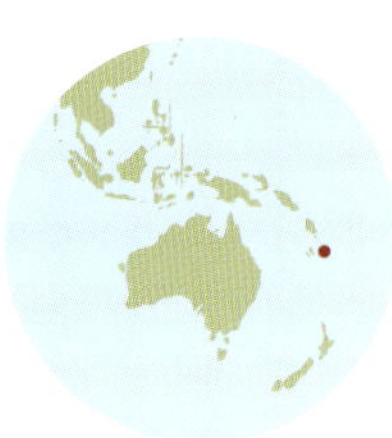

ABOVE White Wall is covered in white corals, making it look like a recent snowfall has covered the reef.

OPPOSITE The coral reef at Fantasea 1 is outstanding.

FIJI IS AN ISLAND NATION which forms an archipelago that is made up of more than 300 islands in Melanesia in the South Pacific. It is famous for its golden beaches fringed by palm trees, its clear blue water and its rugged and lush volcanic hills. It is part of the wider region known as Oceania. Fiji is one of those magical places that can be found on many people's bucket lists of places to visit, and it offers the traveller a host of experiences – whether they want to relax in luxury or be more adventurous, there is something for everyone. Surfers come to ride the waves, you can partake in a natural spa of hot springs and bubbling mud baths which they say will knock 10 years off you (it didn't work for me!), or join in a local kava ceremony which is based around a mildly narcotic drink made from pepper roots that makes your lips tingle and relaxes your body. The landscape and wildlife are as rich and diverse as the local culture and history.

GENERAL DIVING

The diving in Fiji is right up there as one of the top activities to include on any traveller's itinerary. It is probably most famous for its vibrant coral reefs, but many people also want to see the large marine life that inhabits these waters. On the

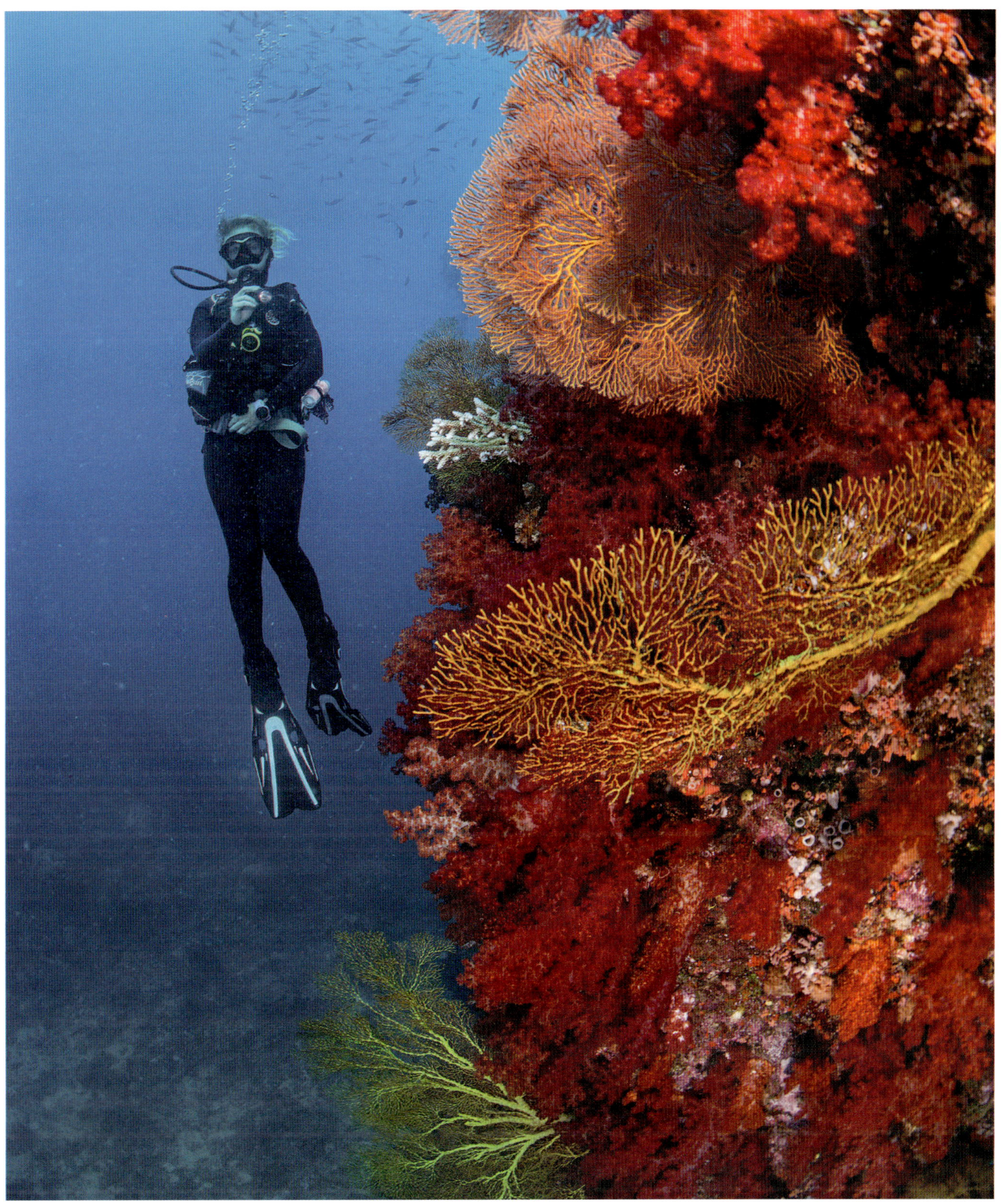

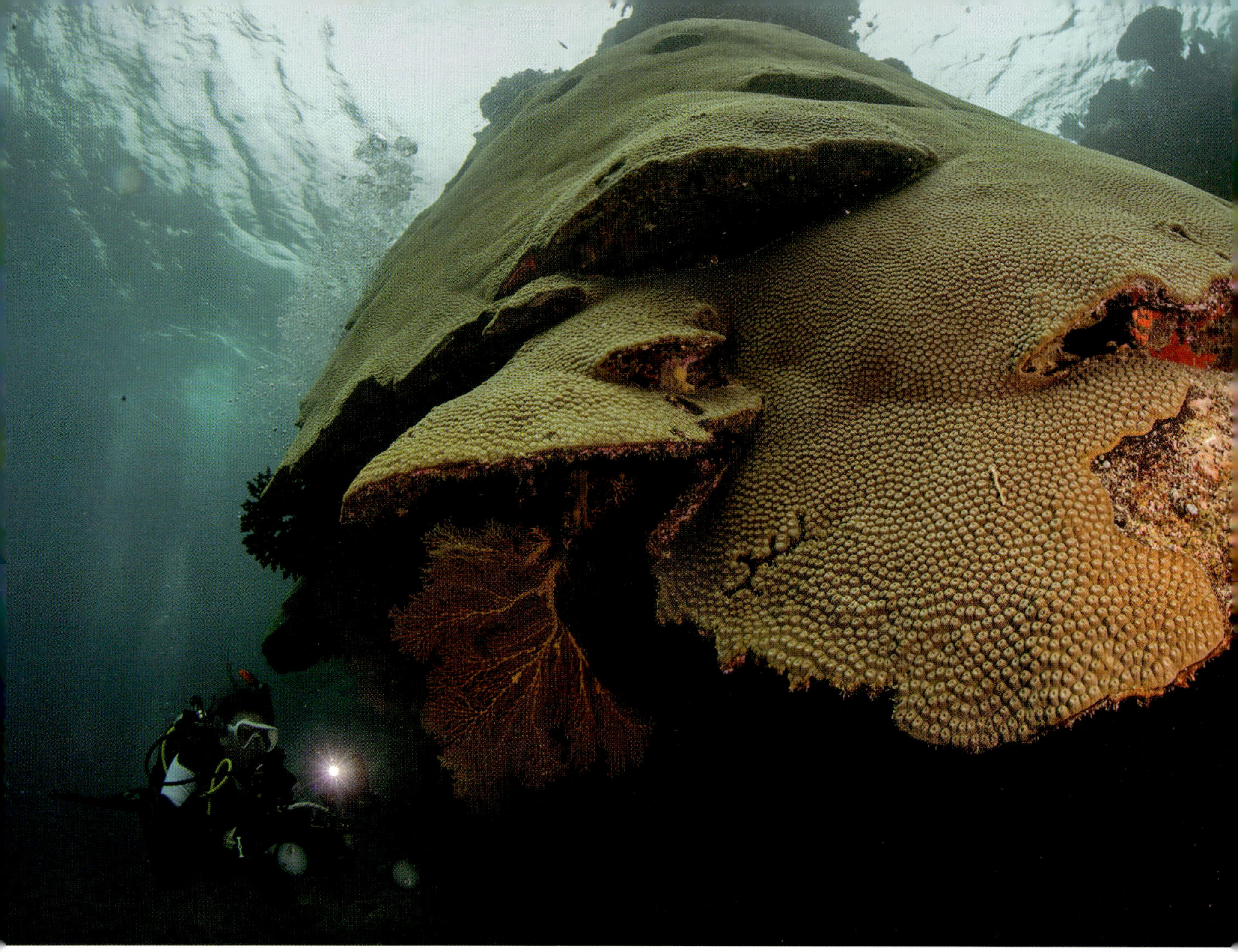

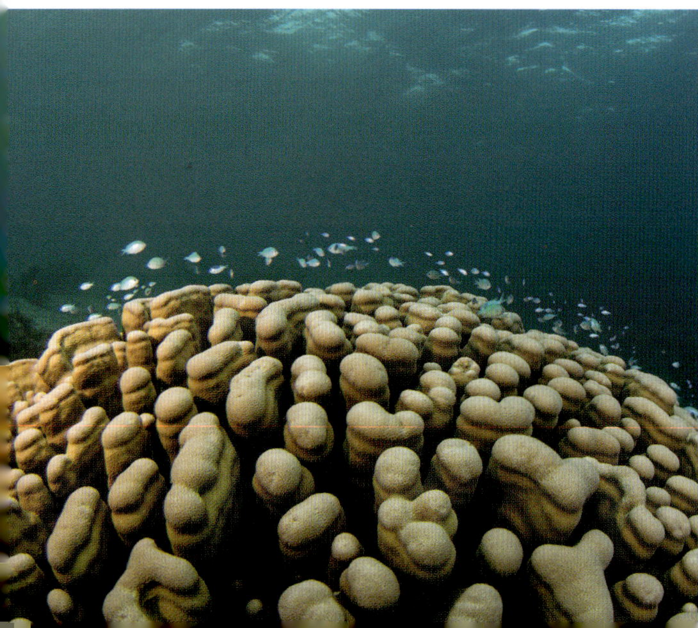

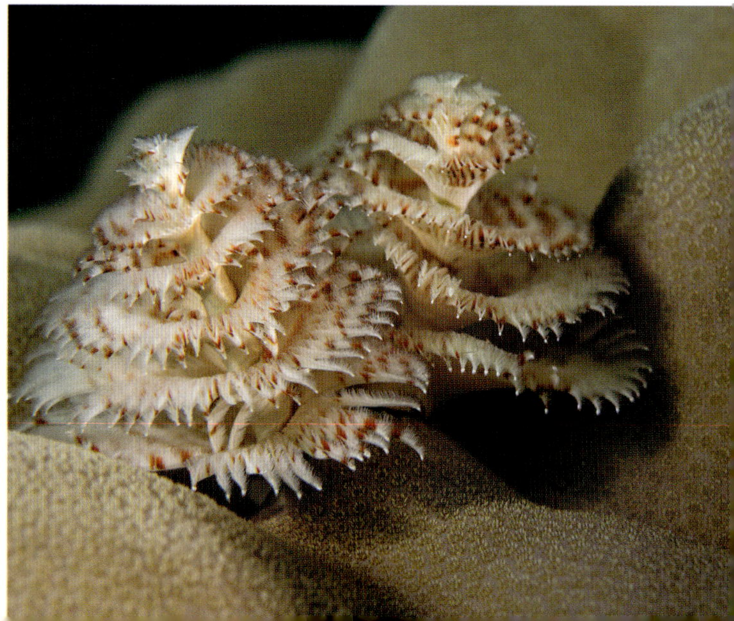

main island of Viti Levu, in a place called Beqa Lagoon (pronounced *benga*), you can dive with dozens of Bull Sharks. The area has been made a marine reserve and the locals have a huge respect and love of the sharks that frequent these waters. Off the island of Kadavu, on the Great Astrolabe Reef, you can snorkel or dive with manta rays, and off Taveuni you can dive the world-famous Rainbow Reef and White Wall. There are so many great dive sites around these islands that it would take many trips or a long stay in order to appreciate them all. On top of this, the region is a hot-spot for Humpback Whales and other impressive marine life.

THE WILD DIVE

We did so much great diving on the three Fijian islands we visited that it still sits as one of our favourite diving destinations. However, coming to the end of our trip we found ourselves on the main island of Viti Levu, staying in the Pacific Harbour area. We arrived at our hotel in the pouring rain, and even though it was just a couple of steps from the taxi to the hotel reception we looked like drowned rats as we dashed in to get some cover. Tropical rainstorms, as anyone who has experienced one knows, can be intense! Once we had got settled into our room and dried off the weather had magically cleared and we sat under a perfect blue sky and then started to relax as the sun set. While enjoying a cool beer with our hosts they asked us where we would like to dive the next morning. We suggested that they should take us to their own favourite location to kick off this part of our trip, and so it was agreed that we would head to a dive site called Fantasea 1 early the next morning.

OPPOSITE TOP Huge coral heads tower over a diver.

OPPOSITE BOTTOM LEFT Tiny fish dash back to the safety of the reef as we approach.

OPPOSITE BOTTOM RIGHT Christmas Tree Worms live in the hard corals.

BELOW The shark dive in Fiji is listed as one of the best in the world.

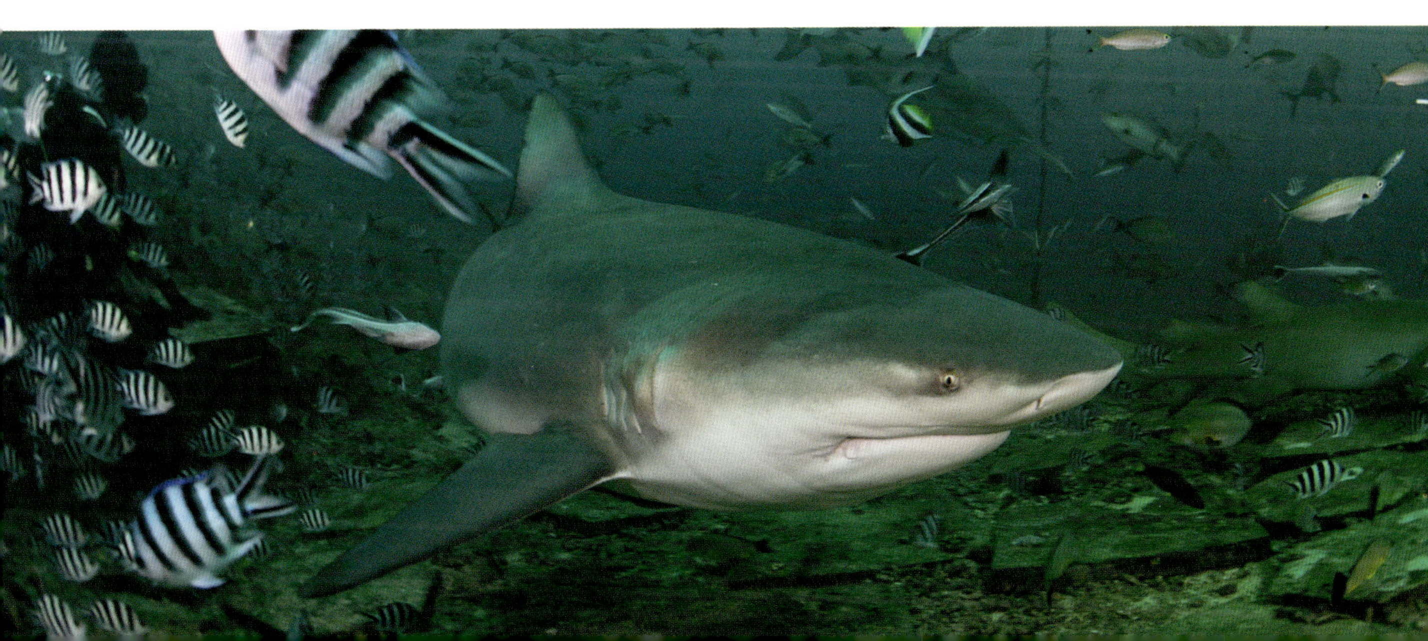

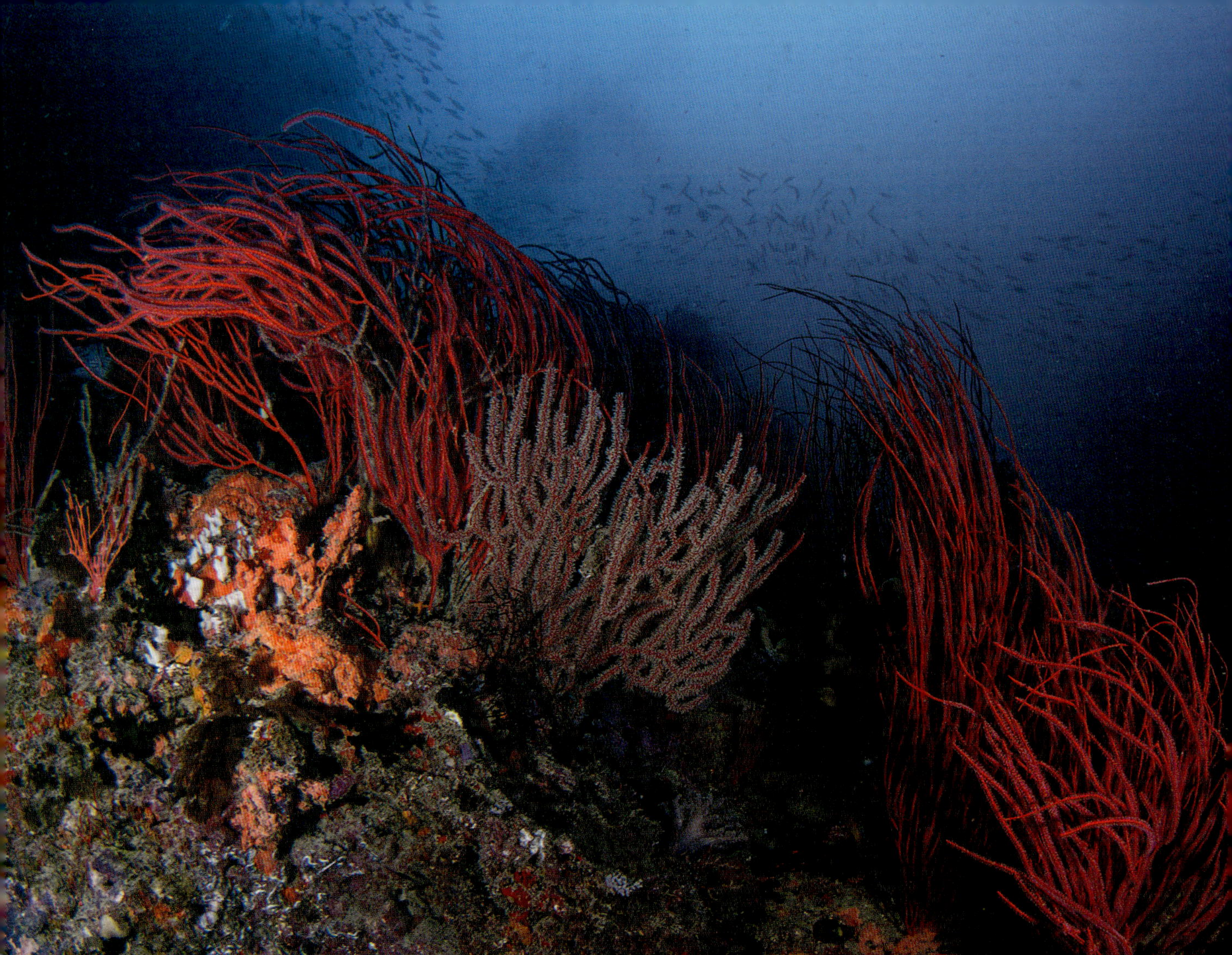

ABOVE Red, pink and purple corals stand out against the beautifully clear blue water.

OPPOSITE Vertical walls have sea fans clinging to every surface.

The following morning we woke to a beautiful day with the sun shining and the seas flat calm. We were excited to be diving a new destination for us. The dive site was adjacent to a small island called Yanuka which would provide us with a place to moor the boat between dives. Our guides eagerly told us that Fantasea 1 was their favourite dive because the coral was like no other dive site they had ever seen, and these guys had worked in a fair number of dive locations. Fiji is famous for its coral and we had already dived several incredible sites, but we were willing to believe them, as this small pinnacle that did not quite reach the surface certainly sounded wonderful. I got into my gear, grabbed my camera, and jumped into the water. As I waited at the surface for my buddy and guide to join me, I dropped my head into the water and looked down at the reef below – the sight was amazing, mind-blowing and

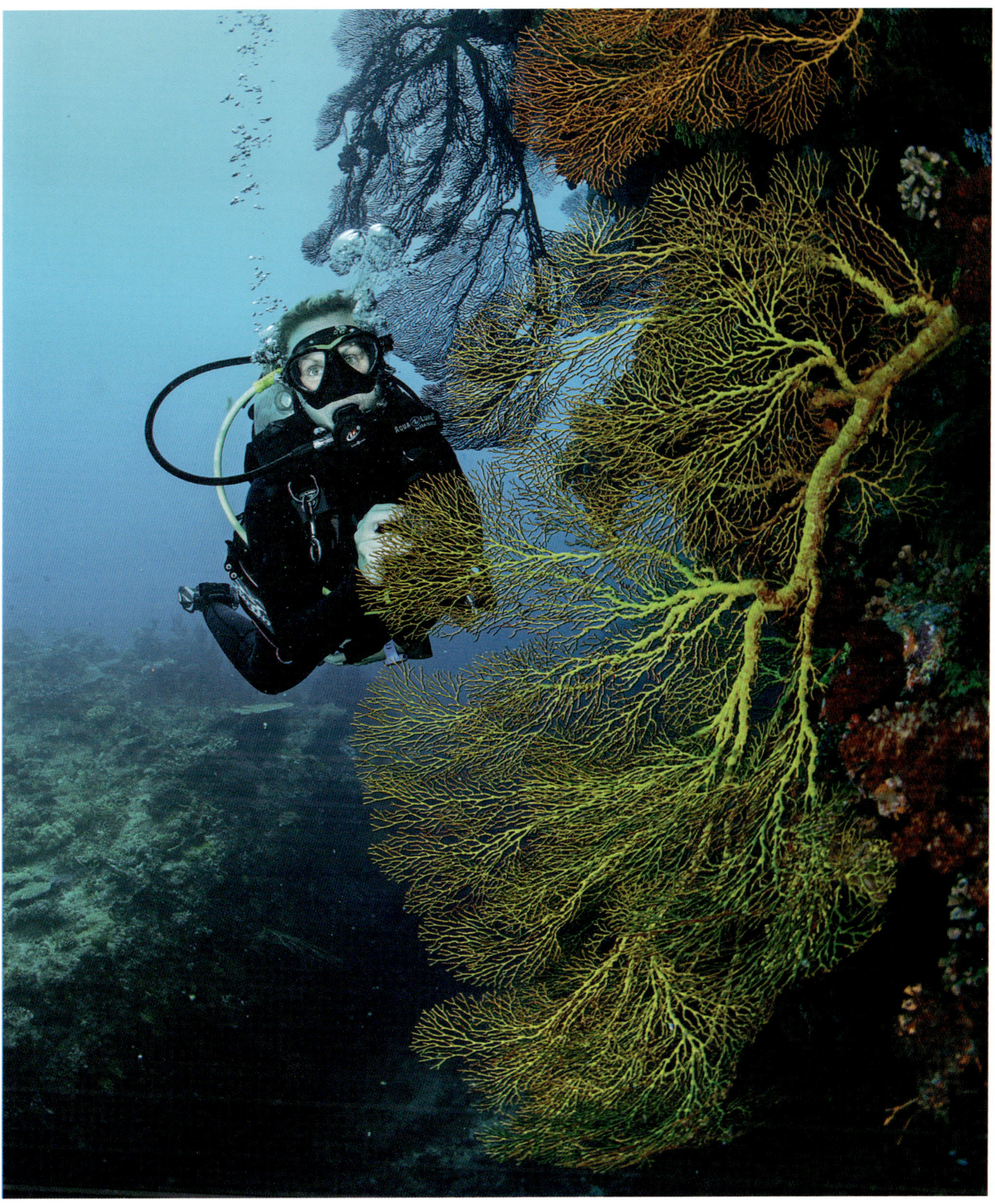

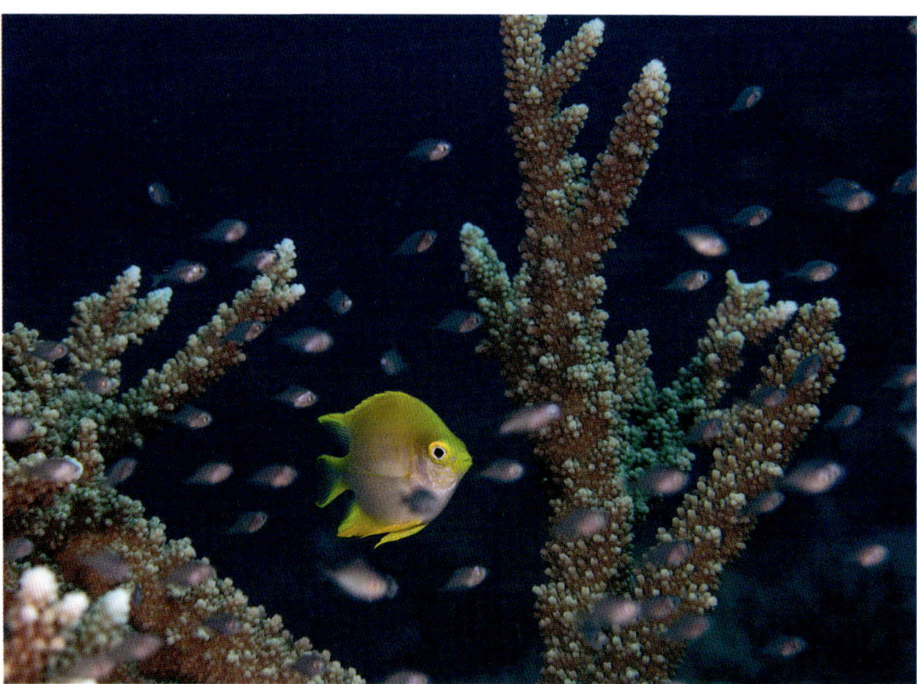

absolutely stunning. Every inch of the top of the pinnacle was covered in soft corals and anemones, all lightly swaying in time with the gentle waves. I was eager to explore more and as we descended, and directed our lights onto it, it just got better and better. As we descended further, the soft coral gave way to huge gorgonians and whip corals. Fish darted in and out of the reef, using the corals as protection. The scene so vivid – with bright pinks, oranges, reds, purples – that it was hard to believe that the reef was real.

As an experienced dive writer and journalist I never thought that coral alone would be enough to blow my mind on a dive, but Fiji is different and it certainly did. It makes your senses leap to attention, and I just wanted to take some time to hover in mid-water and let it all soak in. Back on the boat the dive guides glanced at each other and laughed as we eagerly chattered about what we had seen. For a dive guide, the excited babble of the divers after a dive is always a great indicator of how much they had enjoyed it, and we have seen so many fabulous places that we cannot be the easiest to please. However, in this case they were right all along – this really is a special dive site.

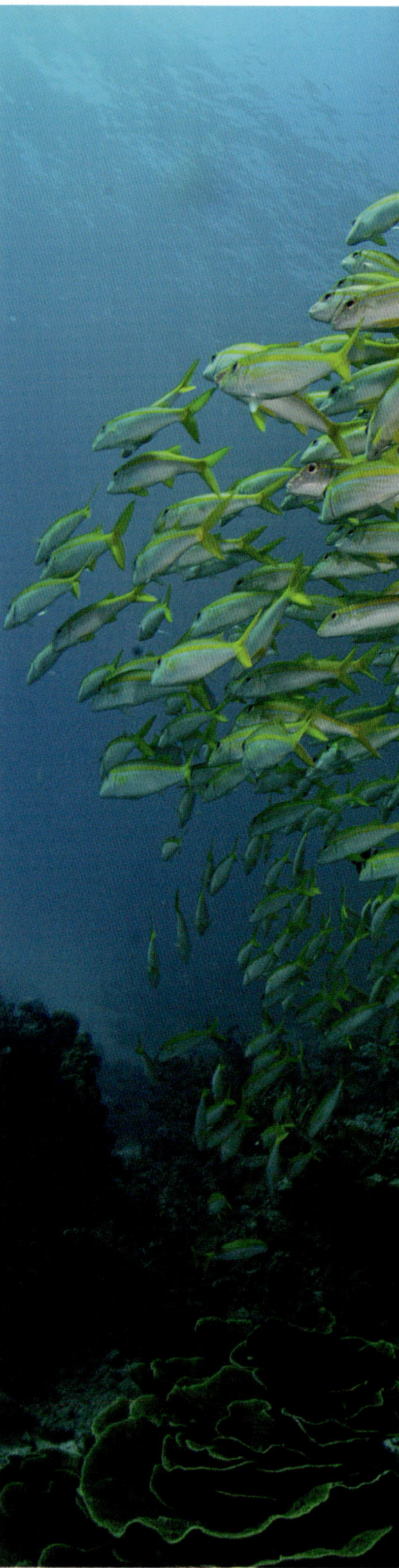

ABOVE Small fish make the coral reefs their home.

RIGHT Schools of fish swim in unison across the reef.

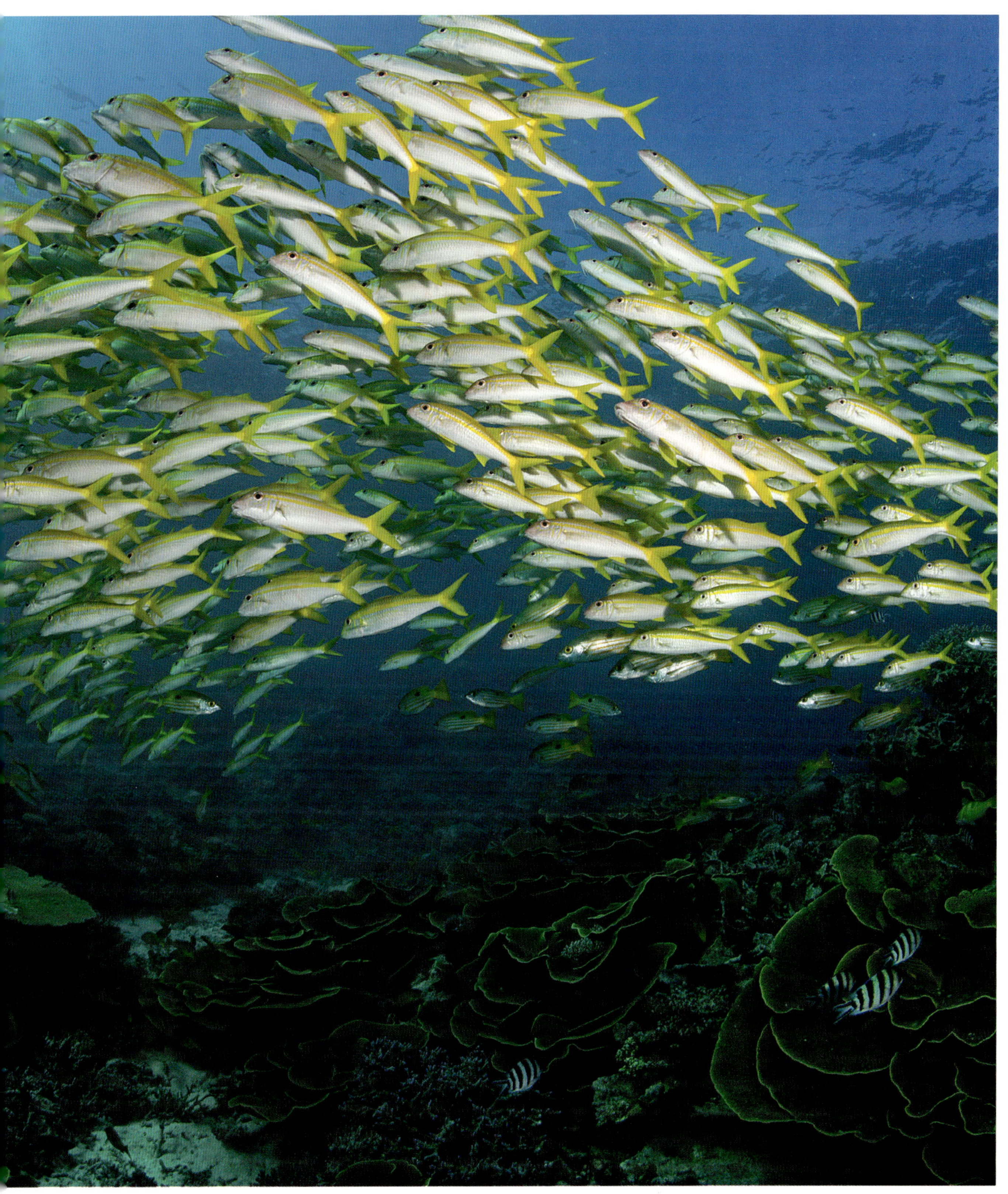

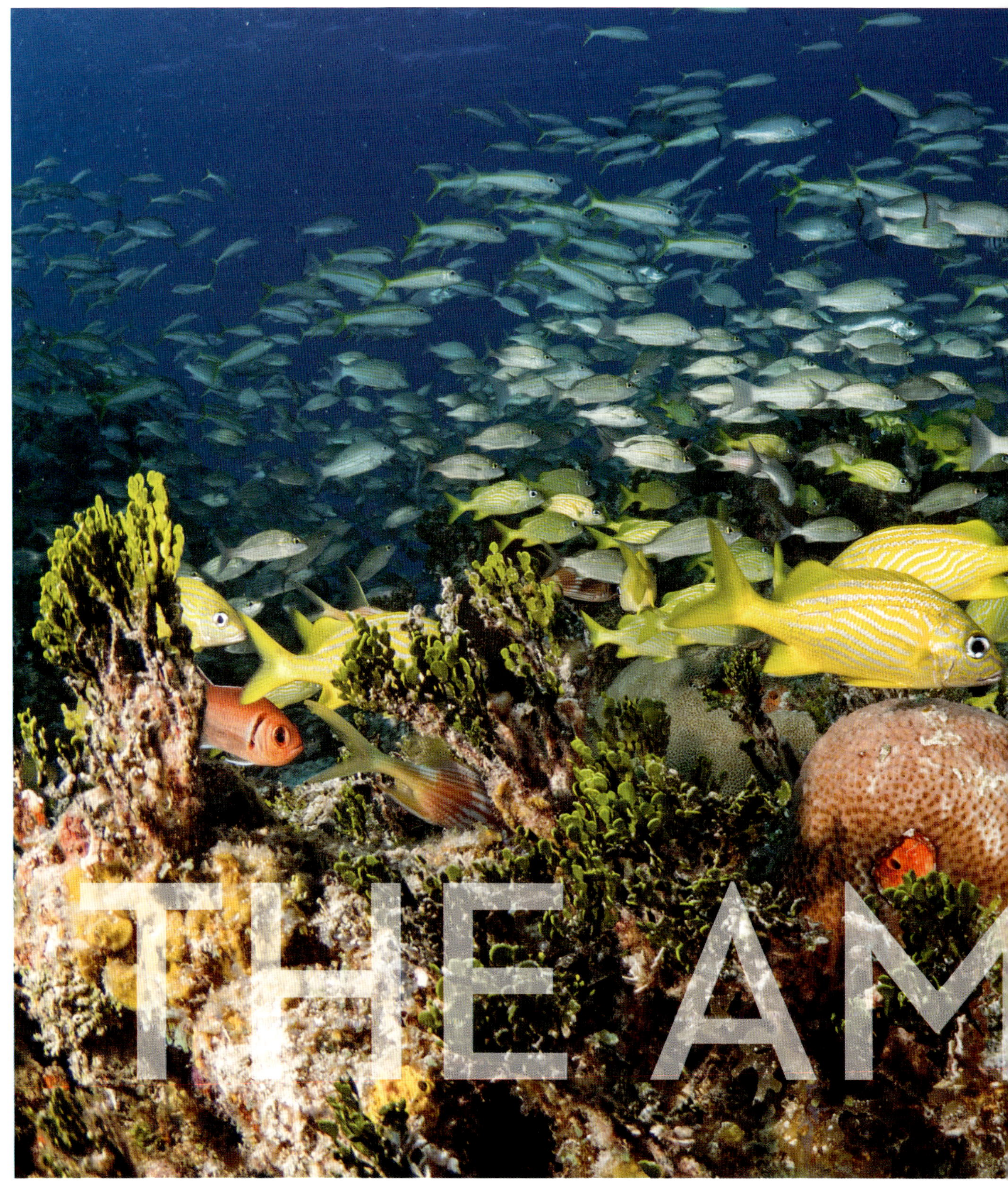
THE AM

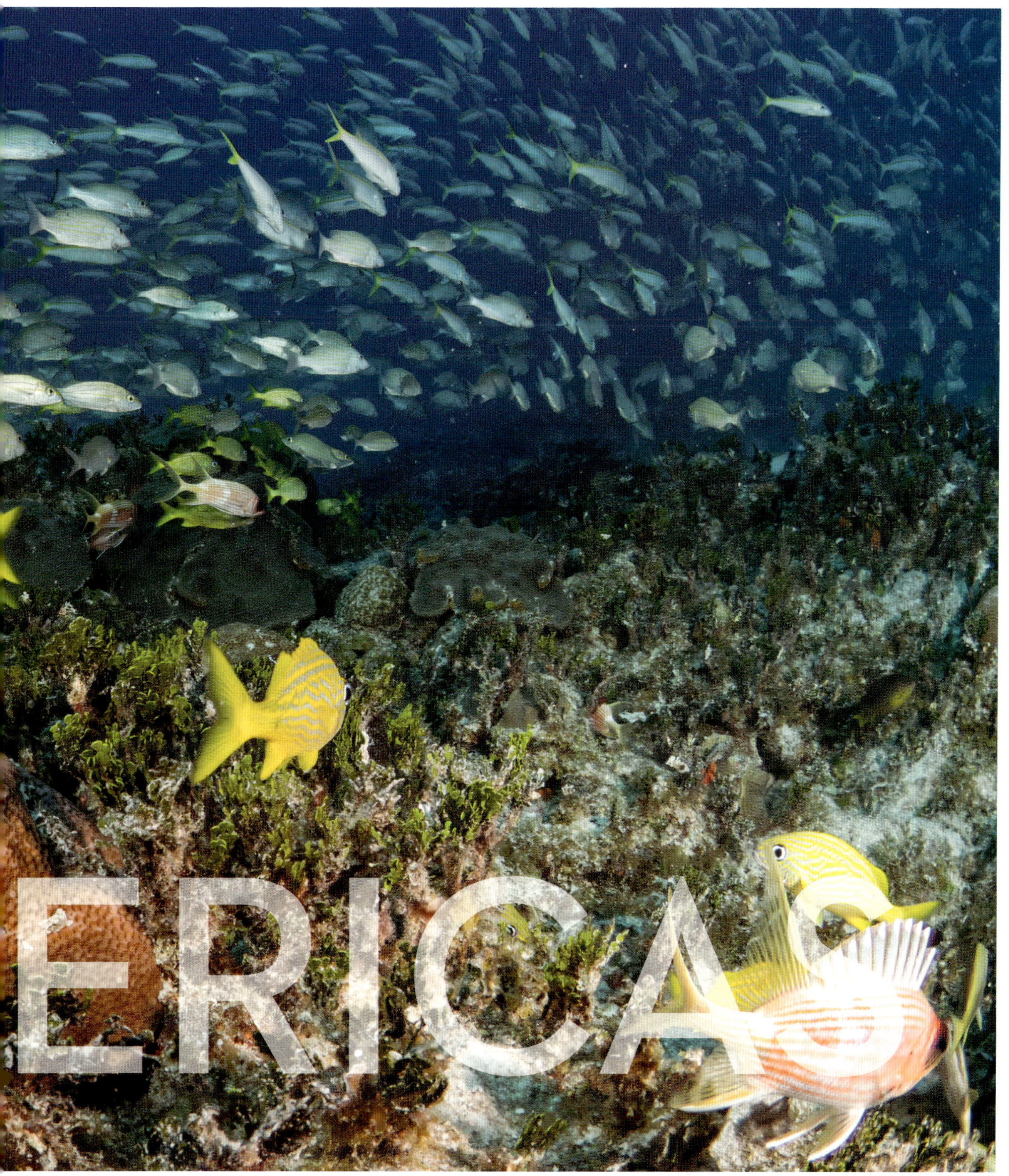
ERICA

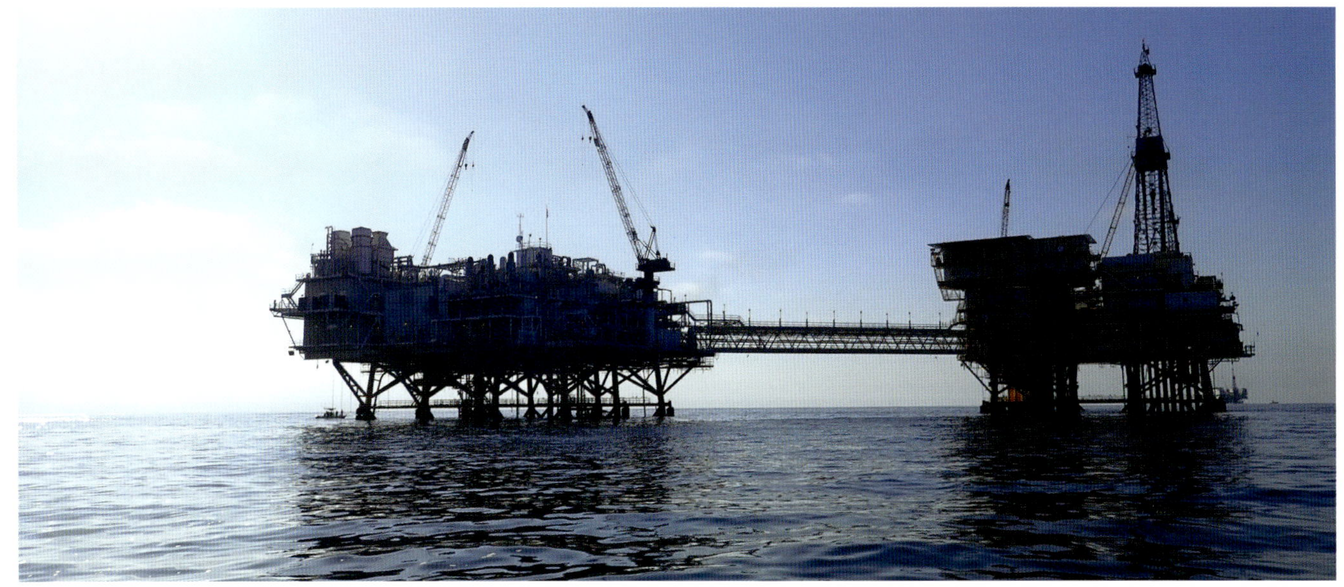

15 DIVING A WORKING OIL RIG IN CALIFORNIA

USA

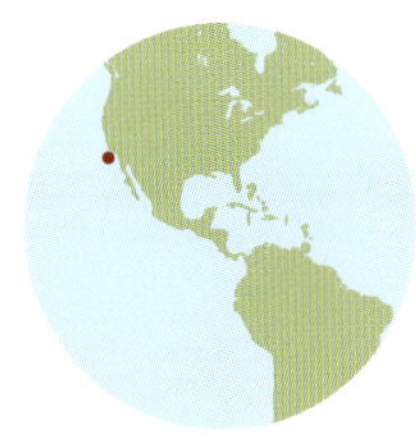

ABOVE Oil Rig Ellen off Long Beach.

OPPOSITE The legs of the oil rig are covered in brittle stars.

CALIFORNIA HAS AN INCREDIBLE COASTLINE that stretches for nearly 1,450km (900 miles), facing west and bordering the Pacific Ocean. From north to south the shores of this state have something to offer everyone, with vineyards, incredible roads, outstanding natural beauty, a rich history and some fantastic cities to take in along the way. It is perfect for an extended road trip. Long Beach is a coastal city and port that sits within the Los Angeles metropolitan area of southern California, and it offers fabulous beaches, great food and drink, sunshine and lots to do.

GENERAL DIVING

The diving in this area is varied, depending on where you dive. From Long Beach, the most popular diving is found at Catalina Island, one of the Channel Islands. Depending on your dive boat you can be here in just over an hour, but can take

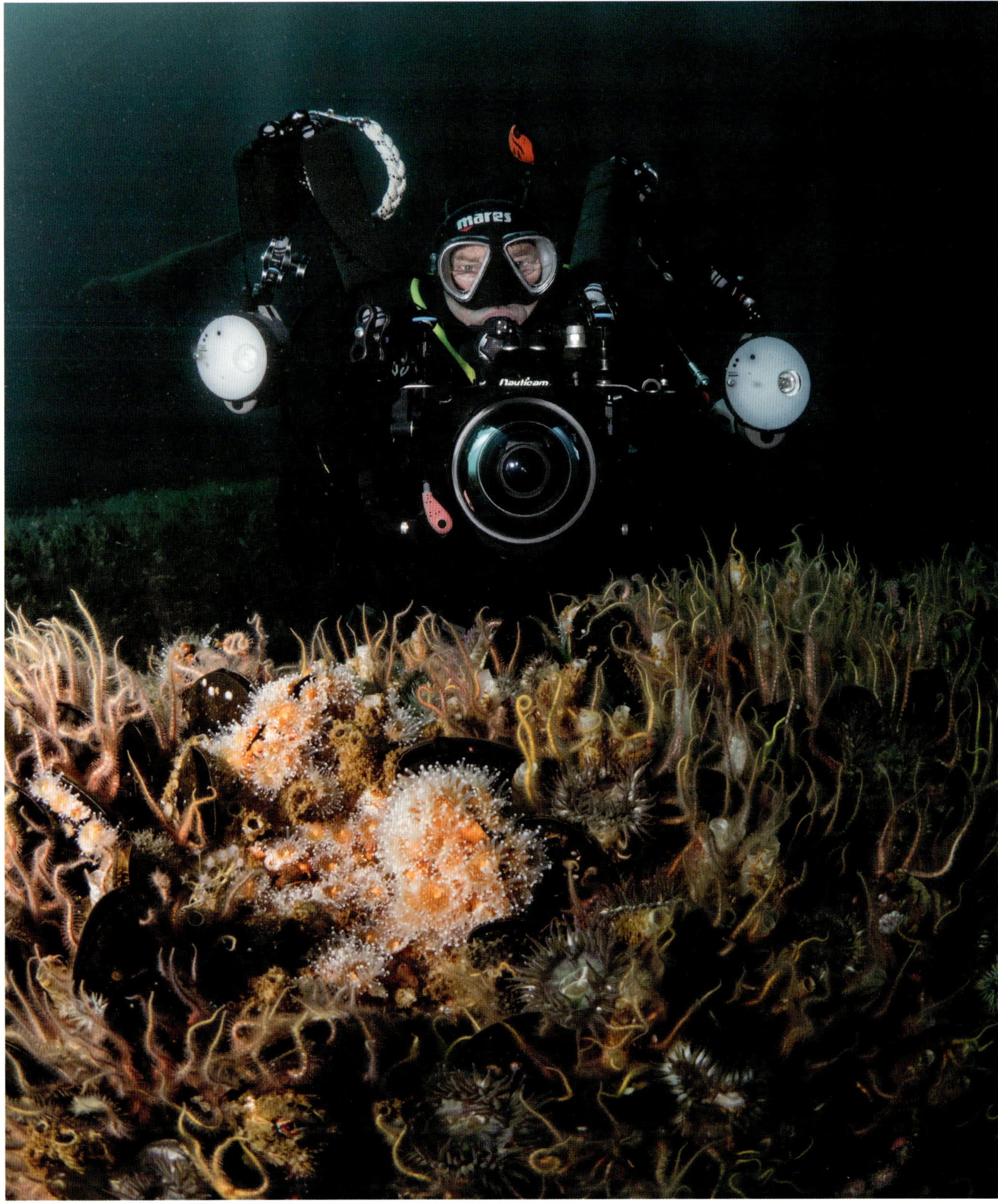
mares
Nauticam

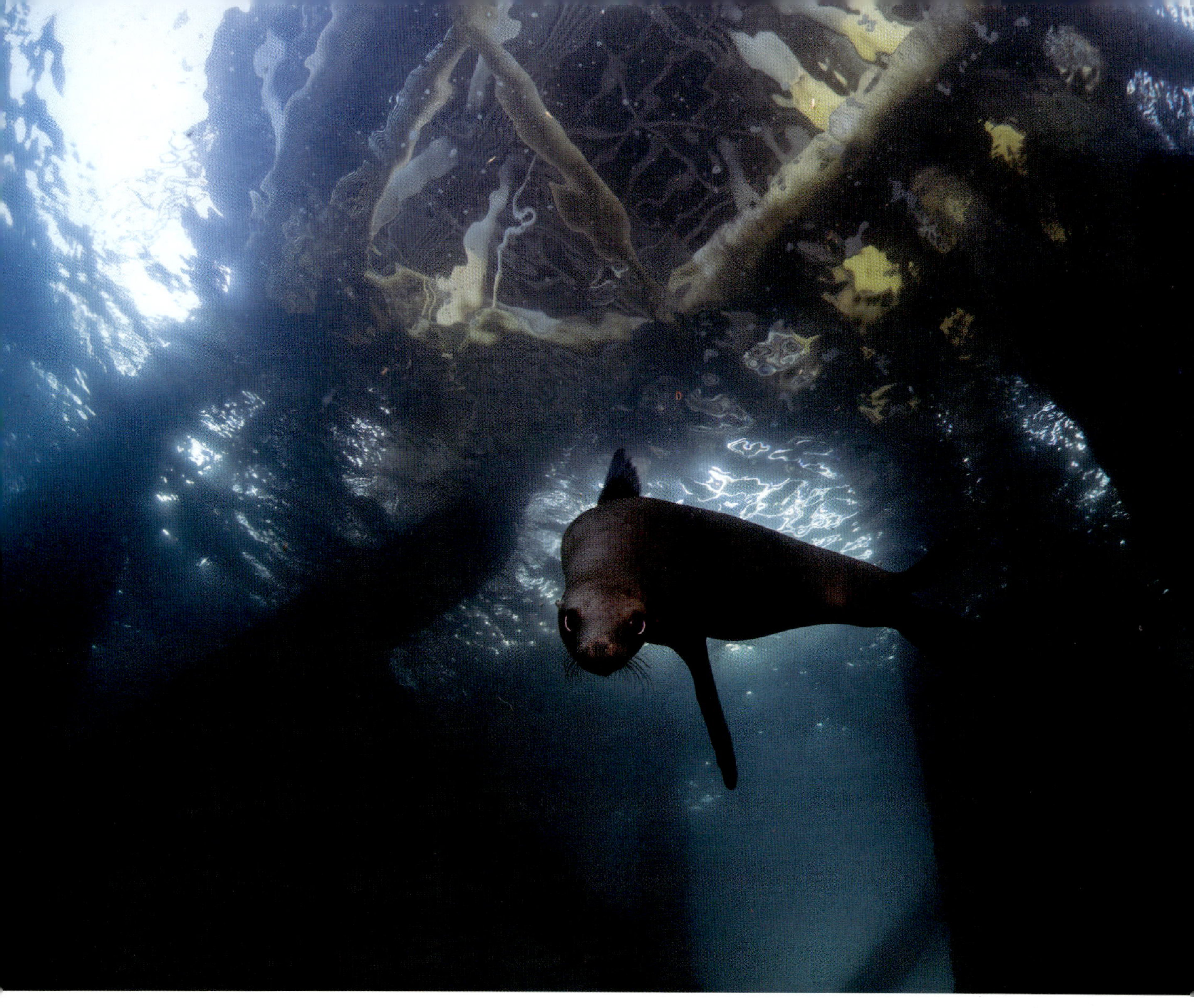

much longer so an early start is well worth considering in order to get the most of a diving day trip. At Catalina Island you can find populations of California Sea Lions and Harbour Seals, playing 'now you see me, now you don't' in the huge dense kelp forests that the Channel Islands are famous for. The kelp grows from the seabed right up to the surface, providing shelter for a whole host of marine life. One of the most famous and sought-after residents of the kelp forest is the Giant Black Sea Bass. Although it was hunted nearly to extinction in the 1950s and 1960s, this huge fish can still be found casually cruising among the kelp fronds. Another fishy highlight is the bright orange Garibaldi, which catches the eye against the green

ABOVE The sea lion put on an acrobatic display.

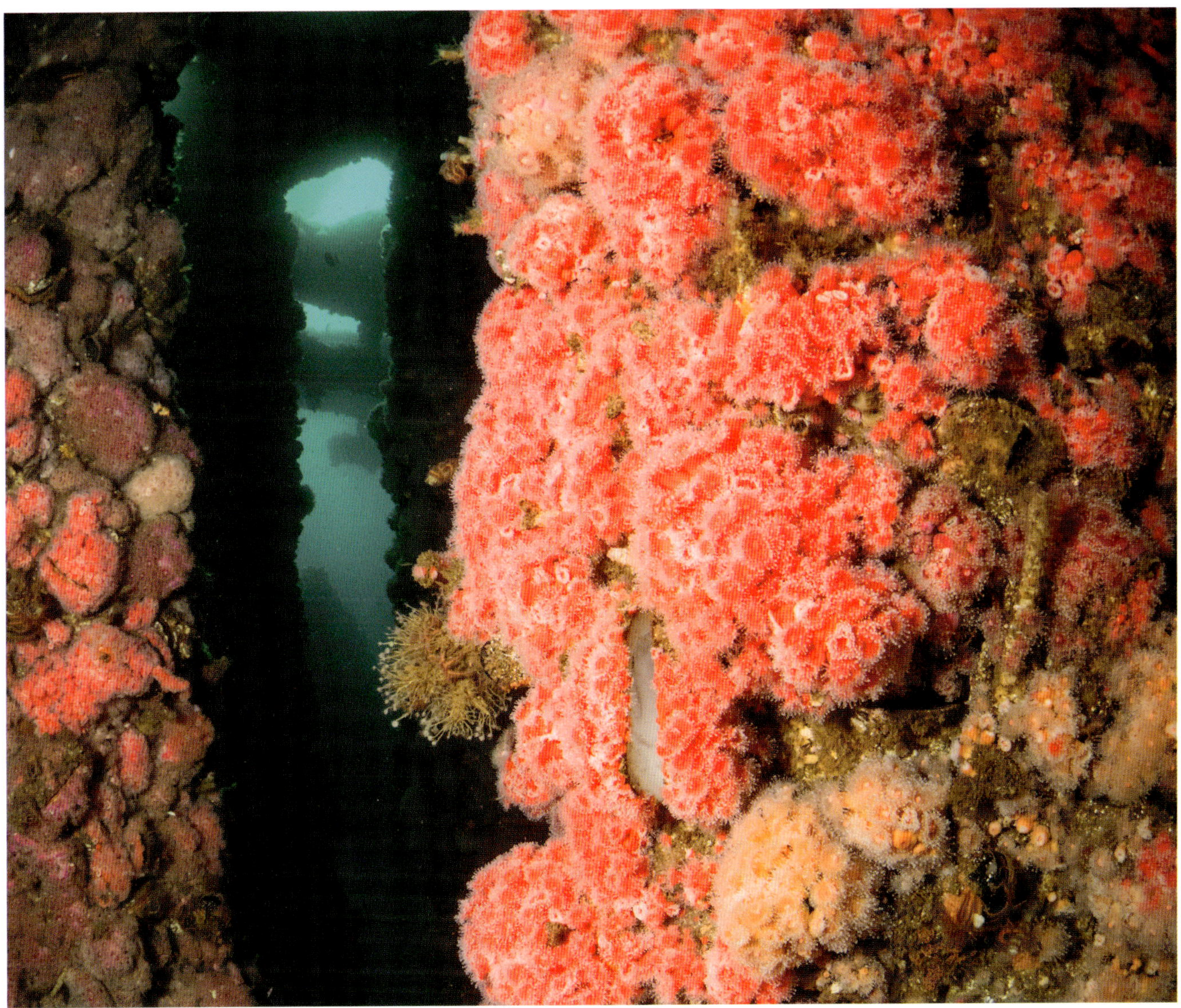

backdrop. The boat ride to and from the island can also provide a real treat, with dolphins and whales common visitors to these waters. If you want to try something a little different you can also take a behind the scenes tour of The Aquarium of The Pacific and dive into the Tropical Reef Habitat.

THE WILD DIVE

Ever since seeing an incredible image of a diving seabird hunting bait fish, I have always wanted to dive one of the oil rigs of southern California. It has always

ABOVE Anemones clinging to the oil rig create a beautiful sight.

appeared to me to be such a marked dichotomy to have such incredible marine life found below what is a less than eco-friendly industry. But that is what the three working oil rigs off Long Beach offer the diver. You do, of course, have to get permission to dive beneath the oil rigs, and the dive boats have to be careful when manoeuvring to drop you in the water and pick you up. As we approached, the noise from the working rig got louder and the smell stronger, and by the time we were in position I could even hear the shouts of the workers above. Luckily the sea was flat calm and so we were given the go-ahead to slip into the cool water of the Pacific Ocean. We were told to swim directly to the oil rig and to then stay close to the structure so as not to drift away into potentially hazardous marine shipping lanes. As it turned out, this was not going to be a difficult command to follow. There was no current and the water was clear. My buddy and I decided to stay close to the surface where there would be more light for our cameras. We dropped just below the surface and started to explore the legs of the oil rig. The very top was covered in mussels, all dark and with sharp edges, and then as we got slightly deeper this changed to a riot of different hues and shades with anemones and brittle stars covering every inch of the structure.

Then something caught my eye. There appeared to be a shadow moving very quickly just at the edge of my vision. As I stared into the blue in an effort to try and see what had attracted my attention, a young inquisitive California Sea Lion darted past me. This is what we had hoped to see by coming to the oil rigs. The structures

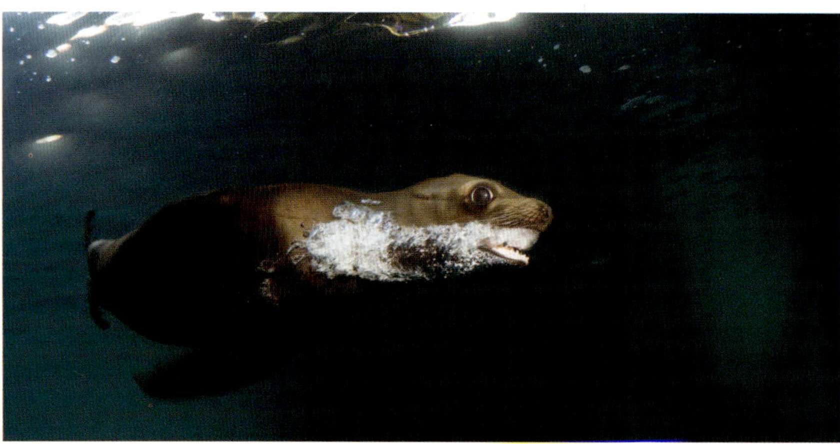

ABOVE A California Sea Lion blowing bubbles at high speed.

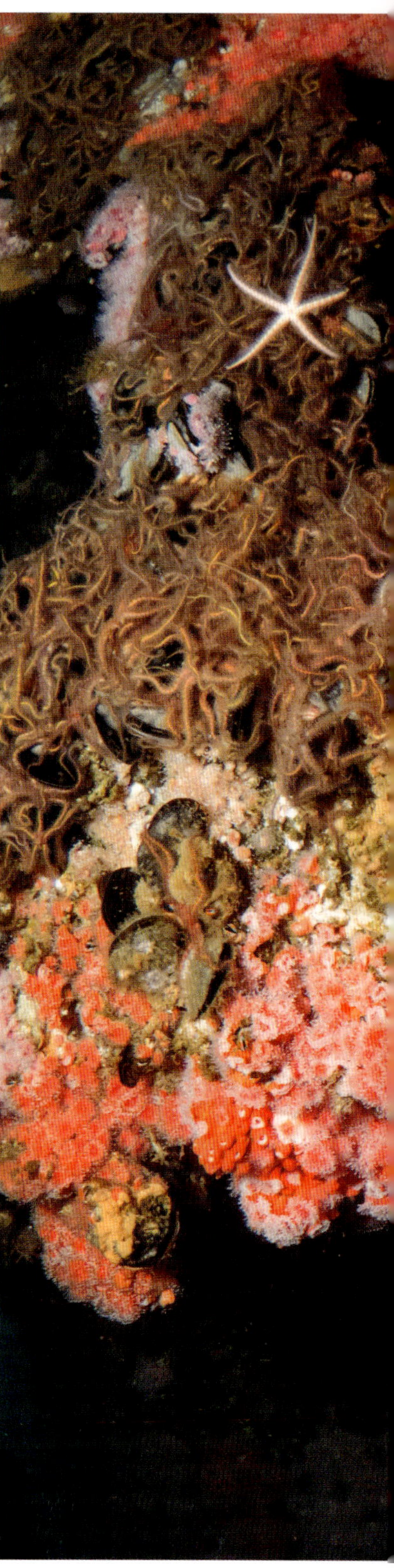

RIGHT It is amazing to see so much life under the oil rig.

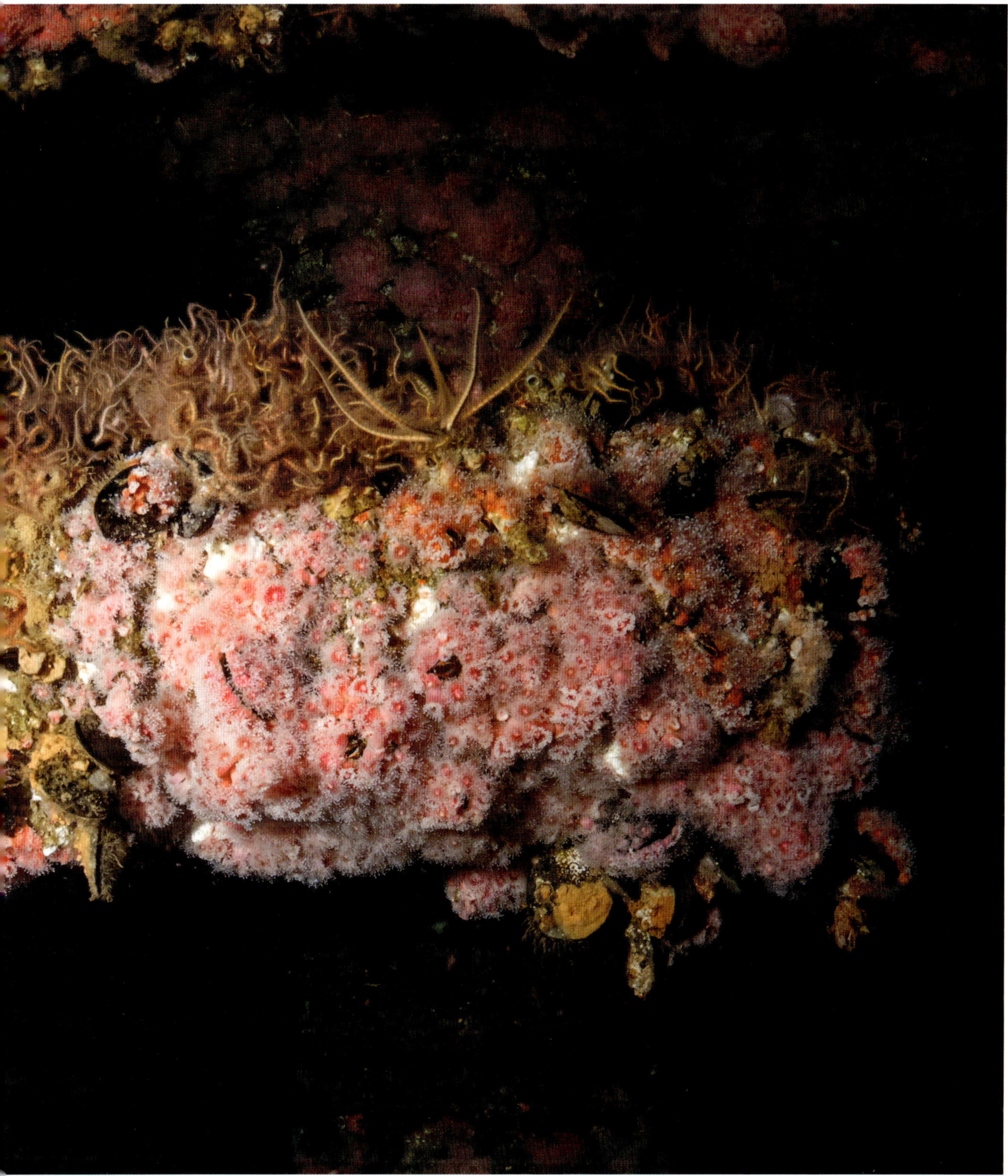

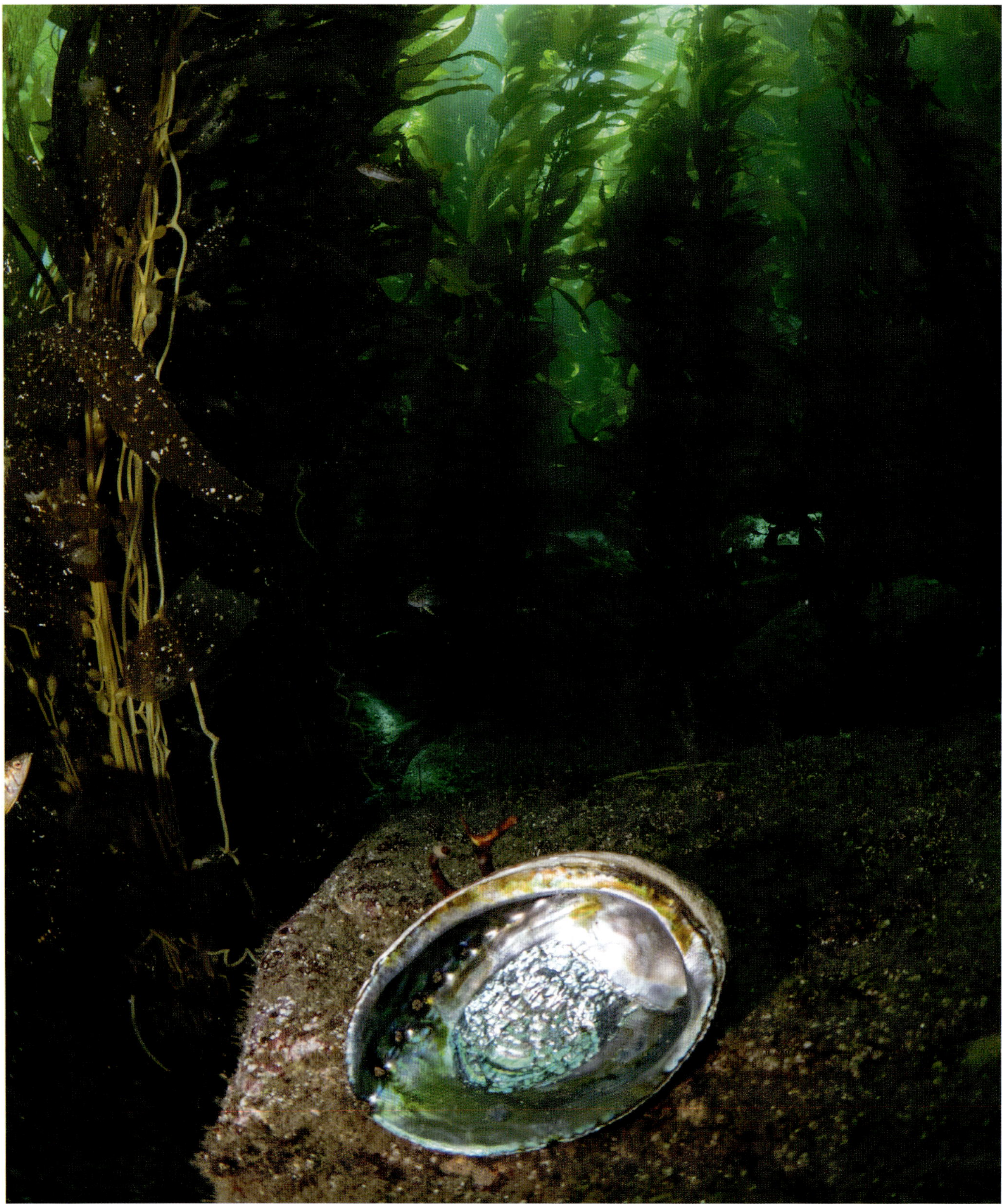

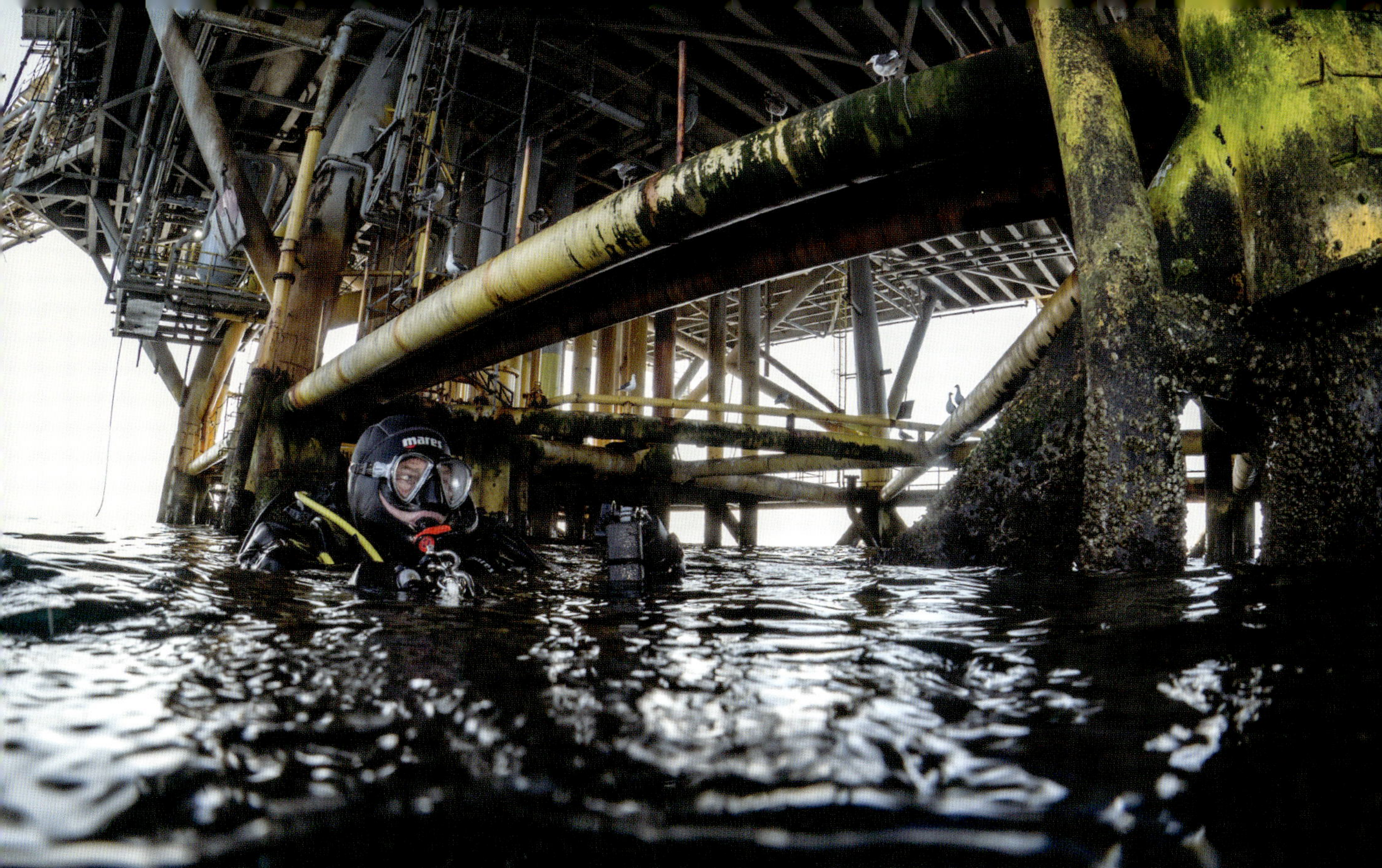

make a superb playground and space for these pinnipeds to spend their day, sometimes resting out of the water by hauling themselves up onto the framework below the working platform and sometimes hunting in the deep water beneath the rigs. However, like seals and sea lions everywhere, the young ones like to play and a group of slow-moving but friendly divers is too good an opportunity to be missed. This sea lion was fast, darting among the steel girders at incredible speeds, turning on a knife edge and putting on an incredible display of underwater acrobatics. Sometimes, as he zoomed past, he would blow bubbles at us or bark. It wasn't long before all six divers in our group had come together close to the surface to marvel at this one individual which had decided to make our dive a truly special occasion.

I tried to get some images of this high-speed display but it was not easy. It was hard enough to keep track of where the young sea lion was, let alone get a shot. In the end I resorted to video as a method of more accurately portraying just what the experience had been like.

I am not sure what I expected but these oil rigs, like so many piers and structures that overhang the sea, offer a haven for wildlife out in the ocean and are certainly well worth diving should you ever get the opportunity.

OPPOSITE An abalone shell in the kelp forest of Catalina Island.

ABOVE About to drop below the oil rig structure.

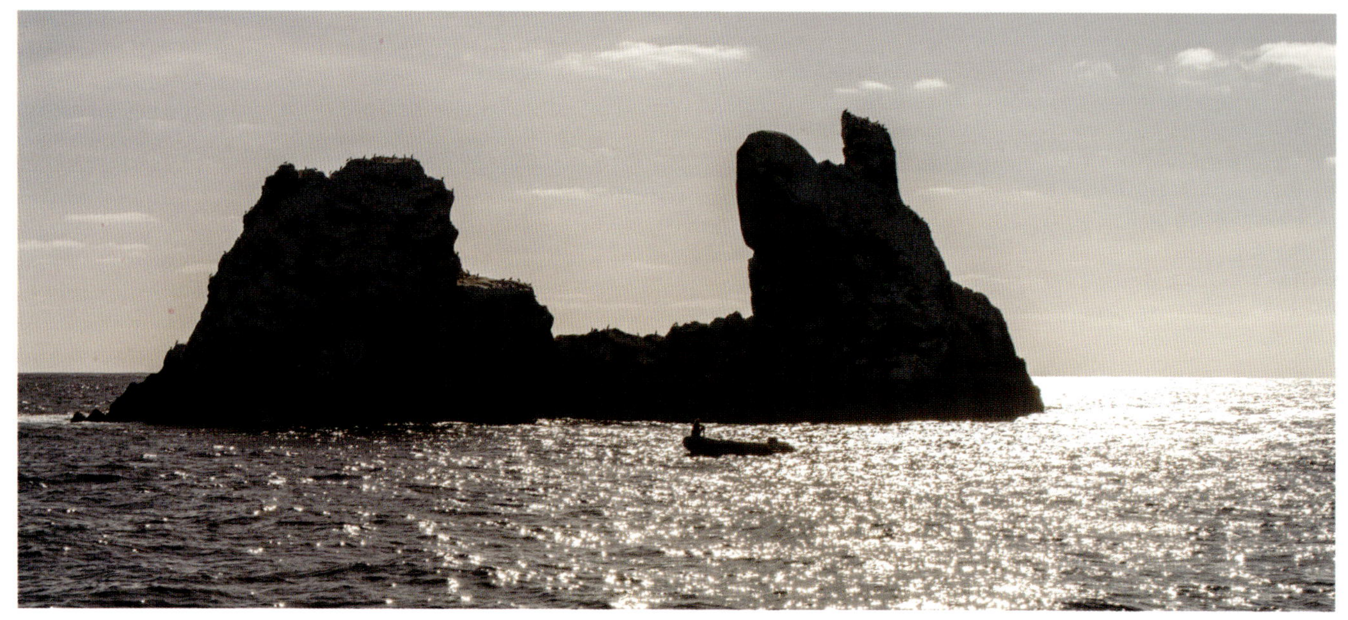

16 PLAYING 'SIMON SAYS' WITH A MANTA RAY IN SOCORRO

MEXICO

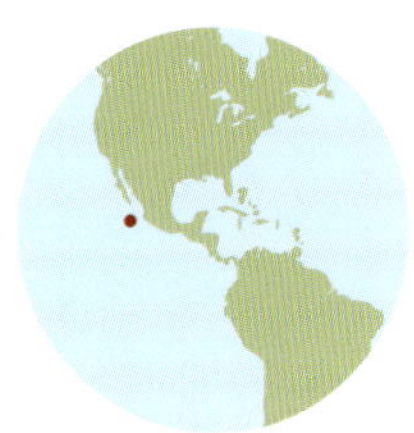

ABOVE A dive boat picks up divers from the last dive of the day at Roca Partida.

SOCORRO IS A VOLCANIC ISLAND that is part of the Revillagigedo Archipelago in the Pacific Ocean. It takes a 36-hour boat ride to reach these islands, which are located some 390km (240 miles) off the coast of mainland Mexico. It is a remote and rugged destination that can only be visited by liveaboard. The islands are a marine reserve and have also been designated as a UNESCO World Heritage Site.

GENERAL DIVING

The diving here can be challenging with strong currents, surge and large waves. As it is also about as remote as you can get, so much so that special procedures are

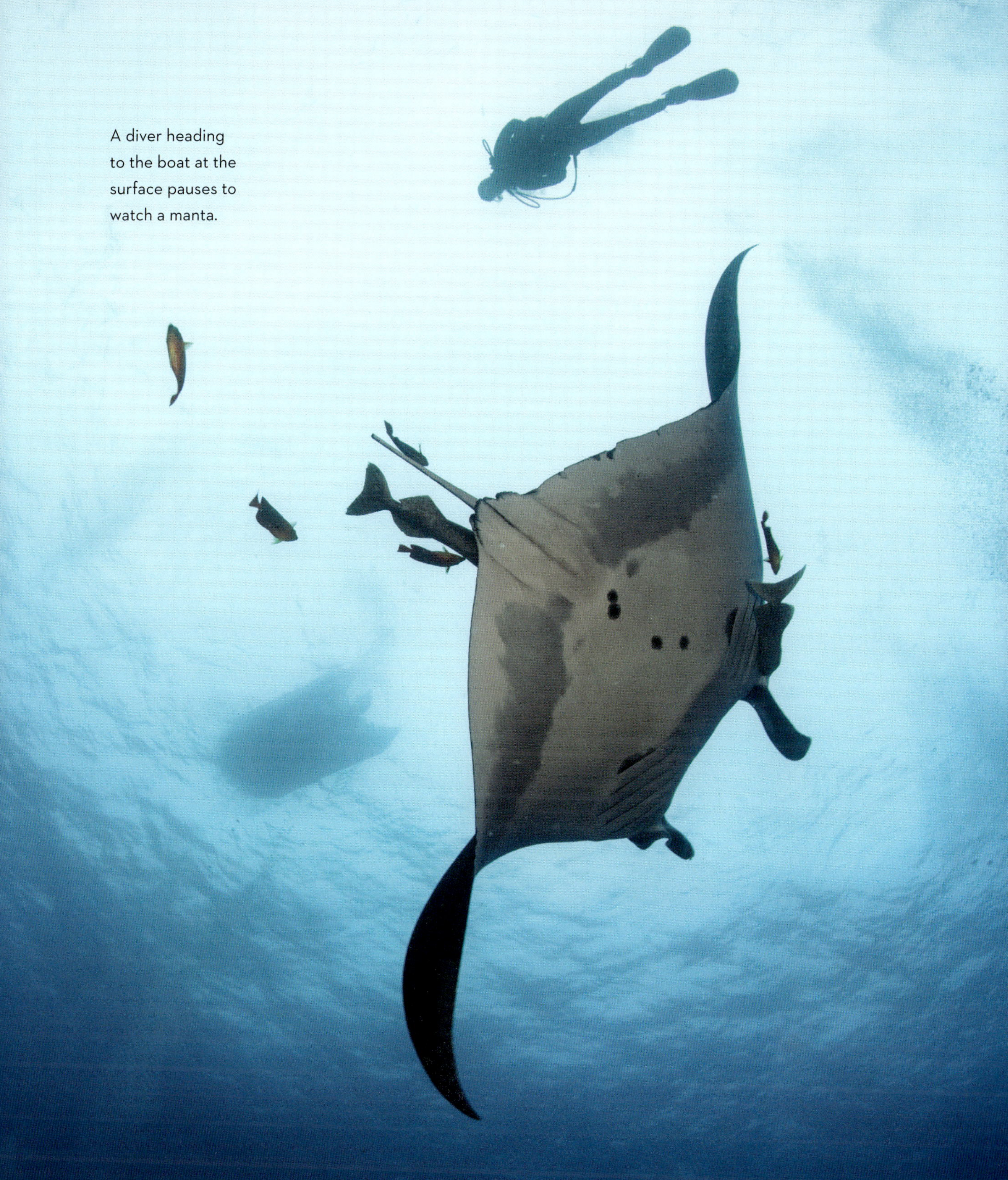

A diver heading to the boat at the surface pauses to watch a manta.

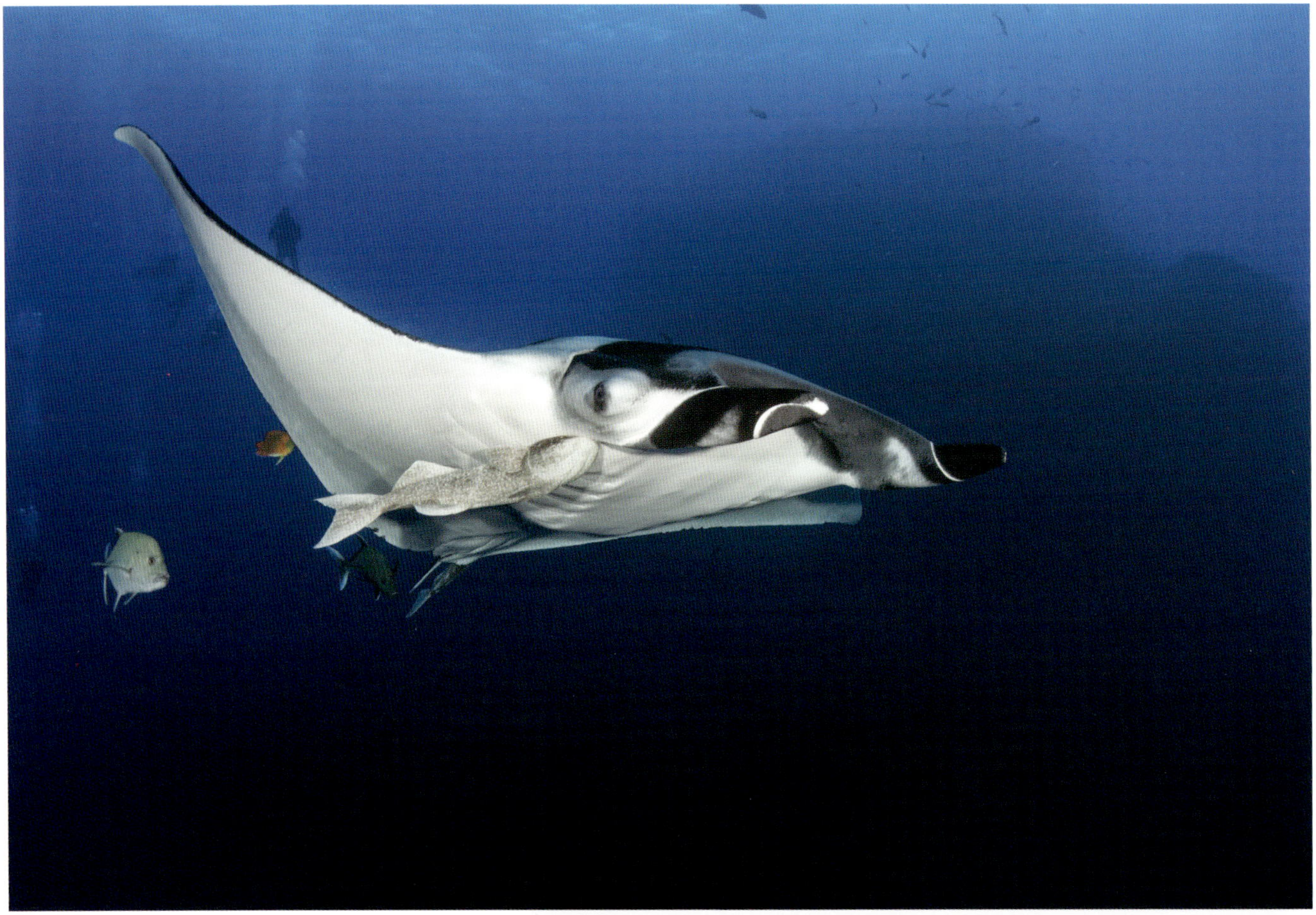

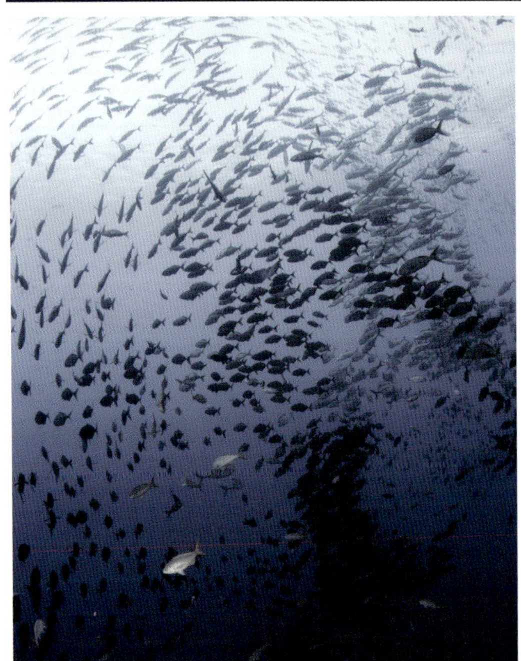

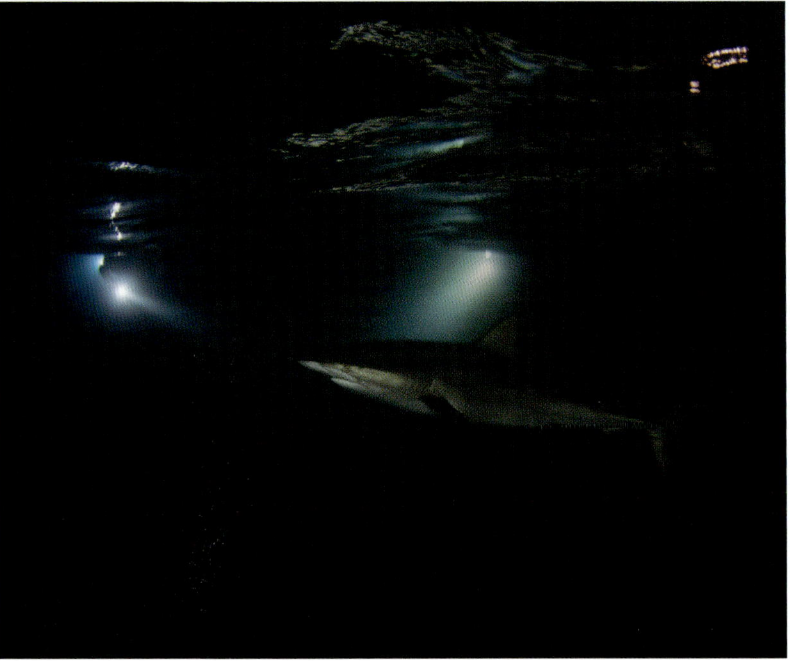

put in place where every diver has to wear a GPS location device in case they drift out of sight of the boat. Underwater, the topography is impressive. These volcanic islands continue their rugged form below the water with huge rock formations that get battered by the waves. Some pinnacles rise up out of the water, and some cannot be seen from the surface, but these islands in the middle of the Pacific Ocean attract an incredible array of pelagic marine life.

On any single dive you might encounter four or five species of shark. Humpback Whales can be heard singing as you dive, and if you are really lucky you might even encounter one underwater. Dolphins are abundant here and often approach divers. Huge schools of fish swirl near to the safety of the rocks, and Orcas have also been known to hunt here. You can even night snorkel with Silky Sharks. The most compelling reason that divers put this remote destination on their wish list, however, is for the Giant Mantas that frequent all the dive sites along the archipelago.

We dived San Benedicto, Socorro and Roca Partida islands on our itinerary. Each one of these islands has amazing diving to offer. On Roca Partida you will find whitetip sharks piled on top of each other trying to get some rest in the small cuttings in the rocks that offer protection from the currents and surge. San Benedicto can offer large numbers of manta rays on every dive, but it is Socorro that really impresses with incredible encounters every single time.

THE WILD DIVE

During the long boat journey to reach the islands various members of our liveaboard crew presented talks in the evenings on the wildlife that we may encounter. Among the specialists on board was a marine biologist whose specialism was in manta rays. On the evening before we arrived at 'El Boiler' he spoke to us at some length and gave us scores of fascinating facts and a mass of information about these incredible creatures.

The Giant Manta Ray is the largest ray in the world and its wing-span can exceed 7m (23ft). They have the largest brain of any fish (relative to their size) and to assist in keeping all parts of the brain active, even in really cold conditions, they have blood vessels that are routed to supply heat to all the principal parts of the brain. Our guide also went on to say that these mantas are truly intelligent and will even make efforts to communicate with us. As a biologist with an MSc in animal behaviour, I was more than a little sceptical about this final part. However, my principal emotion was one of excitement at the prospect of finally getting into the

OPPOSITE TOP Huge manta rays come close to divers off the cliff face of El Boiler.

OPPOSITE BOTTOM LEFT Huge schools of fish create elaborate patterns in the ocean.

OPPOSITE BOTTOM RIGHT At night Silky Sharks hunt by the boat lights.

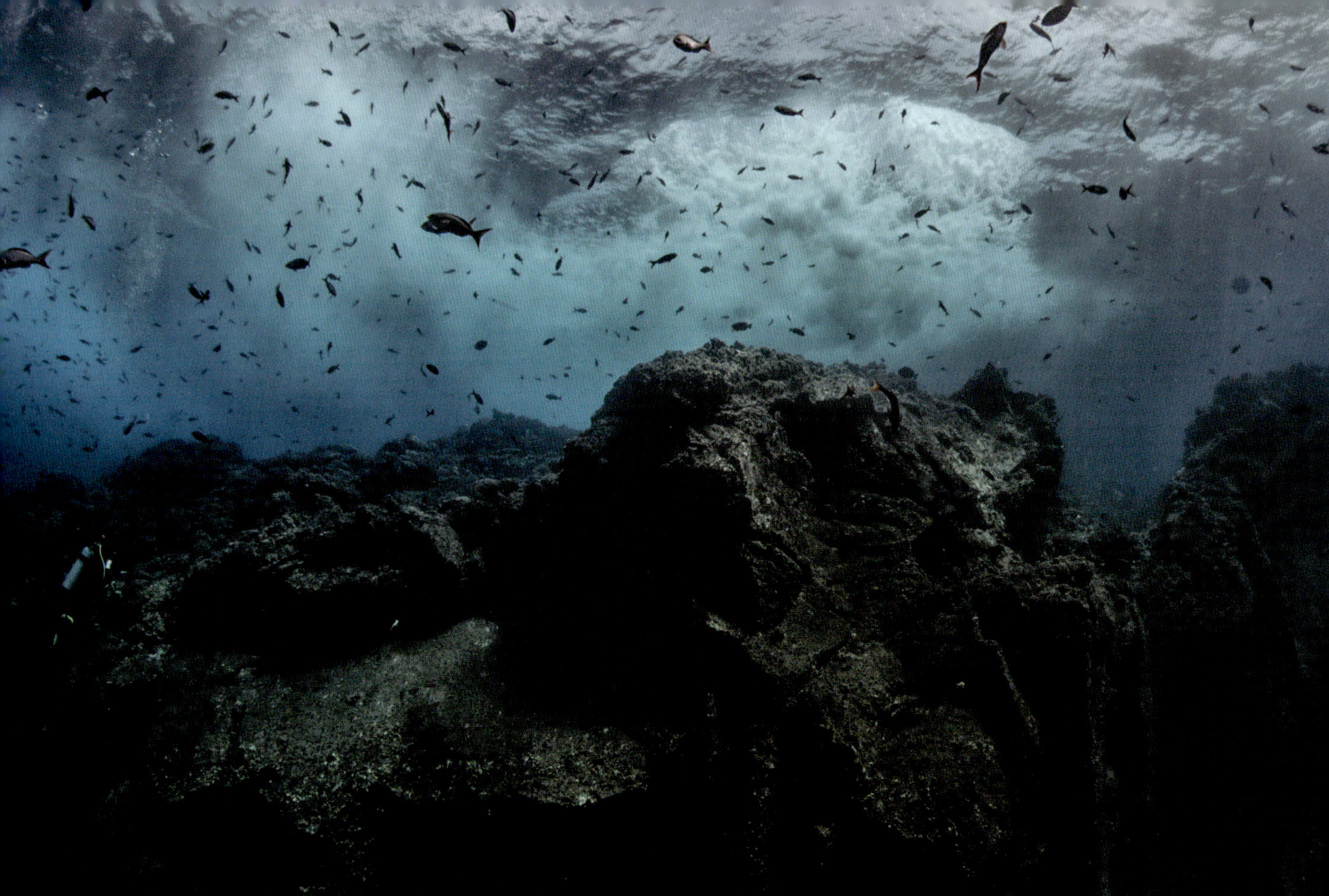

water with these magnificent ocean wanderers and experiencing a close encounter. The giant mantas in this area are known to be familiar with humans and to regularly approach divers. They are inquisitive and docile, and it appears that they like the bubbles that divers exhale brushing up on their undersides.

To reach El Boiler, one of the most famous dive sites of Socorro and the highlight of the whole trip, we had to transfer from our large liveaboard to a small RIB, carrying six divers out to a pinnacle that does not quite break the surface. Once in position our guide gave us the countdown and we all rolled back into the Pacific Ocean together. I looked down and immediately saw a young Whale Shark directly beneath us. It was around 4m (13ft) in length and not at all bothered by our presence in the water. In fact my buddy, who was focussing on getting his camera set up for the dive, did not see it until he was on top of it. Still it seemed happy to swim along with us for a while.

ABOVE Socorro is a rugged island beaten by the powerful waves of the Pacific.

In the distance we spotted two dolphins as they disappeared into the blue. We

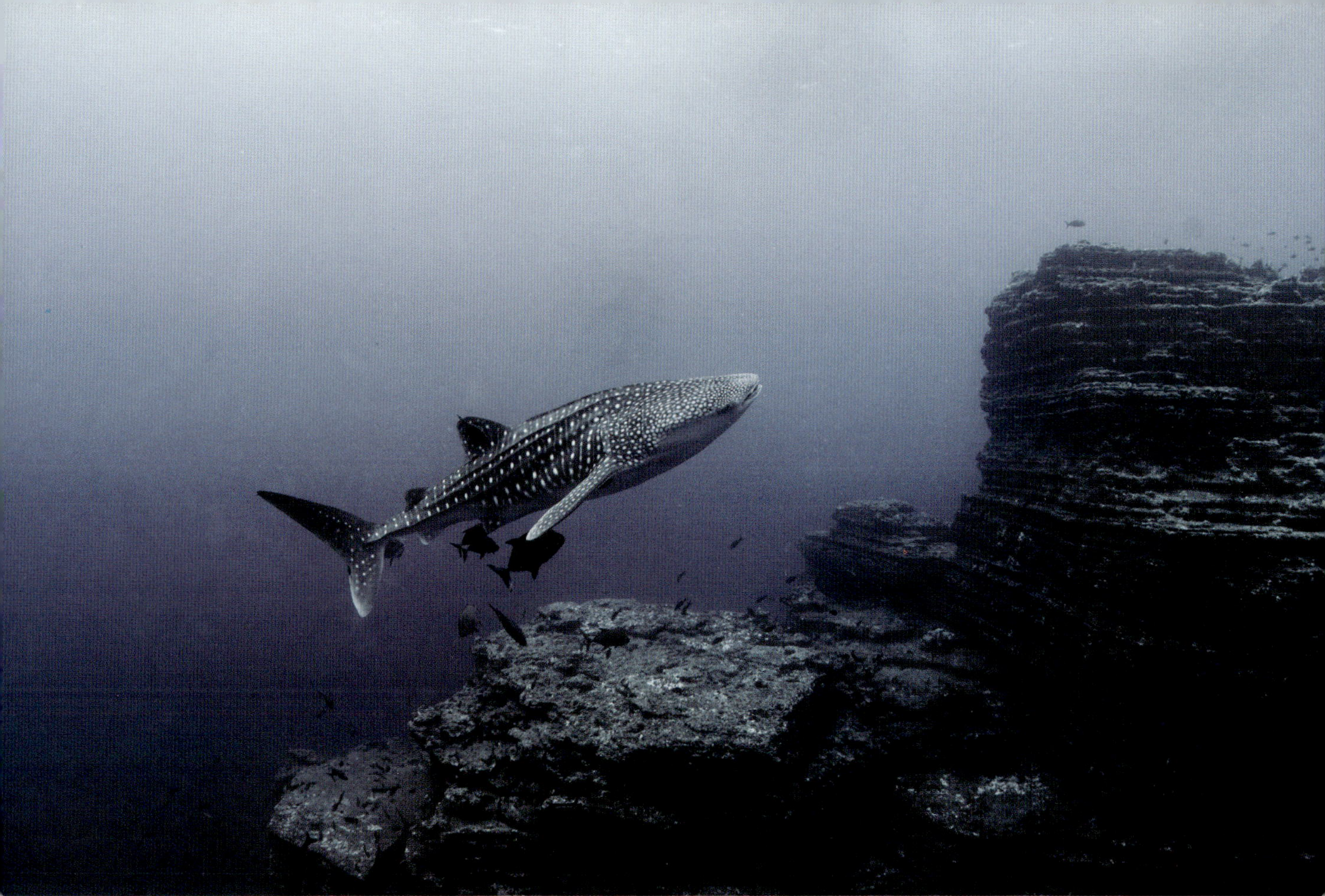

had only been under the water for a matter of minutes and it was already one of the best dives I had ever been on. Divers who ventured a little deeper reported back after the dive that they had also seen a Tiger Shark. We stayed fairly shallow and waited in a sheltered area near the rocky wall. Alongside us there were bright orange Clarionfish waiting for the same event as us – the arrival of the manta rays. We did not have to wait long because very soon two large mantas were soaring towards us. The Clarionfish darted out into the water to meet them, and to begin their task of cleaning them of parasites.

We simply waited, hanging in mid-water, and paused for them to approach us. They were huge, with eyes the size of a softball, and they would slowly glide past looking directly into your eyes as if they were attempting to communicate. One suddenly stopped directly in front of me as I took a few shots before a thought popped into my head: What if I really could communicate with a Giant Manta Ray? How would that work? I did the only thing I could think of at the time and

ABOVE A Whale Shark swims around the submerged rock of El Boiler.

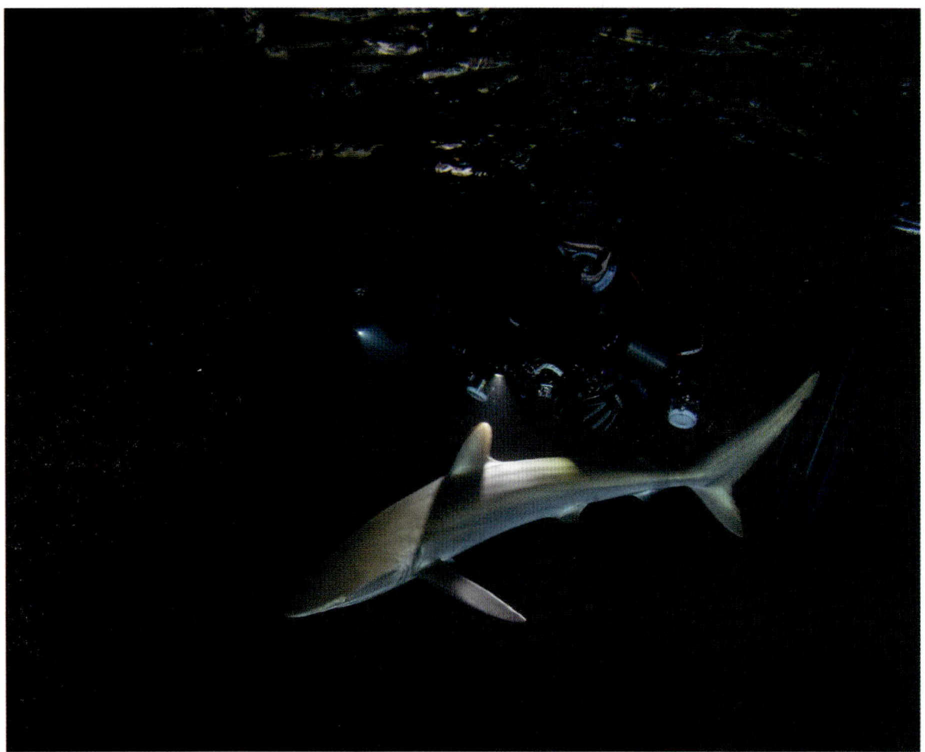

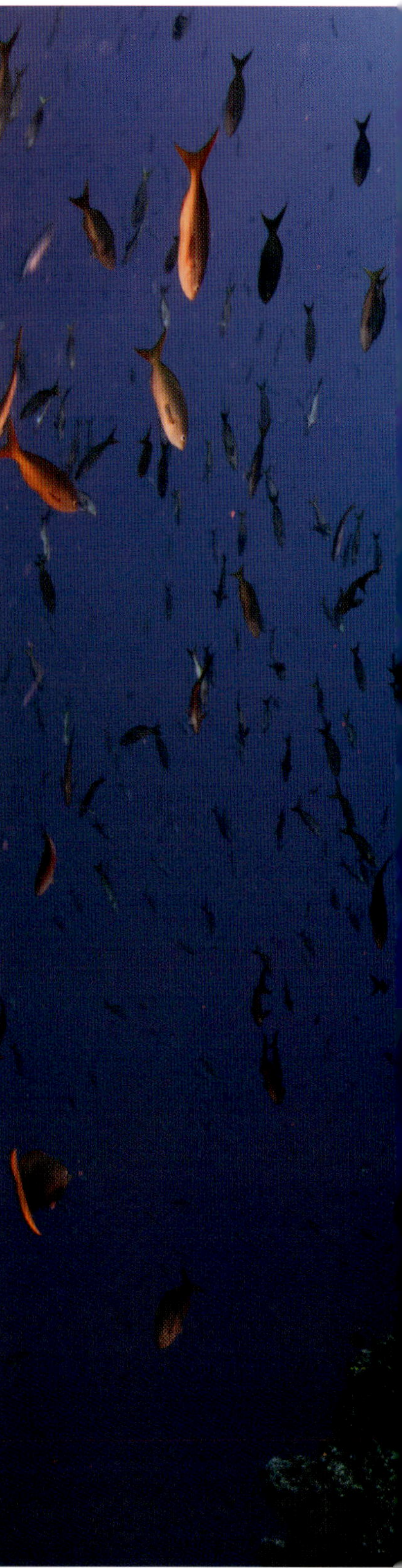

spread my body out, arms and legs as wide as they would go as if I was doing an underwater star-jump. The manta stopped and looked and then reared upwards, manoeuvring so that it too was vertical in the water, its wings stretched out directly in front of me. I could not believe it. Surely this was a fluke? I bent one arm inwards and waited. Sure enough, the manta bent its corresponding wing and waved back. This really was hard to comprehend. Was this manta really playing 'Simon Says' with me? I put my arm in front of my face and waggled it in a wavy motion and paused, barely able to breathe as I waited to see what the manta would do. It did not move its wings, but instead unfurled is cephalic lobe and waggled it back at me. I was overwhelmed and could feel my hairs standing up on end and my eyes filling with tears. I had communicated with a Giant Manta Ray. Obviously my emotions were not of interest to the manta as it got bored and swam off to the next diver to check out their bubbles. Recounting this story still gives me goose bumps and it will always be one of the most amazing dives of my life.

ABOVE Snorkelers get in to see the Silky Sharks hunt.

RIGHT A manta glides over an underwater pinnacle.

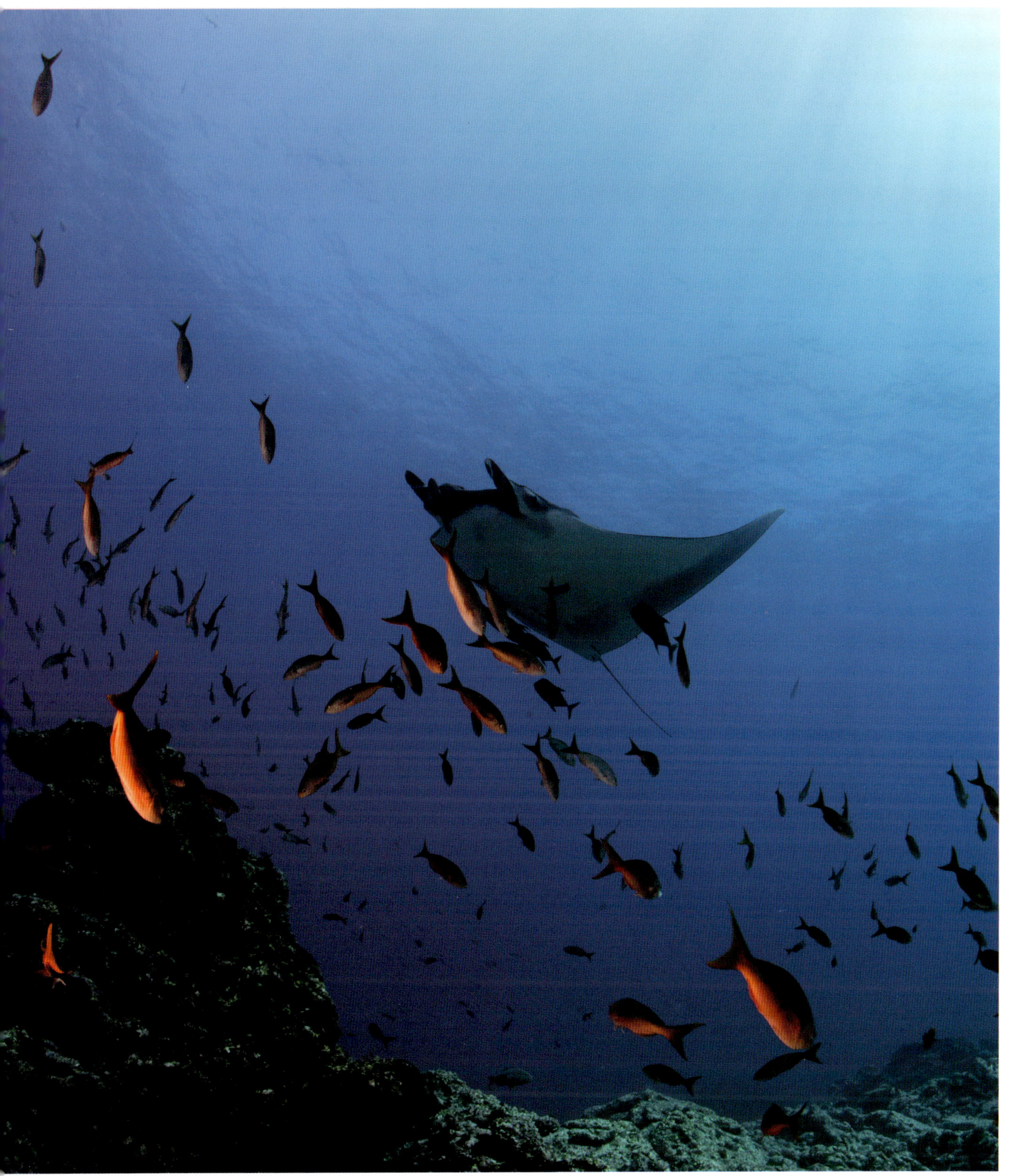

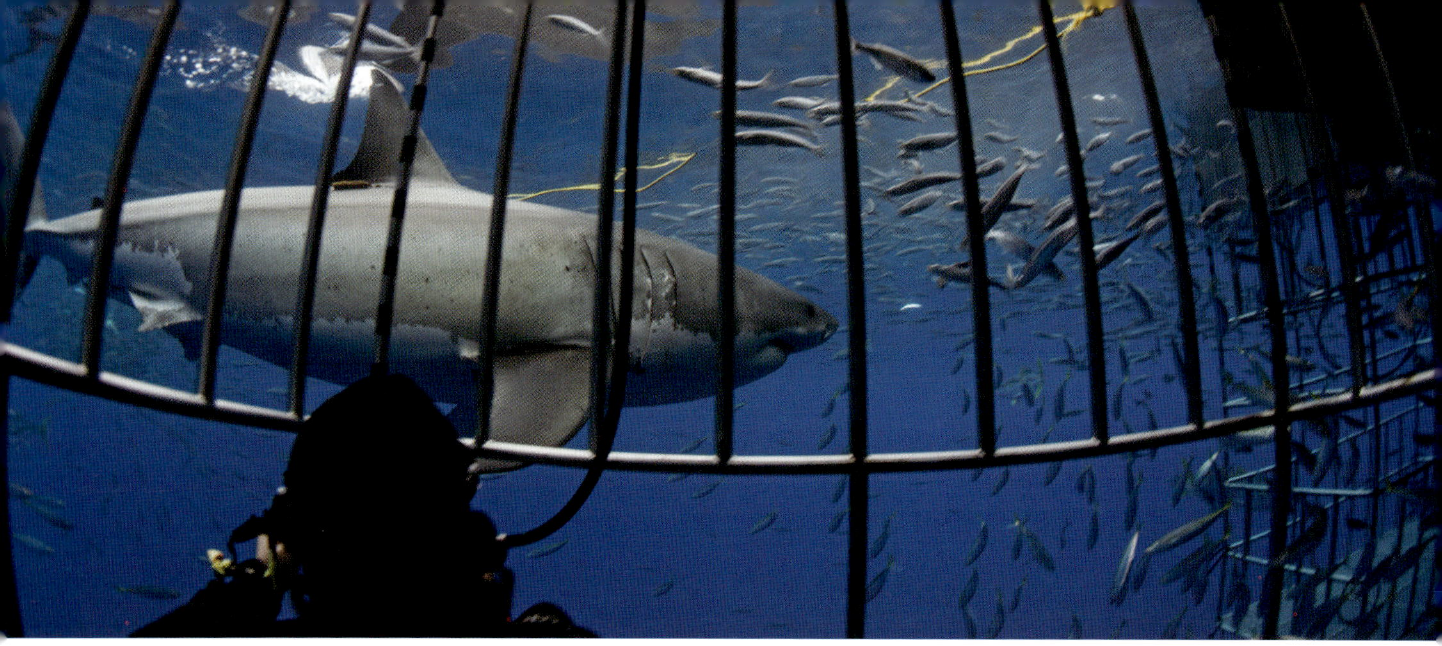

17 THE GREAT WHITES OF GUADALUPE

MEXICO

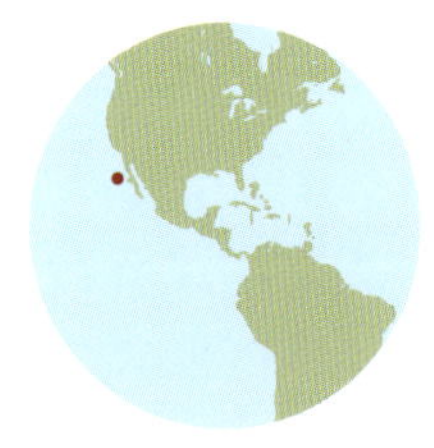

ABOVE A diver looking out of the cage at a Great White Shark.

OPPOSITE Great White Sharks can accelerate at incredible speed.

GUADALUPE ISLAND IS A LARGE VOLCANIC rock in the Pacific Ocean 400km (250 miles) south-west of the town of Ensenada in Mexico. This rugged island is home to a large population of pinnipeds and this is possibly why, during the period from July to November each year, large numbers of Great White Sharks can also be found around its shores. The only way to reach the island is by one of the handful of liveaboard operators licensed to sail here. It takes a full day of motoring to reach Guadalupe, and once there the boat will moor up in the shelter of the barren island, which is 40km (25 miles) long, and prepare for the Great White Shark action to follow.

GENERAL DIVING

There is no scuba diving outside of cages allowed in this area. The male Great White Sharks arrive first, in the earlier part of the season, followed later by the females. Lucky divers might see a dozen sharks on a single dive and more than

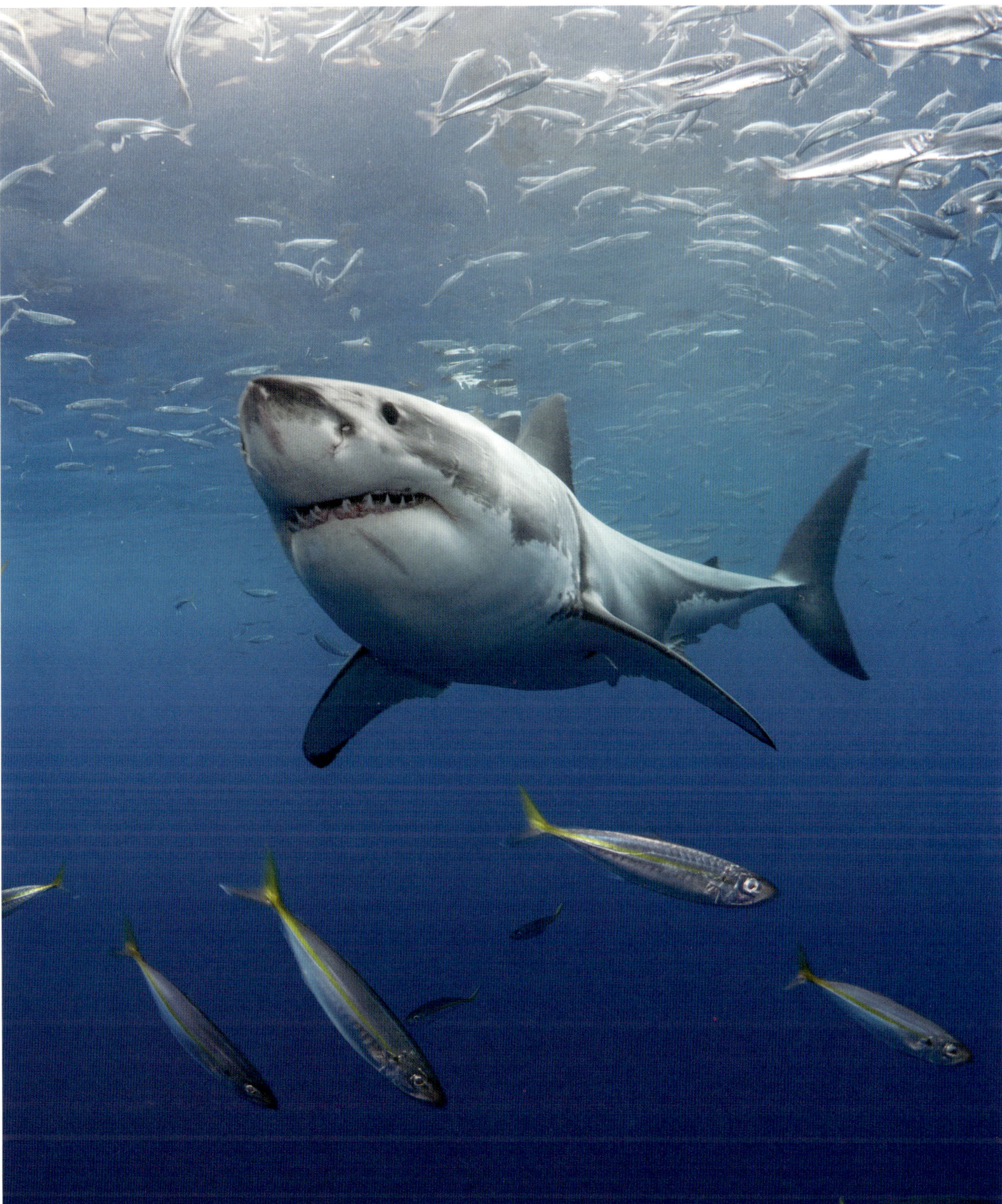

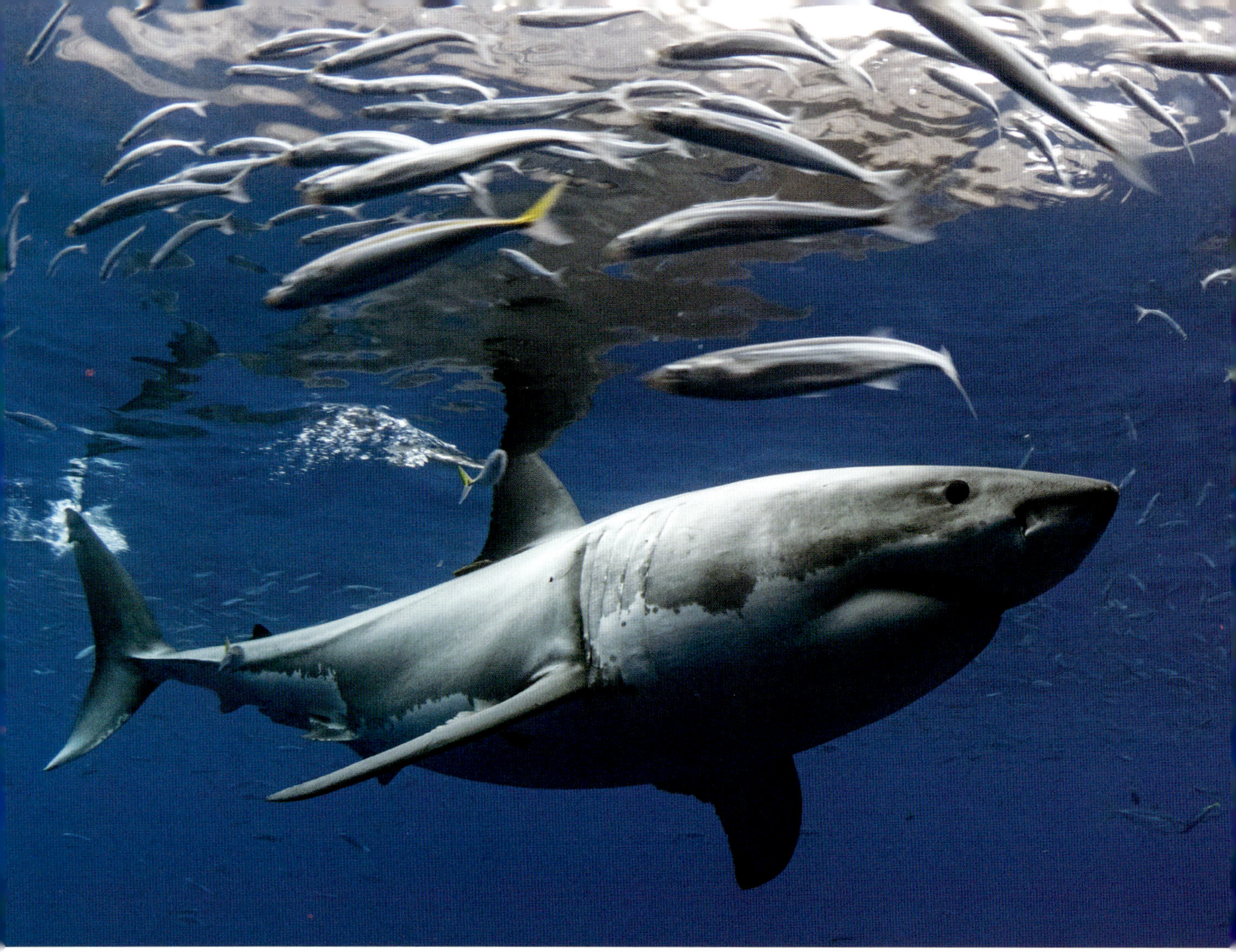

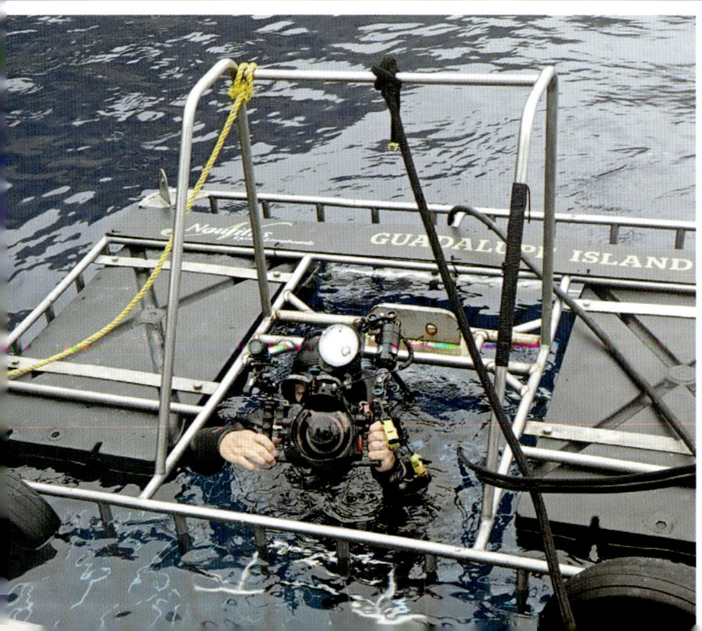
ISLAND

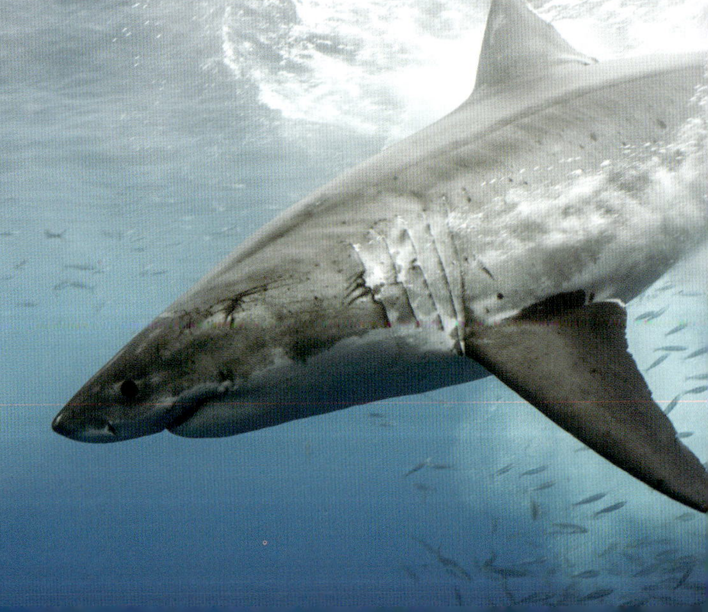

50 individuals on a three-day trip to Guadalupe. Each evening, you can pore over the shark ID books for this area, trying to see if the one you photographed is listed and perhaps, if not, getting the chance to name your own. One of the joys of this trip is the lack of phone signal or internet, making everyone get together in the evenings to share their stories from the day and learn from the on-board experts about the shark research that is done here.

There are strict rules that govern the shark diving around Guadalupe Island. Divers are not allowed outside the cages and the sharks should not be fed. Large hunks of fish are tied to a rope and tossed into the water by the shark wranglers, but their job is to pull the bait clear of the water and try to prevent the sharks from grabbing it. Of course, the sharks are not stupid and sometimes manage to outwit the wrangler and get a prize.

THE WILD DIVE

In the past we have been fortunate enough to have dived with Great White Sharks at both Neptune Island off South Australia and Gansbaai, South Africa, but even so this was a trip that I was really looking forward to since the water is clear and the visibility superb. A trip to Guadalupe Island is almost guaranteed to result in Great White Shark sightings so long as you get the timing right, and that clear blue water and incredible visibility are certain to provide opportunities for great images. We had arrived at the location during the period that the males turn up to wait for the arrival of the females. This meant that there would be lots of boisterous behaviour

OPPOSITE TOP A Great White Sharks swims past the fish at the surface.

OPPOSITE BOTTOM LEFT A diver climbs up to the top of the cage to get back on the boat.

OPPOSITE BOTTOM RIGHT A shark lands in a rush of bubbles after trying to grab the bait.

BELOW A shark heads to the surface to try to grab the bait.

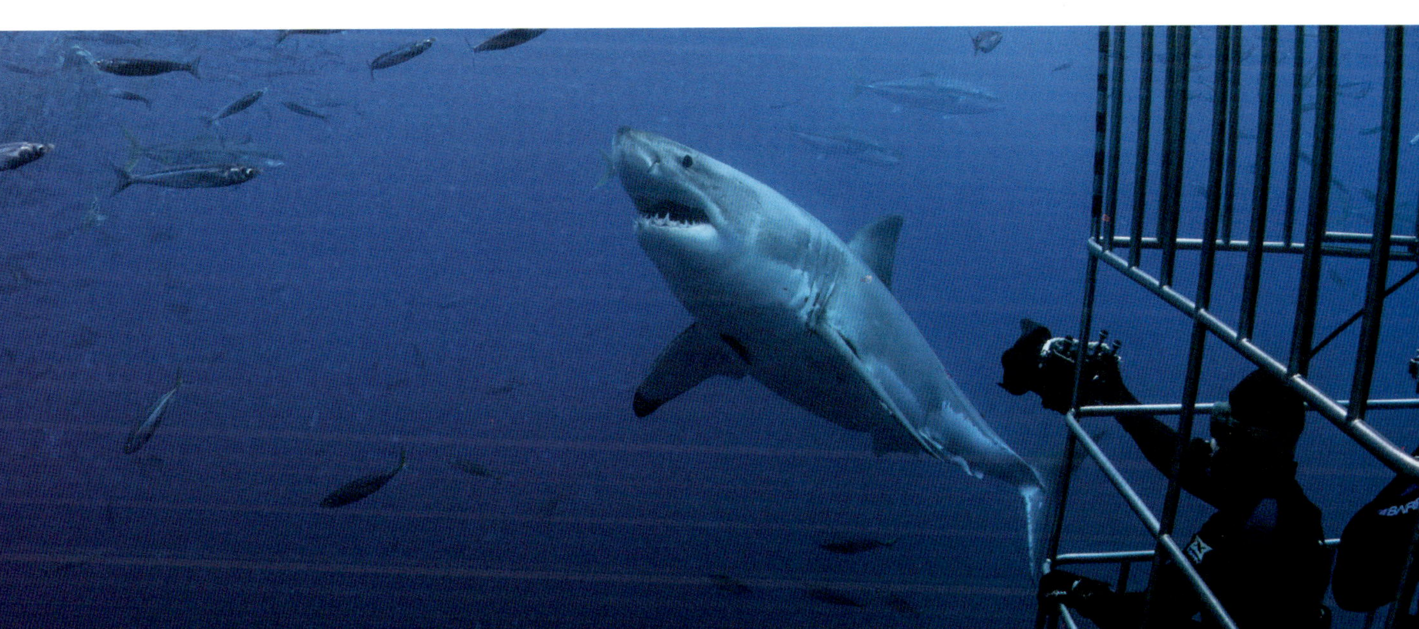

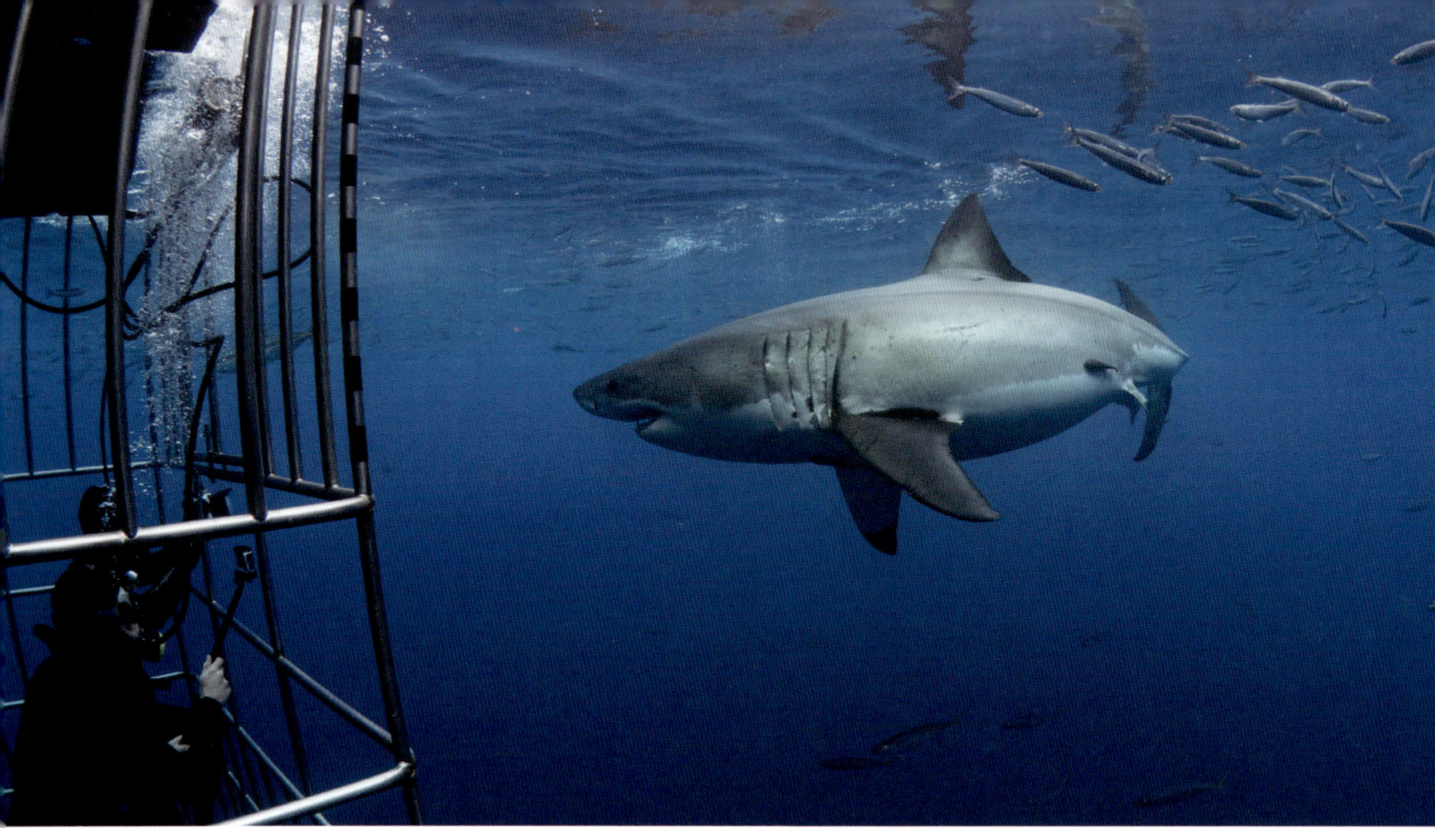

ABOVE Divers film a shark swimming past their cage.

OPPOSITE Pilot fish accompany a shark.

as the 'participants' vied for the position of top dog and first dibs with the females once they arrived in a few weeks.

On the first morning I set the alarm early, excited to be getting into a cage to see the magnificent Great Whites of Guadalupe. I was first on deck and wanting to be first in the water, so I started to haul on the thick wetsuit, hood and gloves that would keep me warm in the chilly waters of the Pacific Ocean. My buddy was, of course, right there behind me, also not wanting to miss out on any of the action. At 6am the shark cages were declared open and I slipped into the cage with my camera and peered out into the gloomy yet crystal clear water. The sun had not yet risen over the top of the island, and I tried to fine-tune my search image and spot my first shark. I was in a surface cage and even though I was completely immersed underwater, and my ears were covered in thick 5mm neoprene, I heard the first shout of 'SHARK' on this voyage. Then I saw it, my eyes focussing entirely on this icon of the oceans as it moved effortlessly and slowly towards me like a shadow. It may sound like a cliché coming from one who has dived with numerous species of sharks across the globe, but really it took my breath away. Even though this was not our first time in the water with a Great White Shark, they never cease to amaze and leave you feeling insignificant and in awe. They are big, they are powerful, and they are all muscle and teeth, but with a big smiley face.

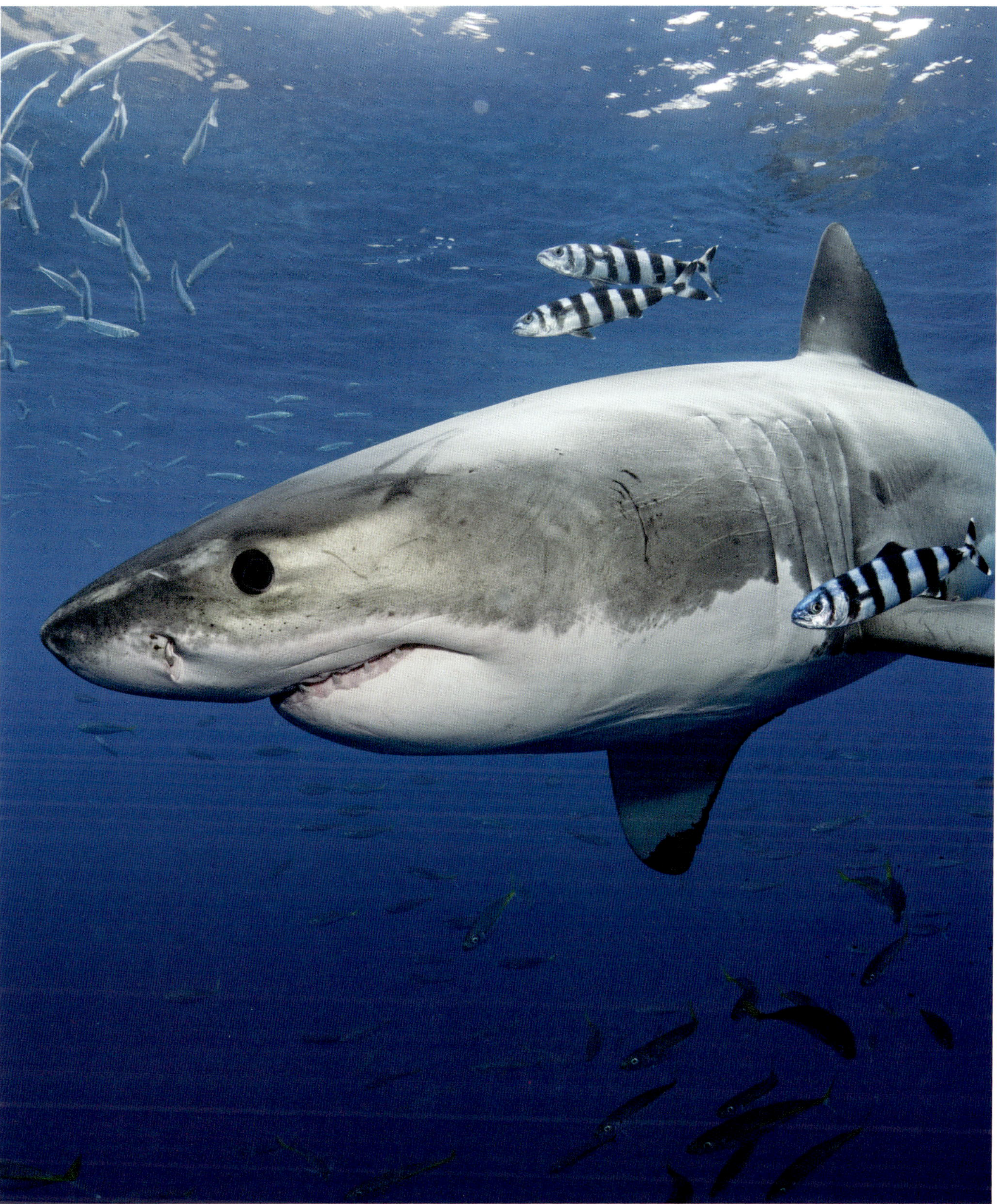

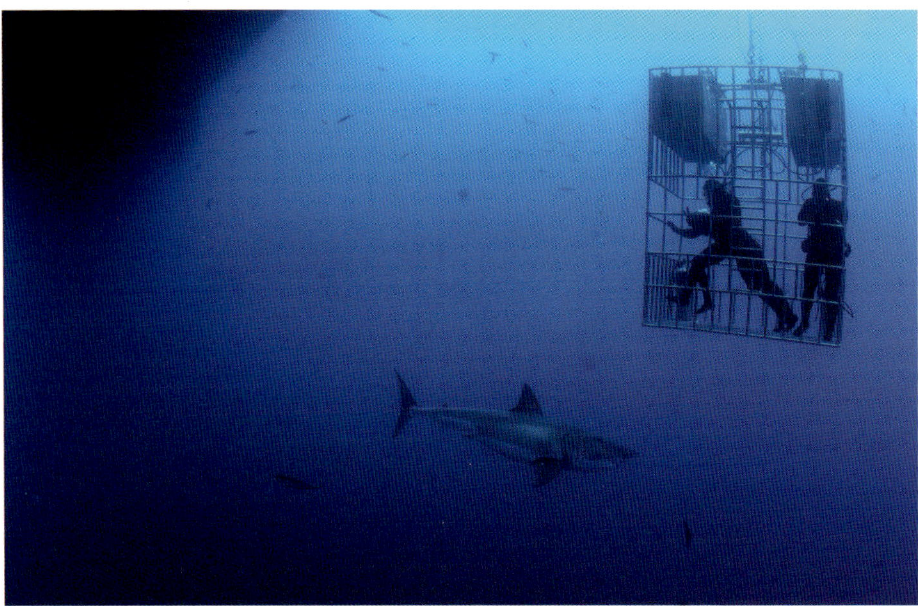

As the day progressed and the sun moved higher into the sky I was still in the water, but by now I had been joined by other divers in the surface cages and also those that were lowered to about 5m (16ft) in depth for an alternative view and experience. There were now five Great White Sharks steadily patrolling around the boat, hoping to fool the wranglers above and try to steal a mouthful of fish being thrown in to tempt them closer. Every once in a while a shark would rise up from beneath, suddenly putting in an intense burst of speed, shooting vertically upwards, breaching at high speed to try to grab the bait that is tied to a rope (no hooks are used). These apex predators are far more intelligent than they look, regularly putting in false moves and then coming out of the sun to try to give themselves the best chance of success.

The divers in the cages around me, as well as those watching from the deck, came and went, either heading towards the galley for meals, to warm up in the hot-tub, or to watch the action from the surface, but I stayed in the water, looking out from my cage for seven and a half hours in constant and total amazement at these magnificent sharks. Having travelled for well over two days to get to this barren and remote outpost, it has to be worth maximising your time in the water as you never know when you may get the chance again.

ABOVE Some boats have submersible cages to give divers a different view.

RIGHT Some of the sharks show off huge battle scars.

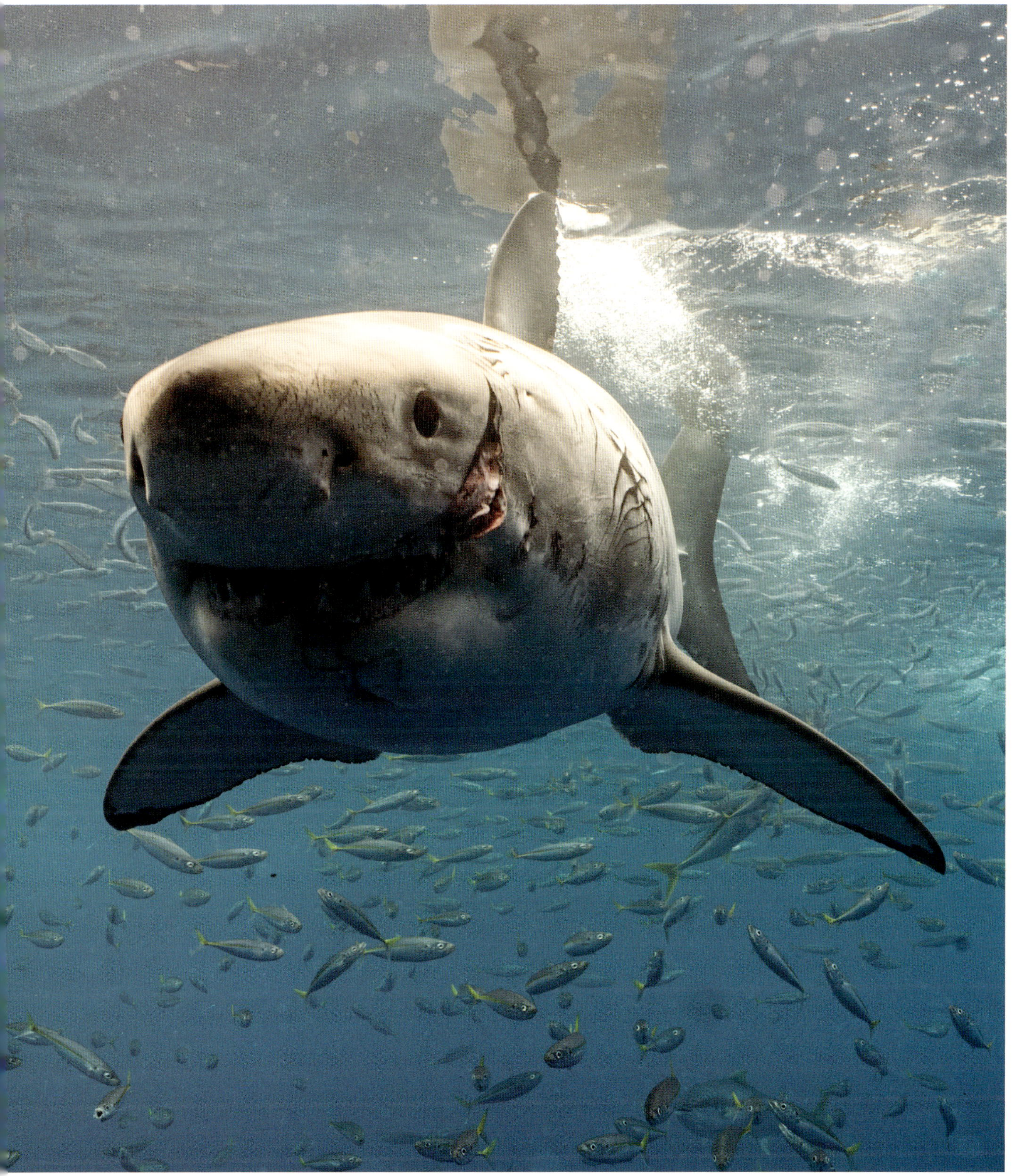

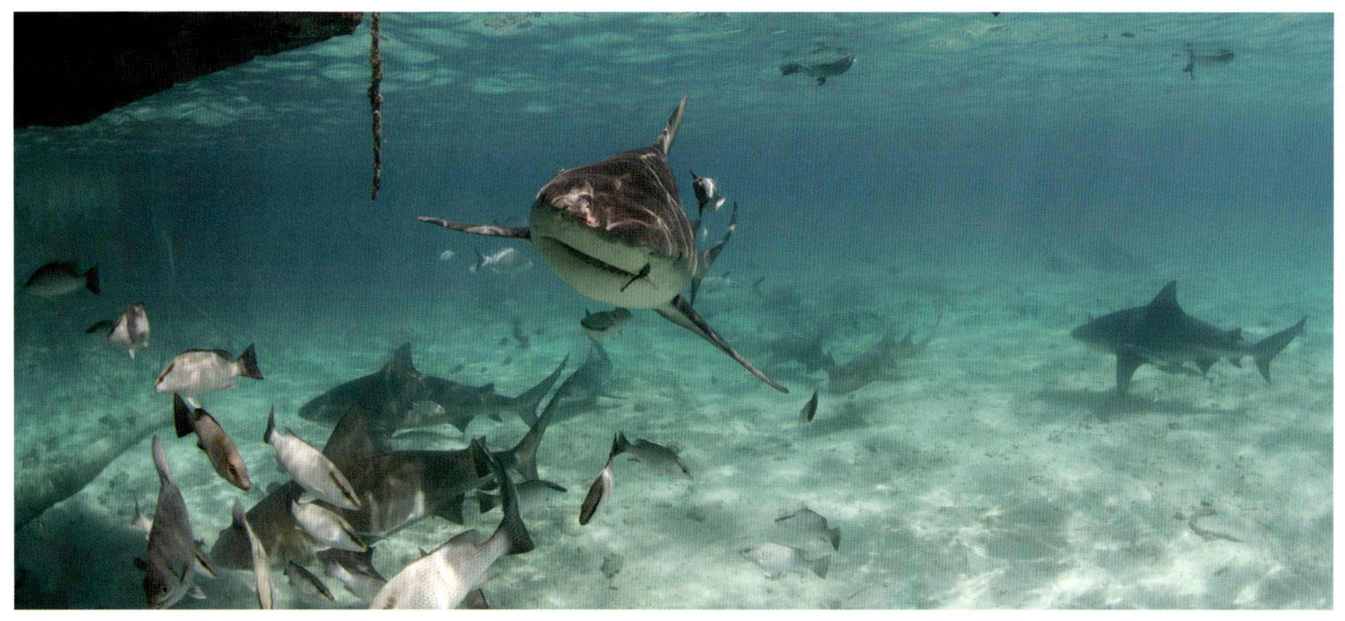

16 BULL SHARKS ON A WINDY DAY

BAHAMAS

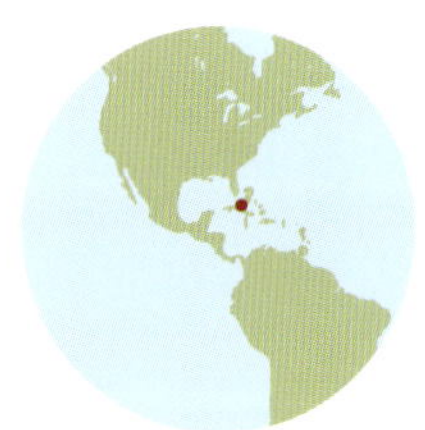

ABOVE Pregnant female Bull Sharks swim around this area for a few months each year.

OPPOSITE A pair of Bull Sharks swim past me.

BIMINI IS PART OF THE BAHAMAS chain of islands that is the closest to the USA, just under 80km (50 miles) off the Florida coast. It has been a popular tourist destination for decades and has seen its fair share of the rich and famous taking advantage of the beautiful weather. Bimini was reputed to be Ernest Hemingway's favourite place to escape and is where he wrote *Islands in the Stream*. Martin Luther King wrote his Nobel Peace Prize acceptance speech whilst on the island and on his final visit also wrote his Sanitation Workers speech – the last speech he would ever give as it happens, as he was assassinated soon after. It is said that he knew he was going to die and also wrote part of his own eulogy during this same visit to Bimini.

GENERAL DIVING

The diving in Bimini is famous for its incredible Great Hammerhead Shark

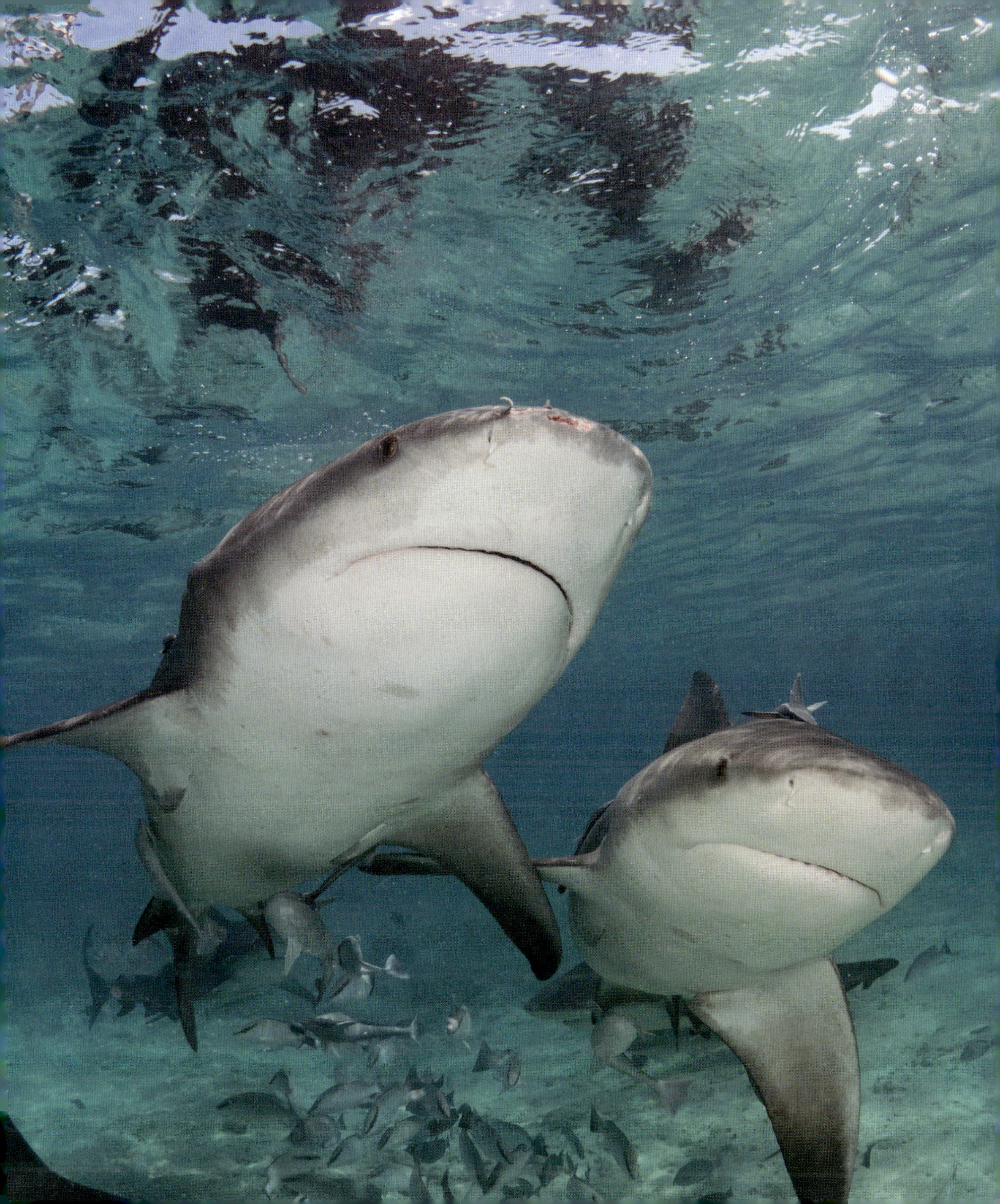

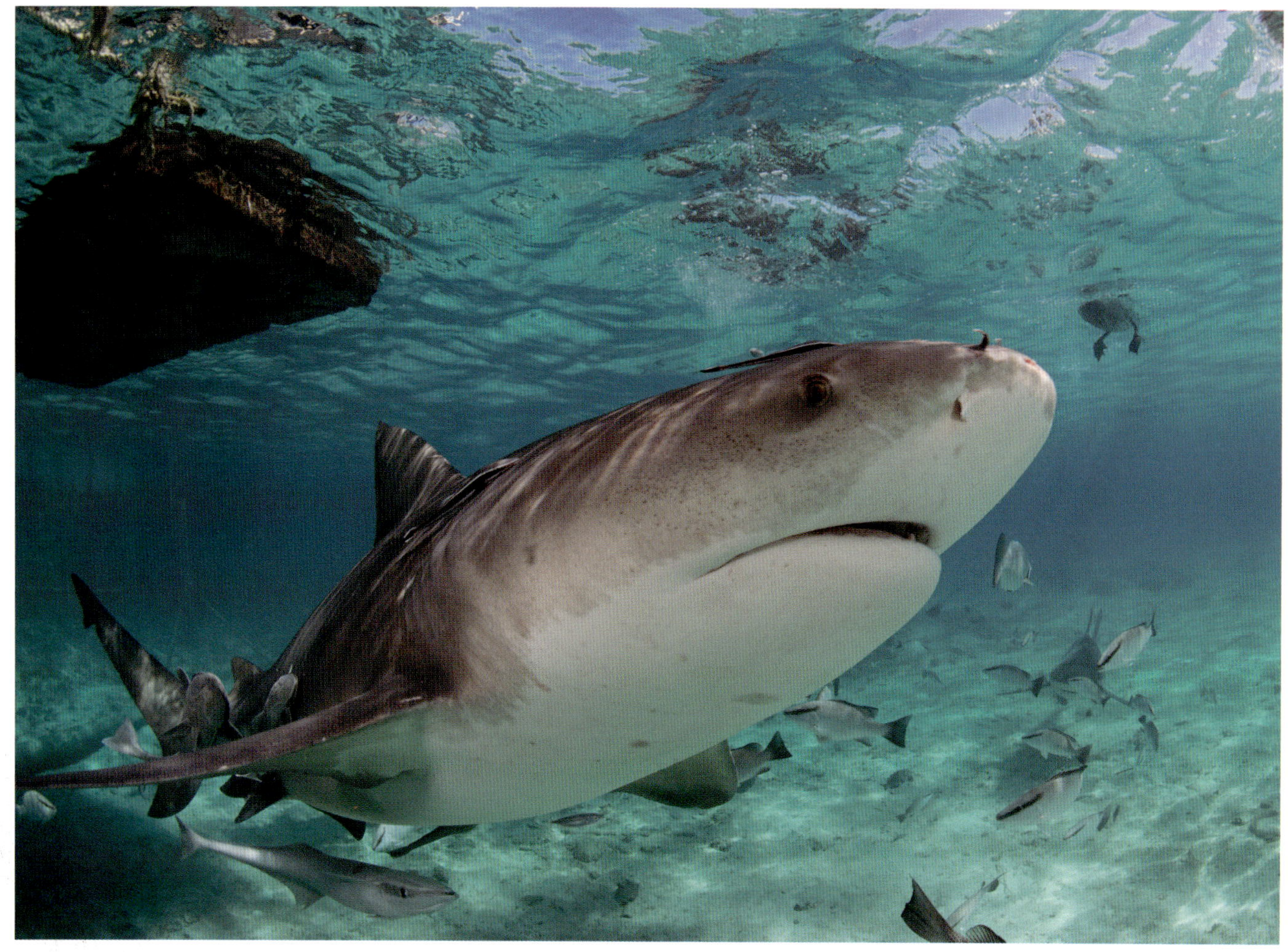

encounters, but it also has a host of good reef diving and a number of wrecks to explore. The clear blue waters here are also home to wild dolphins and in the summer months, when the water is flat and calm, going out to snorkel with them is also a fabulous experience. Spotted Eagle Rays swim in the shallows and can often be seen aggregating in the narrow stretch of water that separates North Bimini from South Bimini. The sand banks between the islands are reputed to regularly give up ancient treasure and the marina area of North Bimini is also famous for its Bull Sharks.

ABOVE A Bull Shark comes in close to the cage.

THE WILD DIVE

As you arrive in Bimini in small plane flying relatively low you get a wonderful bird's-eye view of the low-lying islands of the Bahamas, fringed with white sandy beaches and the pale blue and turquoise shallow water leading into the dark blue

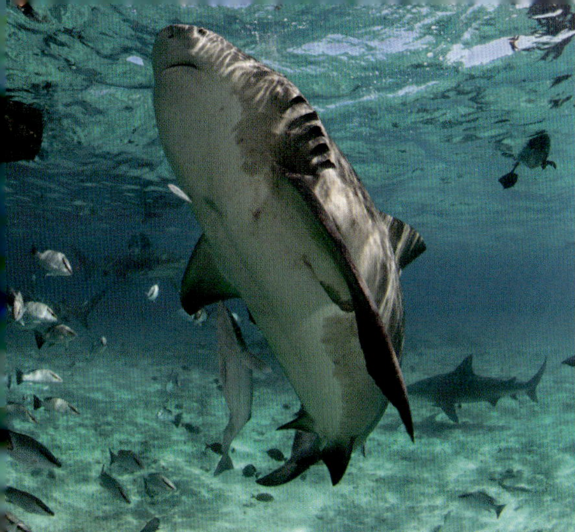

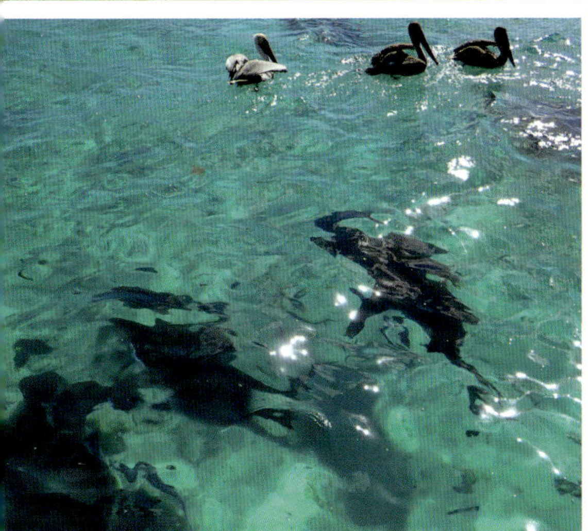

deeper waters. It reminded me of my first visit to these magical islands while working on a dive boat as a divemaster. One of the divers in the group I was taking out asked if his wife could come along for the ride and I agreed that would be fine. As we pulled out of the dock she fished a collection of glass vials out of her handbag and asked me to slow the boat down. As I brought the boat to a stop she dipped the vial in the water, sealed it up and dropped it back in her bag. A minute later she repeated this process and then just as we approached the dive site she asked me to stop once again. The suspense was too much for me and this time I had to ask "What are you doing?"

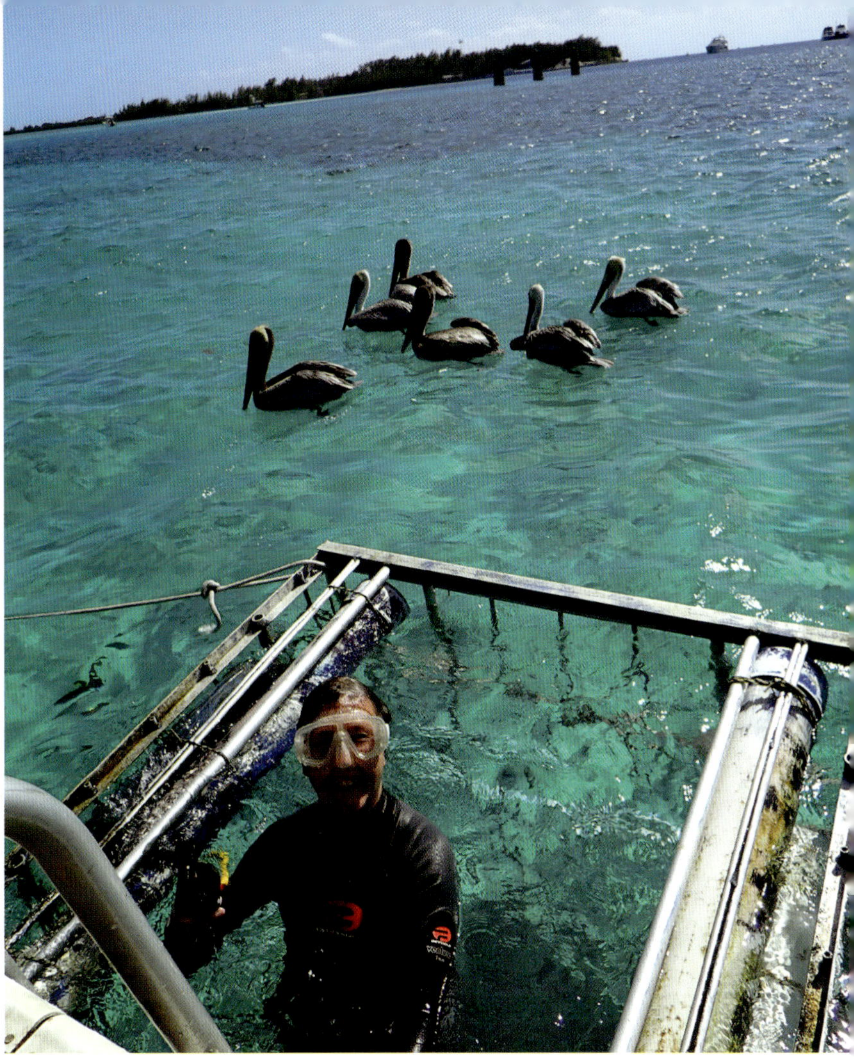

She replied "I just had to show my friends back home all the different shades of blue the sea has here."

I had no idea how to tell her how disappointed she was going to be when she looked at her results back on dry land!

ABOVE TOP LEFT
A shark heads to the surface for food.

ABOVE BOTTOM LEFT
Pelicans sit on the surface whilst Bull Sharks swim below.

ABOVE RIGHT
A happy diver gets out of the cage.

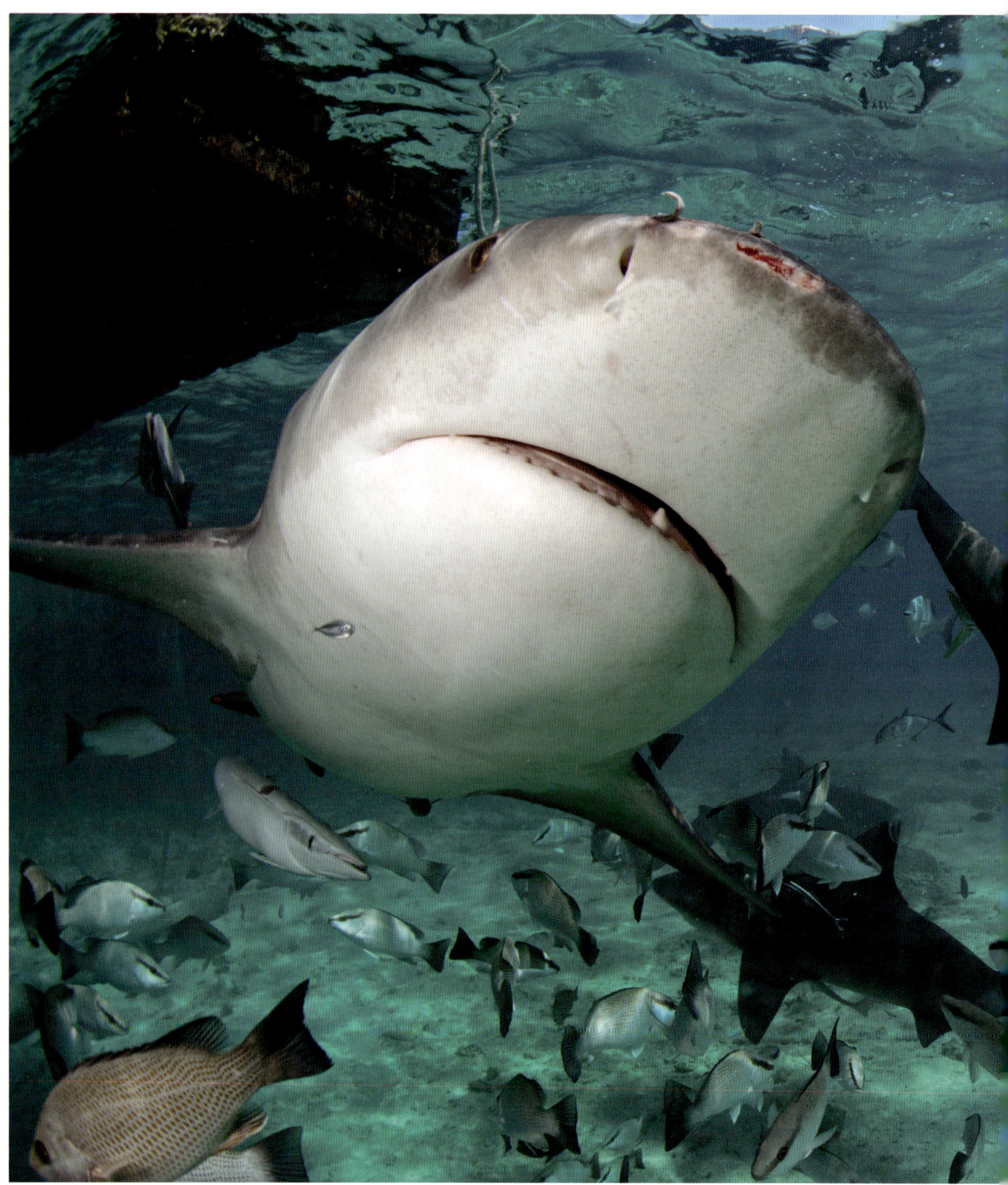

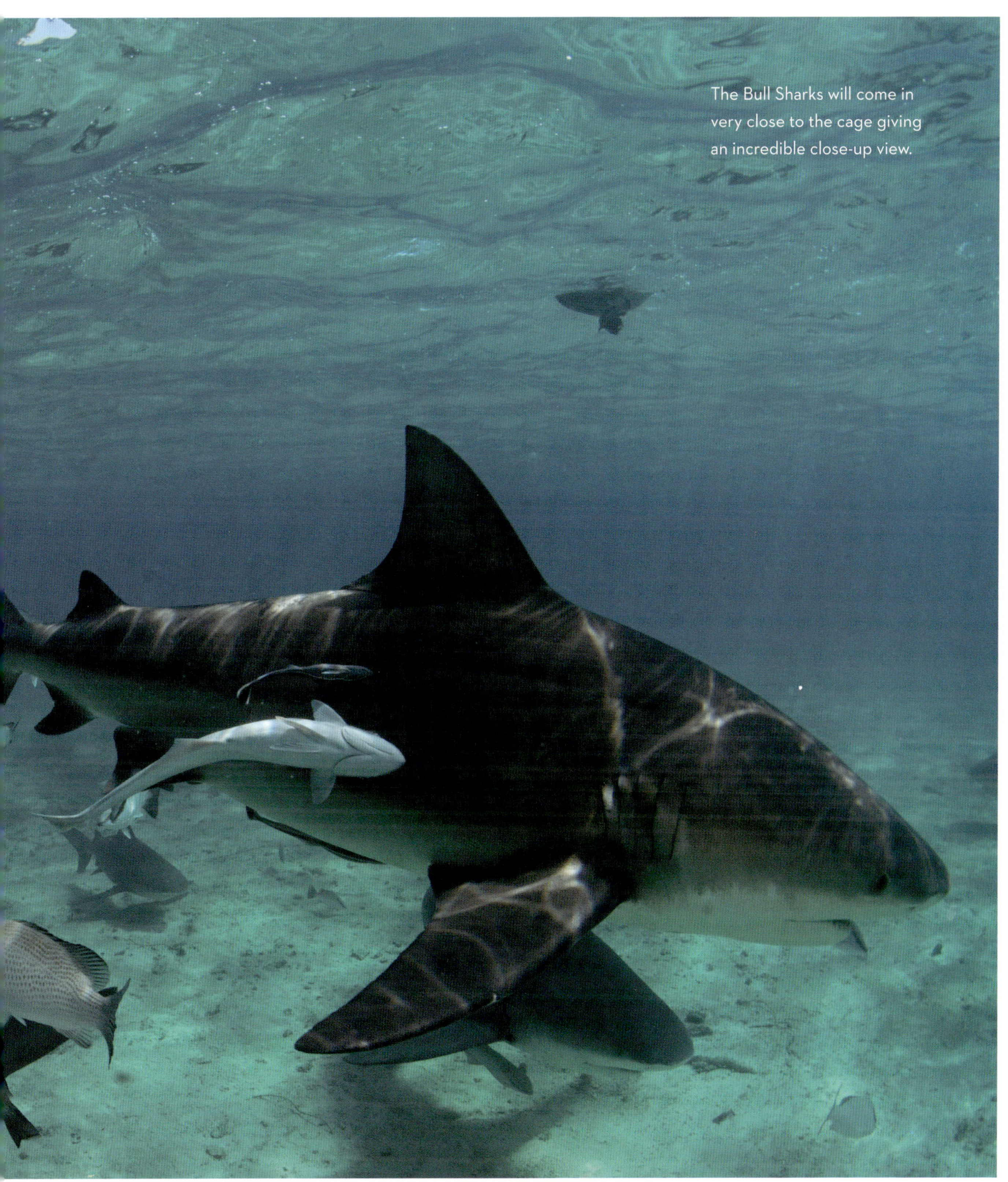

The Bull Sharks will come in very close to the cage giving an incredible close-up view.

On this trip, however, as I hauled my heavy dive and camera kit to the dive shop, I already knew that I was going to be disappointed. I was supposed to be going out to snorkel with Atlantic Spotted Dolphins but the wind had picked up and the seas were starting to get rough – these were not conditions to be snorkelling in. I dropped into the dive shop to check, just in case, but my suspicions were correct and the trip had already been cancelled. However, the owner tried to make it up to me and suggested that I might like try out a Bull Shark cage dive instead.

At the side of the dock, where fishermen often clean their catch and throw the unwanted bits into the sea, was a cage at the surface. Pelicans sat bobbing on the water hoping to get some fish scraps and beneath them I could see the outline of six big Bull Sharks. It was not what I had planned for my day, but sometimes, as any experienced diver knows, you just have to go with the flow. I donned a heavy weight belt, grabbed my mask and camera and slid into the cage. The dive shop owner handed me a long regulator hose attached to a tank that sat at the surface

BELOW Bull Sharks are bulky and powerful.

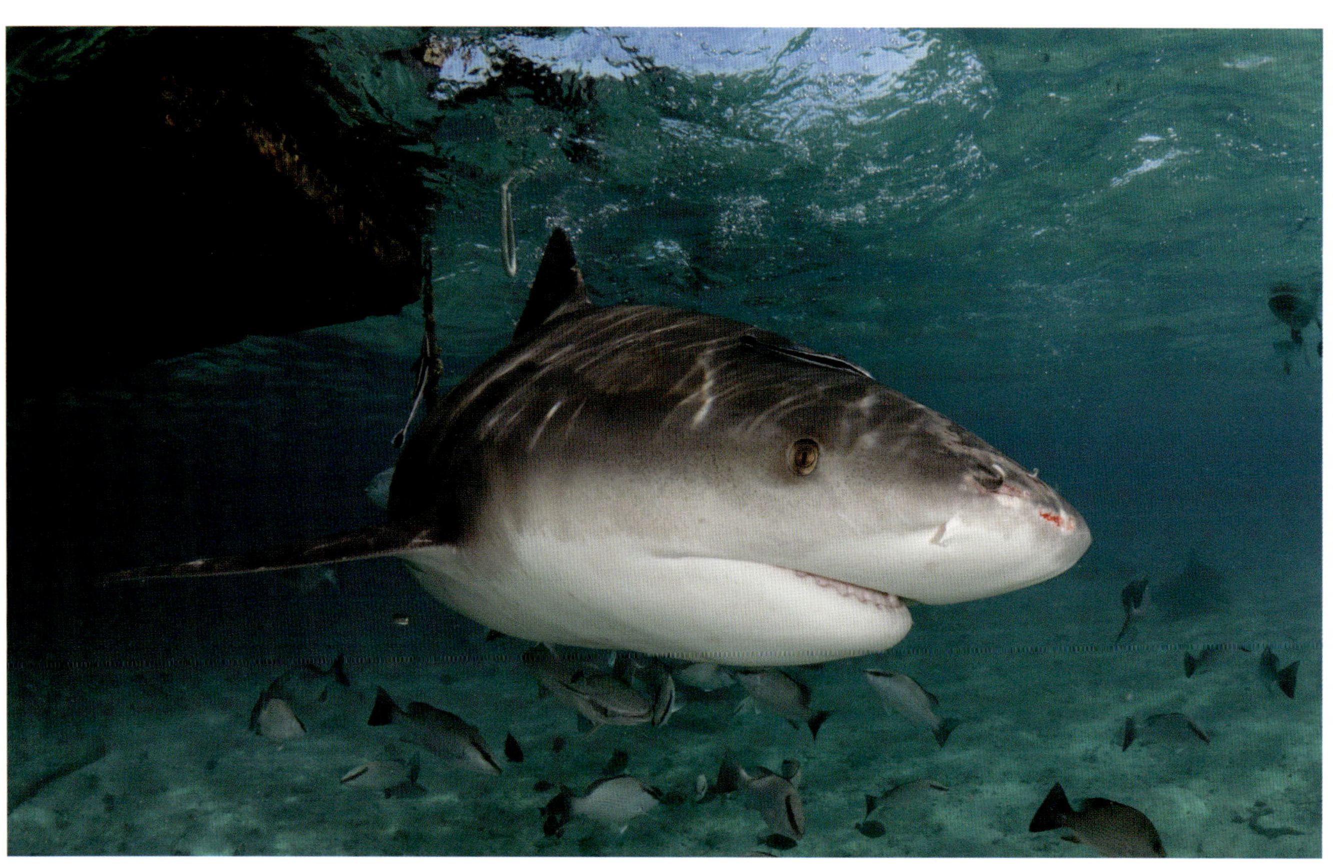

and I dropped down to the cage floor. The water was clear and in front of me I could see the sharks circling and the feet of the pelicans paddling above me. One of the Bull Sharks had some pretty impressive scars on its nose and this one had all the attitude of the alpha, the other sharks deferring and moving out of its way. Scarface – the name I gave this shark in my head – also liked to come in close to the cage, perhaps checking out its reflection in the glass dome of my underwater camera. Bull Sharks are always impressive to see up close. They are bulky and all muscle, built for fast bursts of speed. As a shark approached the surface the pelicans would scatter, but then would soon be back again hoping to snatch a free snack and I wondered why the Bull Sharks did not attempt to eat them.

There was a queue of divers waiting to take their turn in the cage, due to the weather having stopped all the other diving and snorkelling activities Bimini has to offer and so soon I was being called out. It was not the way I had planned to spend my afternoon, but getting to see these jaw-droppingly impressive sharks up close in clear water was an exhilarating alternative.

BELOW A perfect palm-tree-lined beach

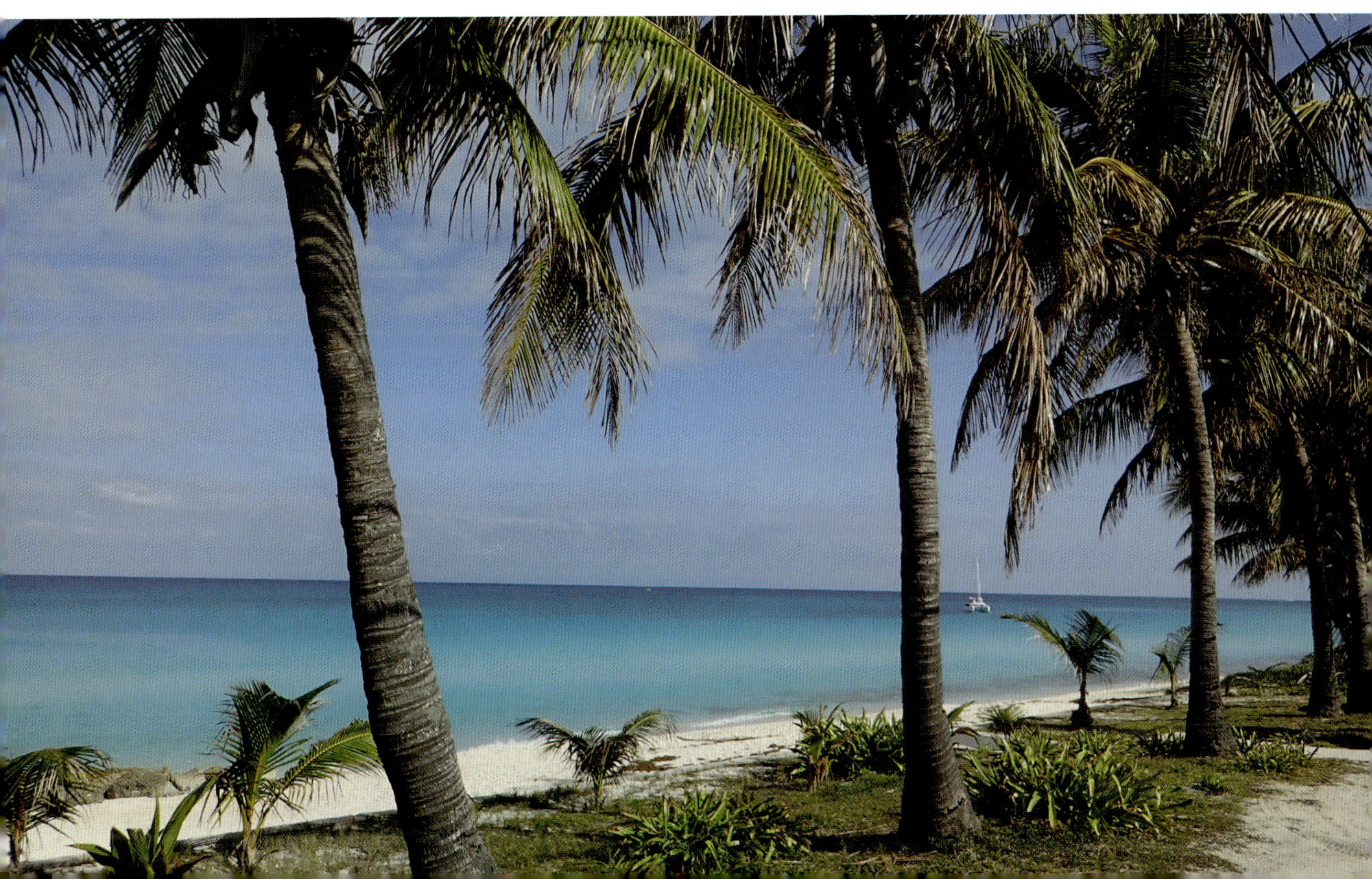

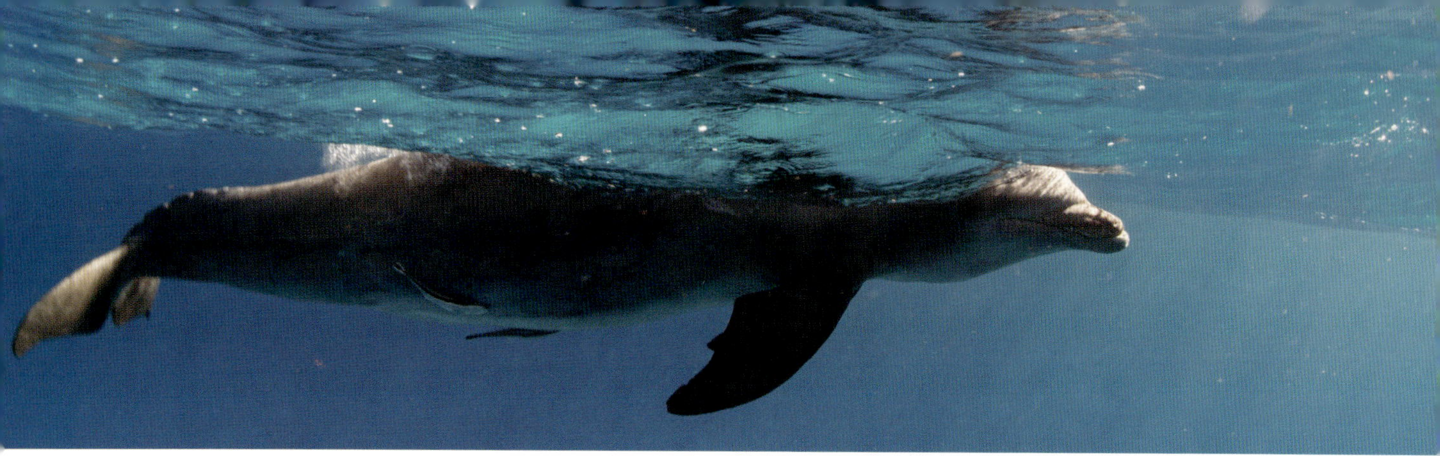

17 BIMINI'S WILD DOLPHINS

BAHAMAS

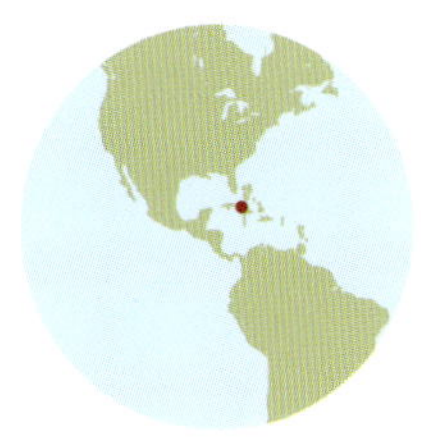

ABOVE A Bottlenose Dolphin peeks under a wave.

OPPOSITE A pair of dolphins investigate the snorkellers at the surface.

LOCATED IN THE FAR WEST of the archipelago, Bimini is closest group of Bahamian islands to the Florida coastline. Bimini has two main islands, North and South, in addition to a series of small cays. North Bimini in particular is famous for being Hemingway's favourite getaway, but less well known is the fact that the final scene from the film *Silence of the Lambs* was shot here – the one where Hannibal Lecter (Anthony Hopkins) talks on the phone to Clarice Starling (Jodie Foster). "I have to go now Clarice, I'm having an old friend for dinner."

It is also believed that Bimini Road, a series of artificial blocks off the coast of North Bimini, could be the remains of the road to the lost city of Atlantis. What is certain is that Bimini is famous for its sharks and is home to the world-famous Bimini Sharklab that conducts invaluable research into shark behaviour and biology.

GENERAL DIVING

Among the diving fraternity Bimini is most famous for its Great Hammerhead Sharks, which visit the islands for a few months each winter. There are plenty of other species of shark that reside here permanently, and this is why one of the leading global shark research facilities is based on South Bimini. Divers flock here to

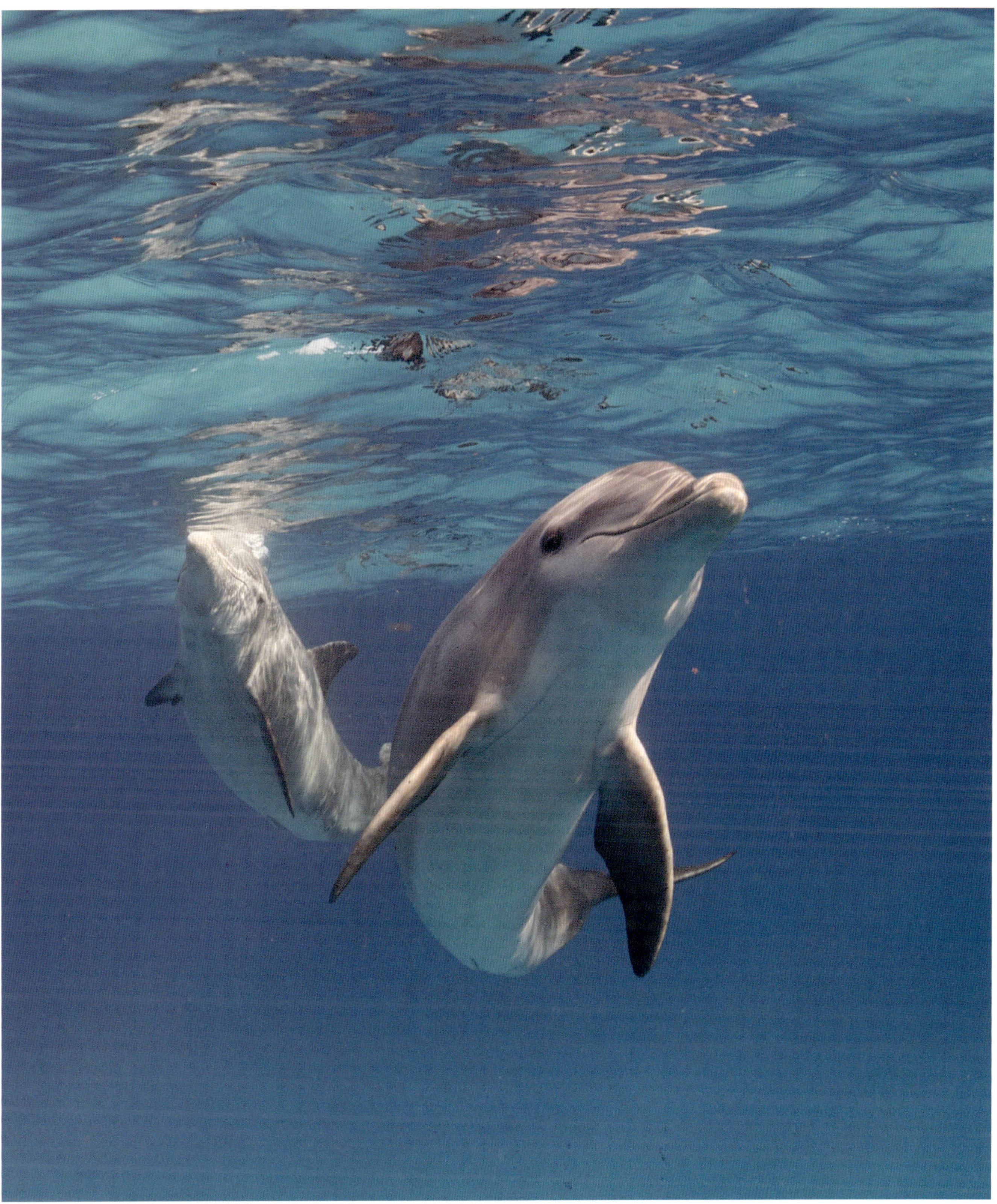

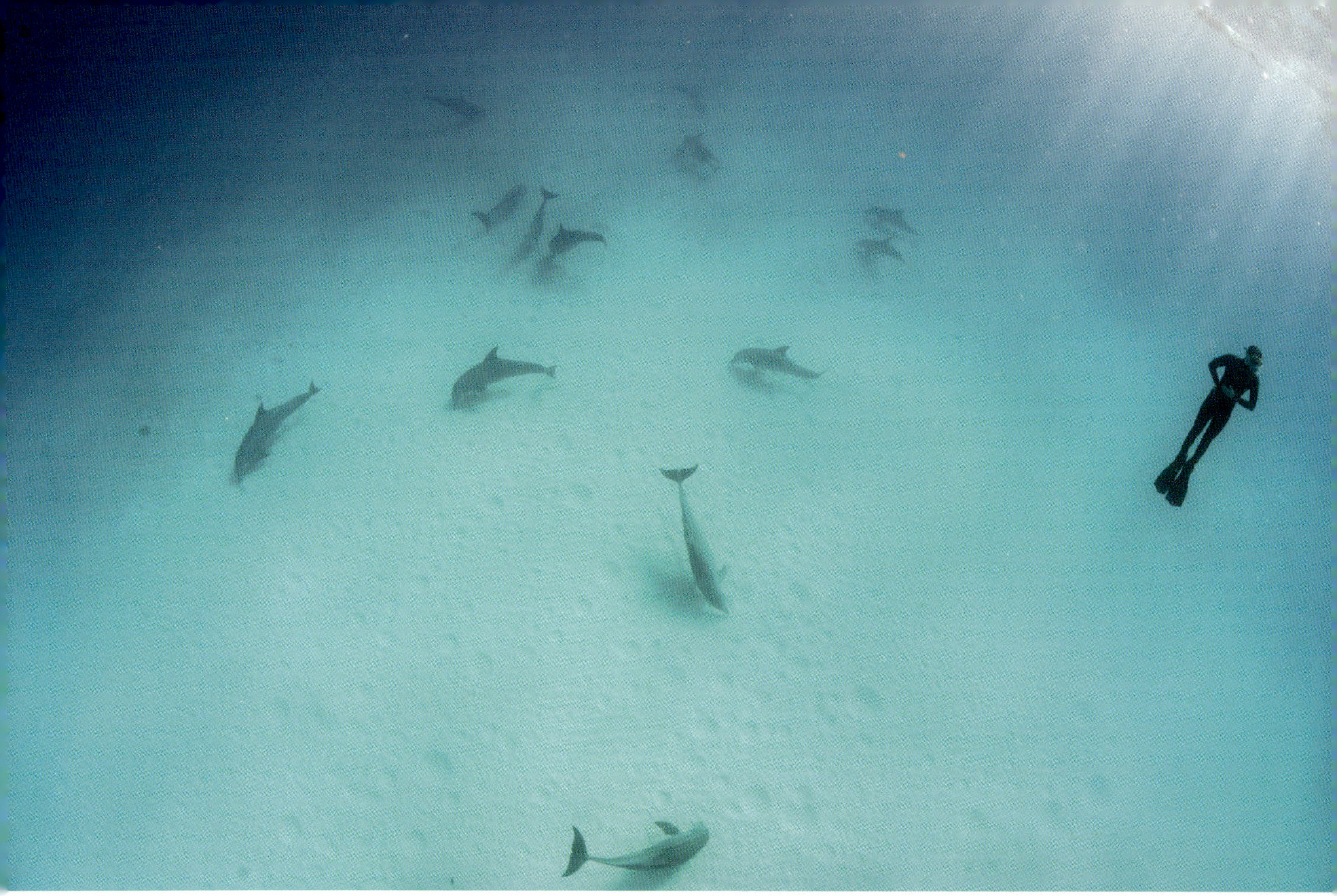

ABOVE A free-diver ascends from the seabed.

RIGHT Even though the dolphins were busy hunting they still came to visit the snorkels when coming up for air.

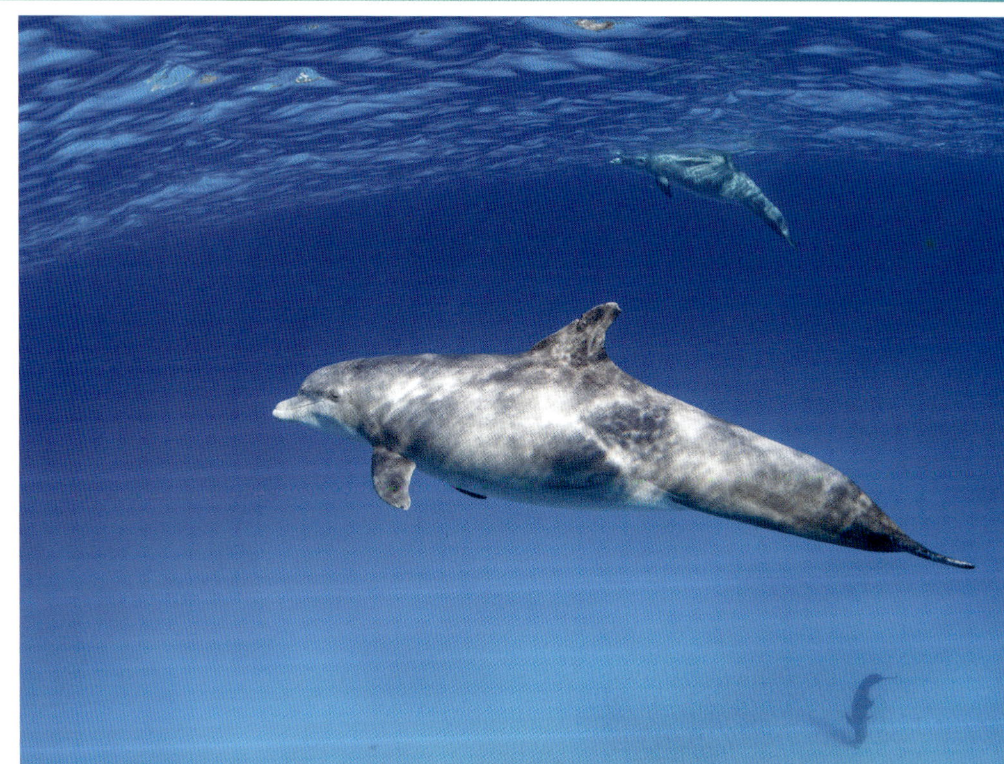

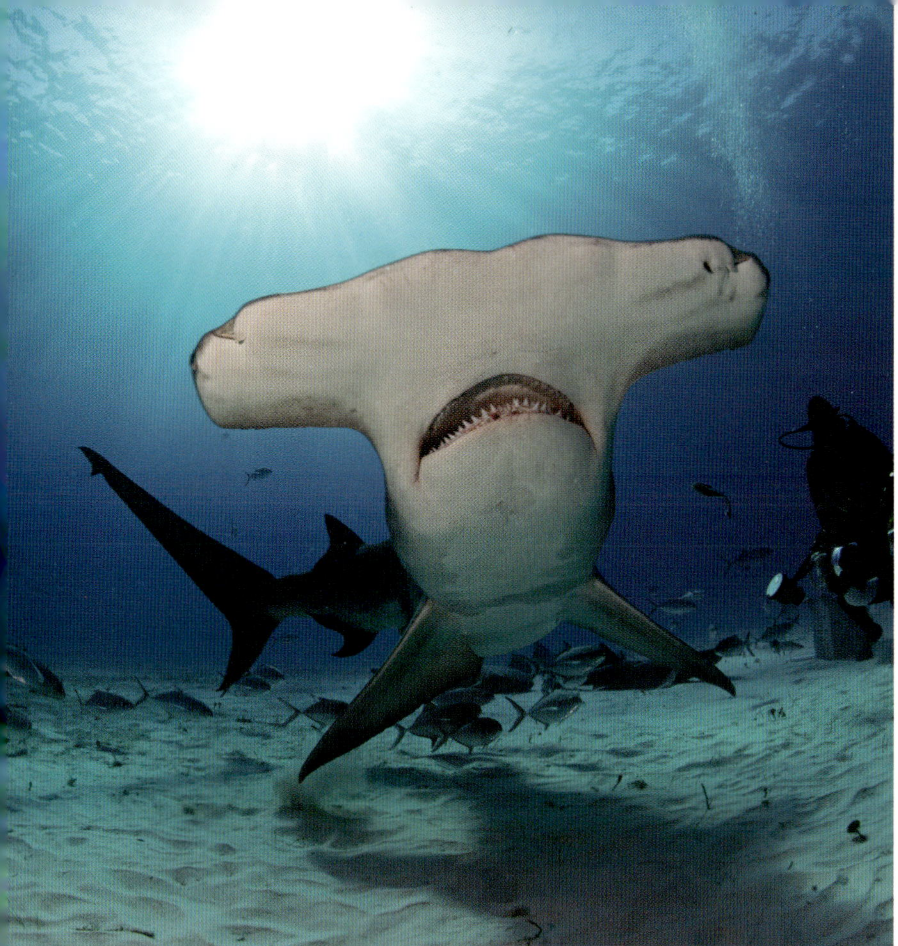

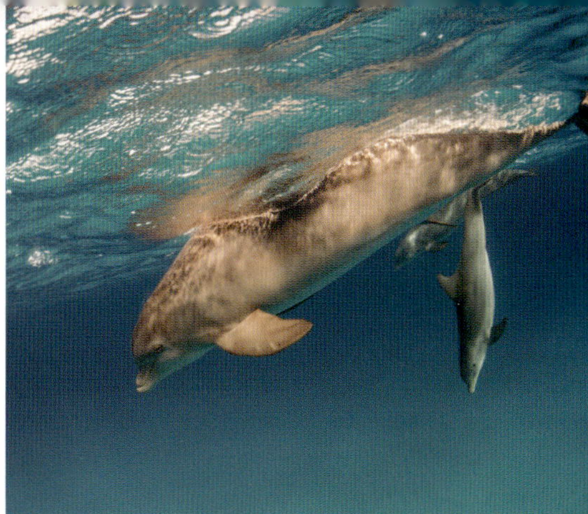

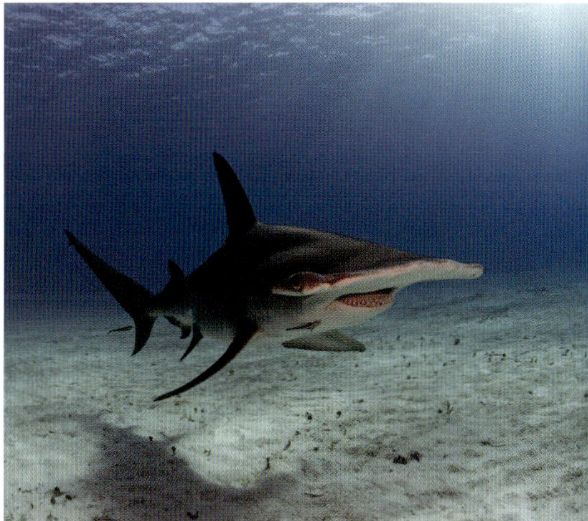

dive with the sharks in warm shallow clear blue water, and 'shark photography' is a major form of income to the Bahamas. Shark fishing has been banned throughout the islands of the Bahamas since 2011 and hefty fines are imposed on anyone found landing a shark of any species.

There are shallow coral reefs around the islands of Bimini, as well as an impressive wall dive that plummets more than 1,200m (4,000ft) below you. These islands are also home to important mangrove areas which are nurseries for reef fish and sharks, as well as providing the islands with protection from storms. A number of shipwrecks, both deliberately and accidentally sunk, have created havens for marine life.

THE WILD DIVE

Sometimes the most simple of experiences can bring the most pleasure and this was certainly the case on a sunny afternoon in Bimini. We were asked if we

ABOVE LEFT Bimini is also famous for its Great Hammerhead Shark diving.

TOP RIGHT The baby dolphin heads to the seabed.

ABOVE Great Hammerhead Sharks are winter visitors to Bimini.

would like to join a boat going out to see if they could find any wild dolphins. The waters around Bimini have resident populations of both Bottlenose and Atlantic Spotted Dolphins and the latter are renowned for interacting with snorkellers and free-divers in the water. There wasn't much to think about really, so we ditched our diving equipment, grabbed our masks, snorkels, fins and cameras, and jumped aboard one of the waiting dive boats. We travelled north past the Bimini Islands until we arrived at the western edge of the Great Bahama Bank. Our captain told us to start looking for any signs of dolphins leaping out of the water, and as we peered into the sunlit waves it was not long before we caught our first glimpse.

At first our captain did not appear to be as keen as we were, and this was because the dolphin pod we had spotted were Bottlenose Dolphins which are generally not as curious as their Spotted cousins. However, this particular group seemed to be staying in one place and everyone on the boat was keen to jump in to have a look. Bowing to pressure, he cut the engines, brought the boat to a stop and declared the 'pool' to be open. We did not need any further encouragement and were finned and masked-up and in the water in a matter of seconds. Around 20 dolphins were hunting using echo-location on the sandy seabed to find their hidden meal. You could hear and feel the dolphins' sonar clicking away as they searched, and then you would see one plunge its head, or rostrum, into the sand, flapping its tail fins to push deeper. Eventually it would come away with a small fish as its prize. It was mesmerising to watch. Dolphins need to come to the surface to breathe and so, as we floated at the surface watching, every now and then a member of the pod would come up towards us and, rather than avoiding us, would swim close around us for a few minutes before returning to the job of feeding.

This experience would have been special enough, but then we spotted a baby dolphin. It could only have been a few weeks old, perhaps even just a few days. It was tiny! Always accompanied by an adult, with the adult positioned between us and the baby, it would come up to breathe and swim at the surface. As the pod became more tolerant of us the adults and youngster got closer and closer until we had an incredible close-up swim past with baby and escorts swimming right past our masks. It was incredible. Far too soon our skipper was encouraging us to come back on to the boat and we were amazed to find we have been in the water for hours.

RIGHT Four dolphins surface together.

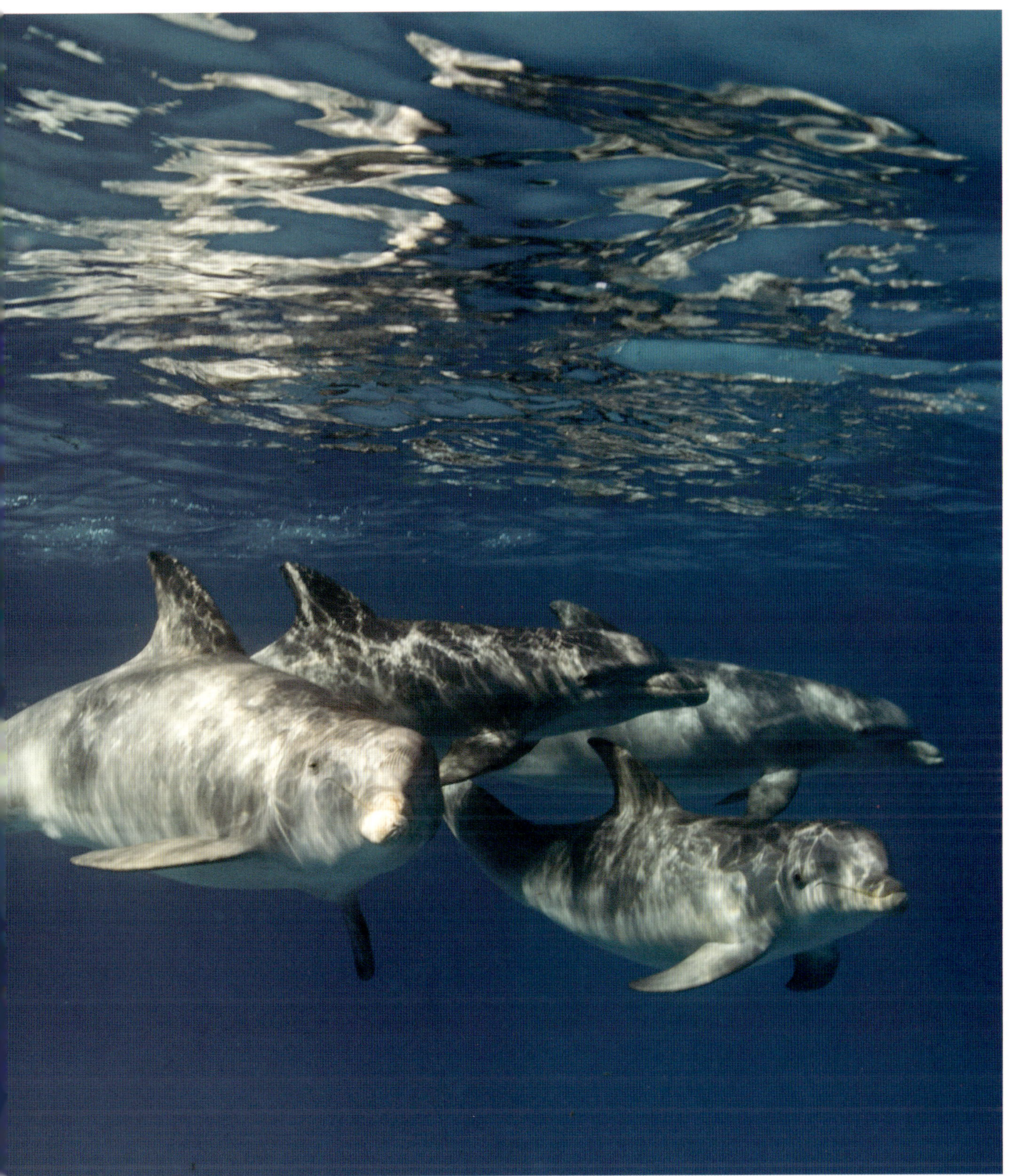

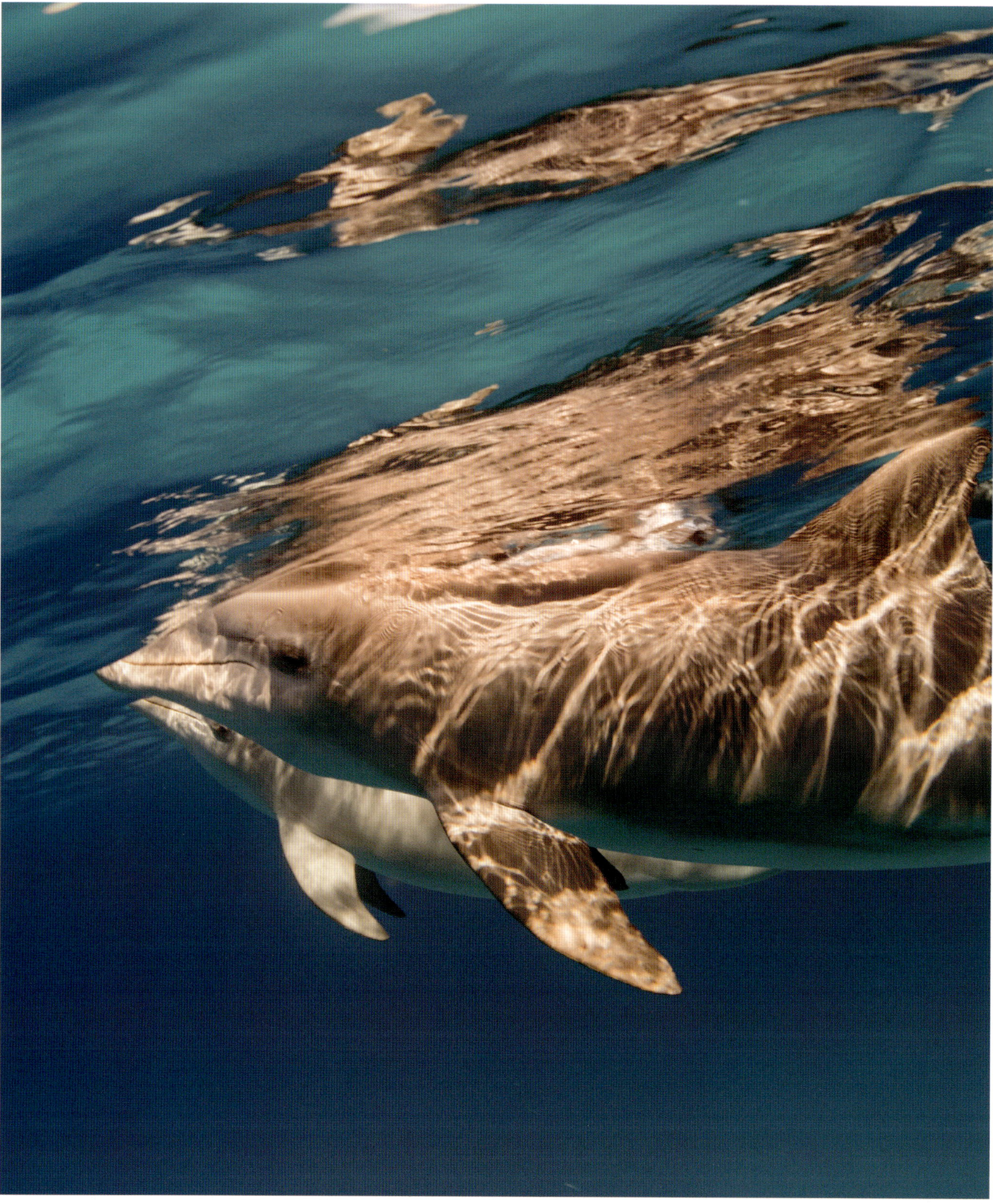

This is how dolphins should be experienced. Wild and free, not in captivity where they have been ripped from their families in order to 'entertain' humans. We did not get to see the Atlantic Spotted Dolphins we had hoped for, and yet we had still had an incredible encounter. We will just have to go back and do it all again some time.

OPPOSITE The dolphins surface together, keeping a watchful eye on the baby.

BELOW One of the more recent wrecks sunk deliberately to make an artificial reef.

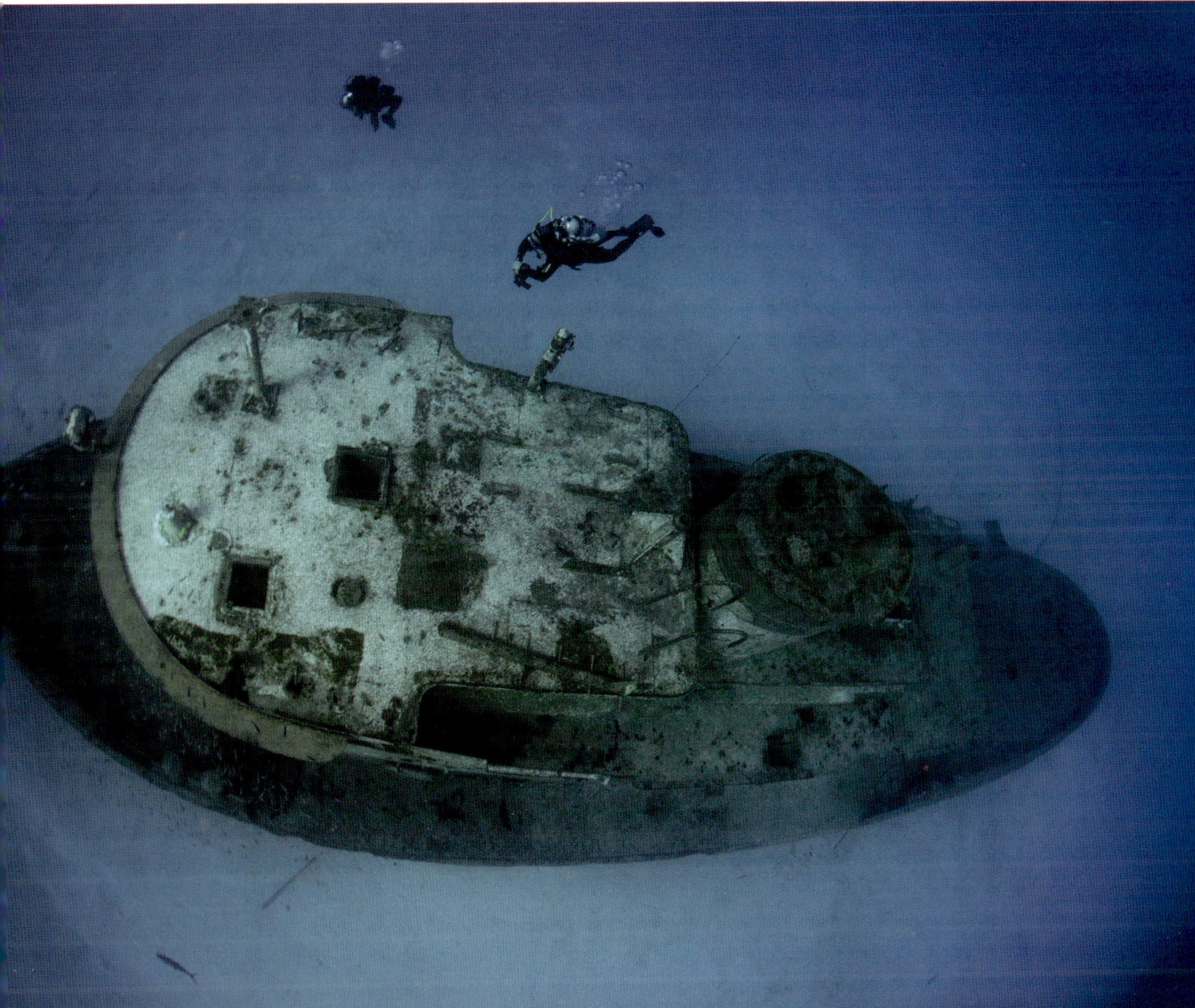

17 FINALLY GETTING TO DIVE WITH TIGER SHARKS

BAHAMAS

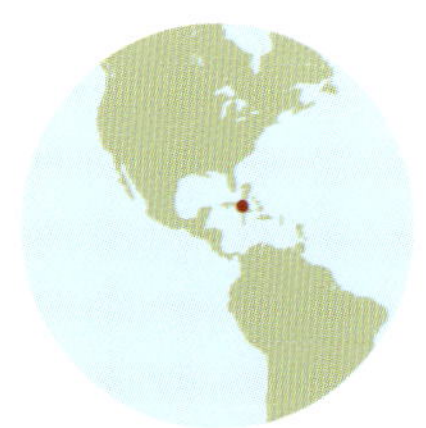

OPPOSITE Tiger Sharks are big and beautiful.

BELOW Three beautiful Tiger Sharks approach Nick in formation.

GRAND BAHAMA IS THE NORTHERNMOST island in the Bahamas. It is a popular tourist destination, with typical white sandy beaches, plenty of night life and lots to do for those that want more than just to lounge in the sunshine by the pool or on the beach. It is surrounded by rich mangroves that help to protect the island from the worst of the hurricanes.

GENERAL DIVING

Grand Bahama has a host of excellent dive sites with varied experiences that include diving with sharks, exploring wrecks and enjoying reefs that sit just off the shoreline. One of the signature dives on Grand Bahama is Shark Alley, a site where several dive operators have been performing shark-feeding dives for decades. Now,

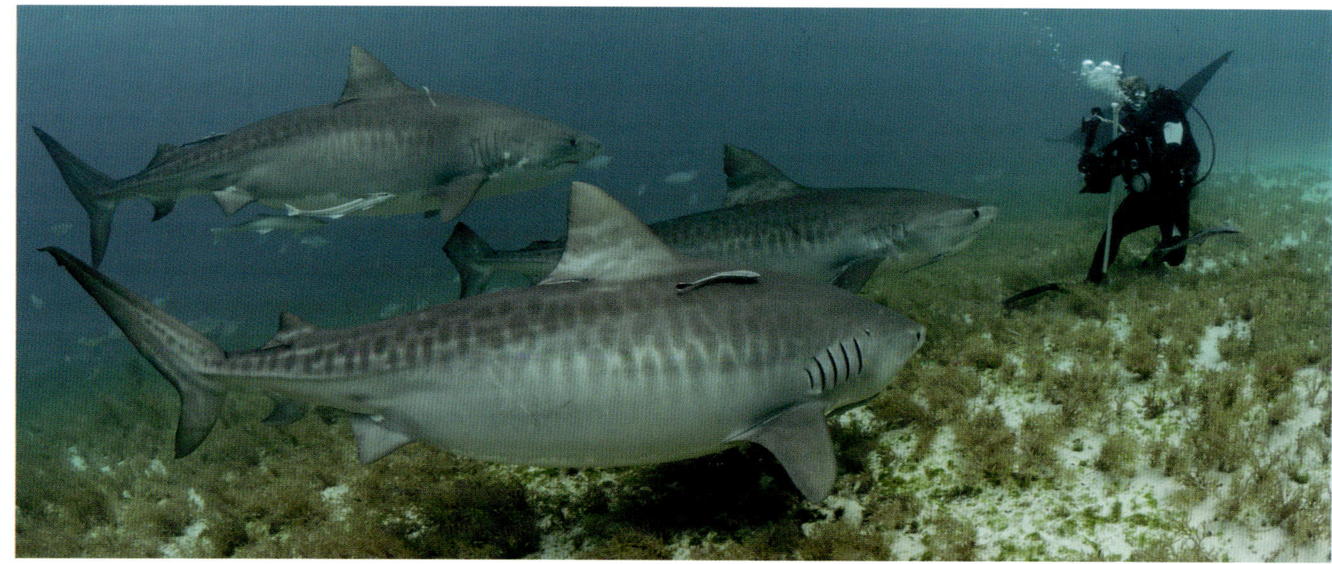

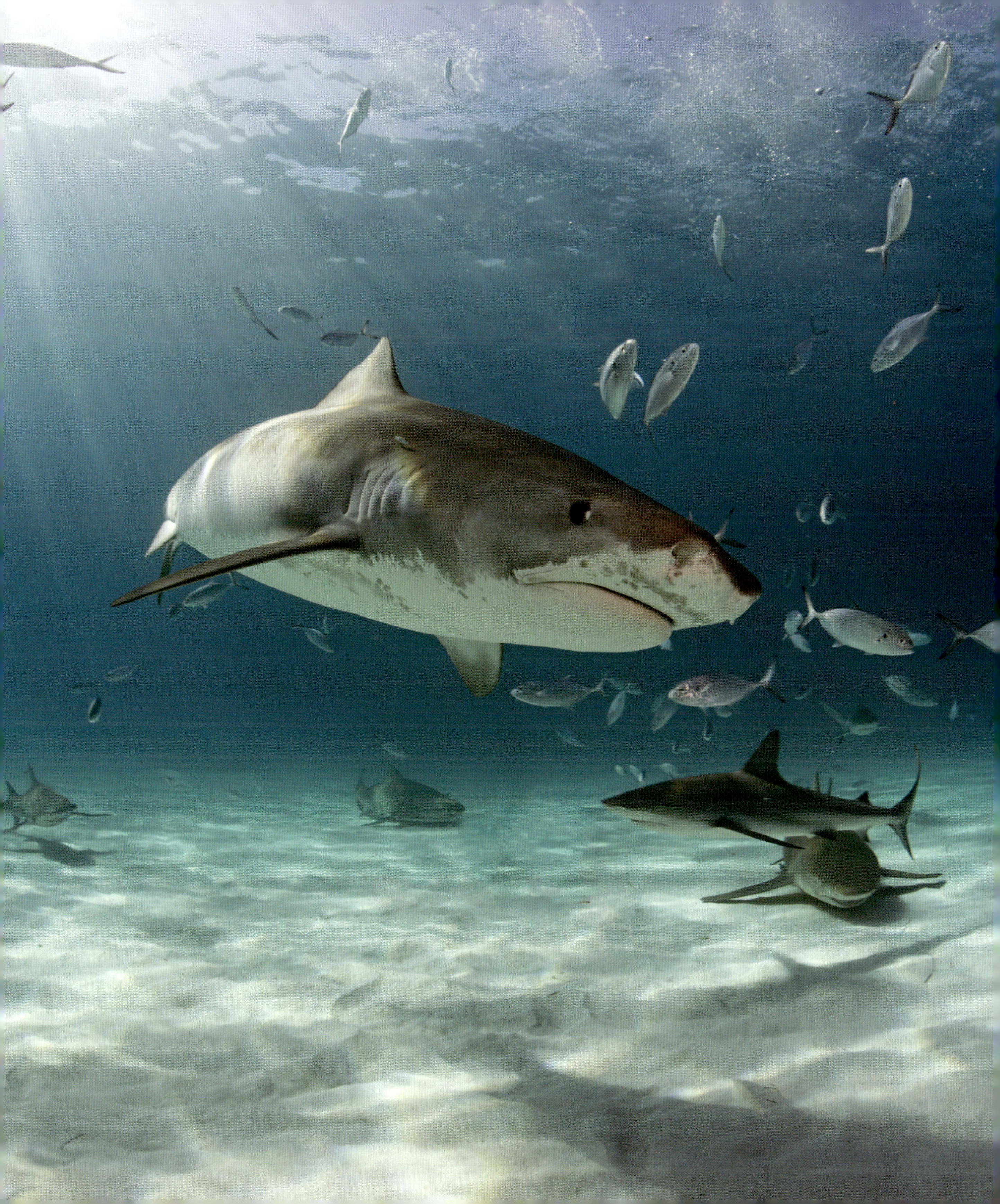

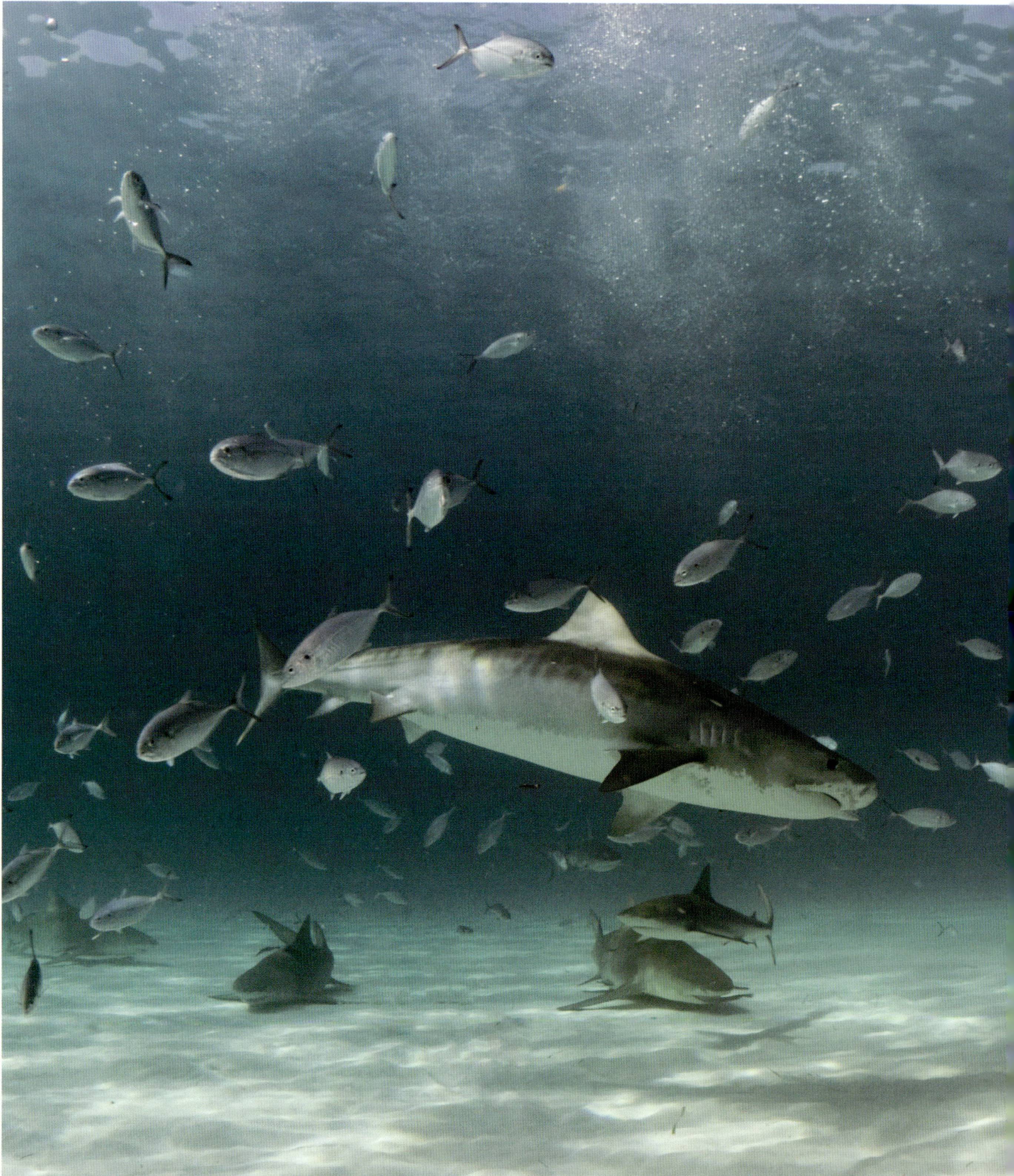

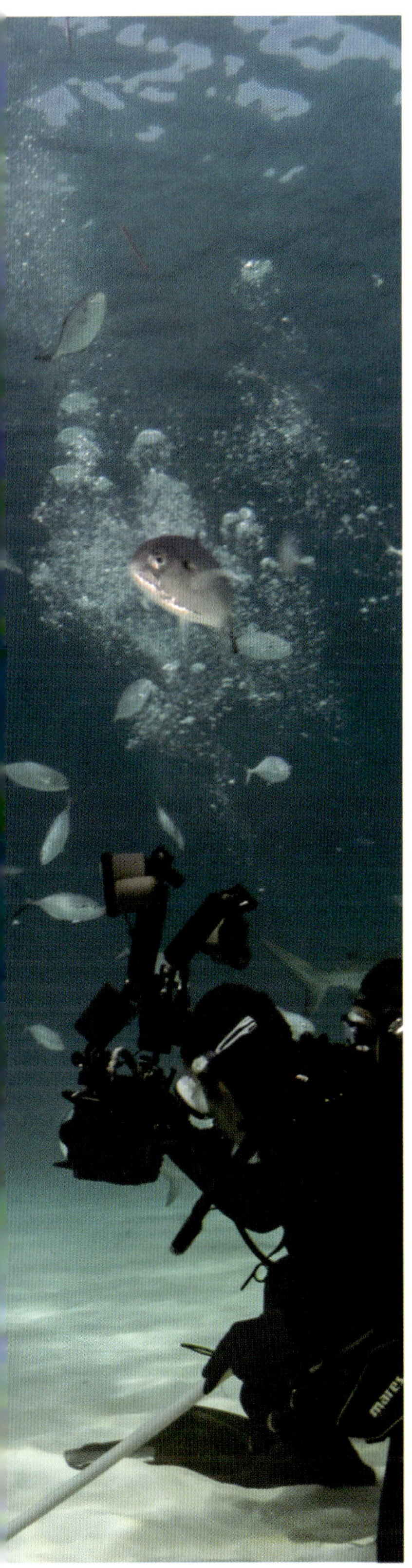

if you dive in this area, regardless of whether there is any bait in the water, you will be sure to see plenty of Caribbean Reef Sharks patrolling the reef. Stingrays and large green moray eels also frequent this area, along with a large school of Horse-eyed Jacks. The shallow dive in clear blue warm water is high on many people's lists of favourite dive sites.

The reefs that surround the island form long fingers of rock formations covered in coral, with sandy channels in between. Here marine life thrives with delightfully varied creatures living on the reef including weird arrow crabs hiding in sponges, beautiful and delicate angelfish darting along the reef, bright parrotfish audibly chomping on the coral, and the oddly named Flamingo Tongue Snail eating soft corals.

Another very famous dive off the West End of Grand Bahama is Tiger Beach, which is named after the Tiger Sharks that frequent this area. Along with Tiger Sharks you may also encounter Caribbean Reef Sharks, Nurse Sharks, Bull Sharks, Great Hammerhead Sharks and Lemon Sharks here, so it is a perfect dive for those that love their sharks!

THE WILD DIVE

There are very few places in the world where you stand a good chance of diving with Tiger Sharks. It has been a dream of ours since we started diving over 20 years ago to be able to see these magnificent sharks up close and one that proved very hard to bring to fruition with the weather and animals proving less than cooperative. So it was with some excitement that we woke up to an early alarm that indicated today was the day.

The boat ride was going to take between one and two hours depending on conditions, so we focused on getting our dive and camera gear ready and loaded onto the boat and then went to find coffee and breakfast for the journey. Our group consisted of us, another diver we had met at our hotel, our captain and two dive guides, so we knew that if we got the chance this was the perfect small group for a great day out. As we arrived it was hard not to get overexcited. As I looked

LEFT A Tiger Shark approaches a diver at Tiger Beach.

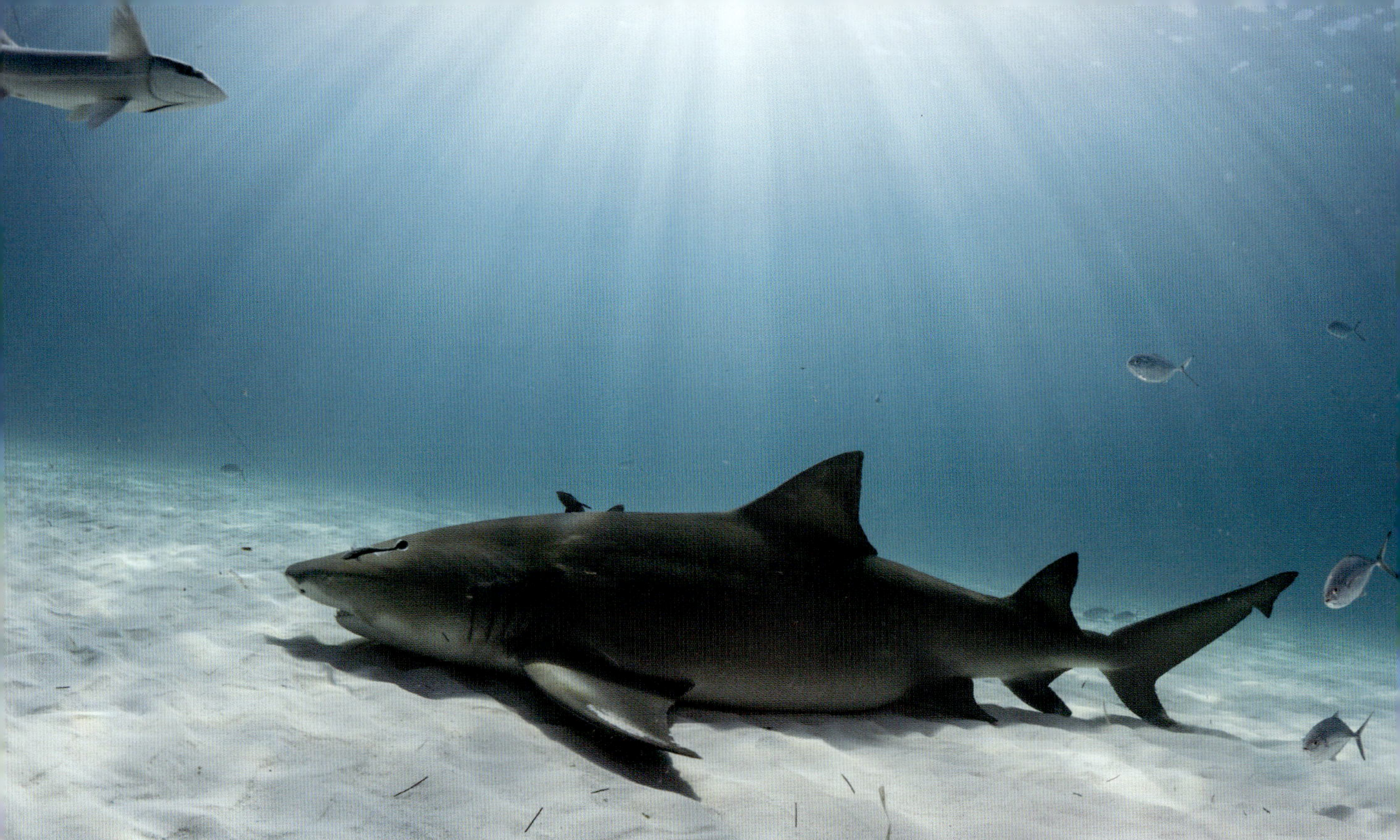

ABOVE Lemon Sharks lie on the sandy seabed.

OPPOSITE A Tiger Shark gets in very close to the bait box and my camera, almost grinning for the shot.

down through the clear blue shallow water I could already see Caribbean Reef Sharks and Lemon Sharks swimming under the boat. As the crew start to chum the water it was time for us to set up our dive and prepare our underwater photo and video equipment. Soon I heard one of our dive guides ask "Is that one there?" while pointing to a larger shark swimming below. The first Tiger Shark had arrived.

I slid into the water and made my way down to the pre-agreed spot on the sandy seabed. The view was incredible. Lemon Sharks joined us, lying on the sand and swimming lazily around our legs. Nurse Sharks stayed right next to the bait box. Caribbean Reef Sharks swam in circles, never straying too far from the alluring smell of the bait box. Then as I got my eye in I spotted my very first Tiger Shark in the distance. I could also see a Bull Shark, but this shark stayed away from us and was too nervous to come any closer as the first big Tiger Shark approached.

Tiger Sharks are extremely beautiful. They get their name from the patterns on their sides that are similar to those of the eponymous big cat species, but these patterns fade as the Tiger Shark gets older and so the biggest of the individuals seen on this dive had very few of their striped markings still obviously visible. They are big sharks, with the largest reaching 5m (16ft) in length. They are broad and imposing and on this dive they came very close indeed to the divers.

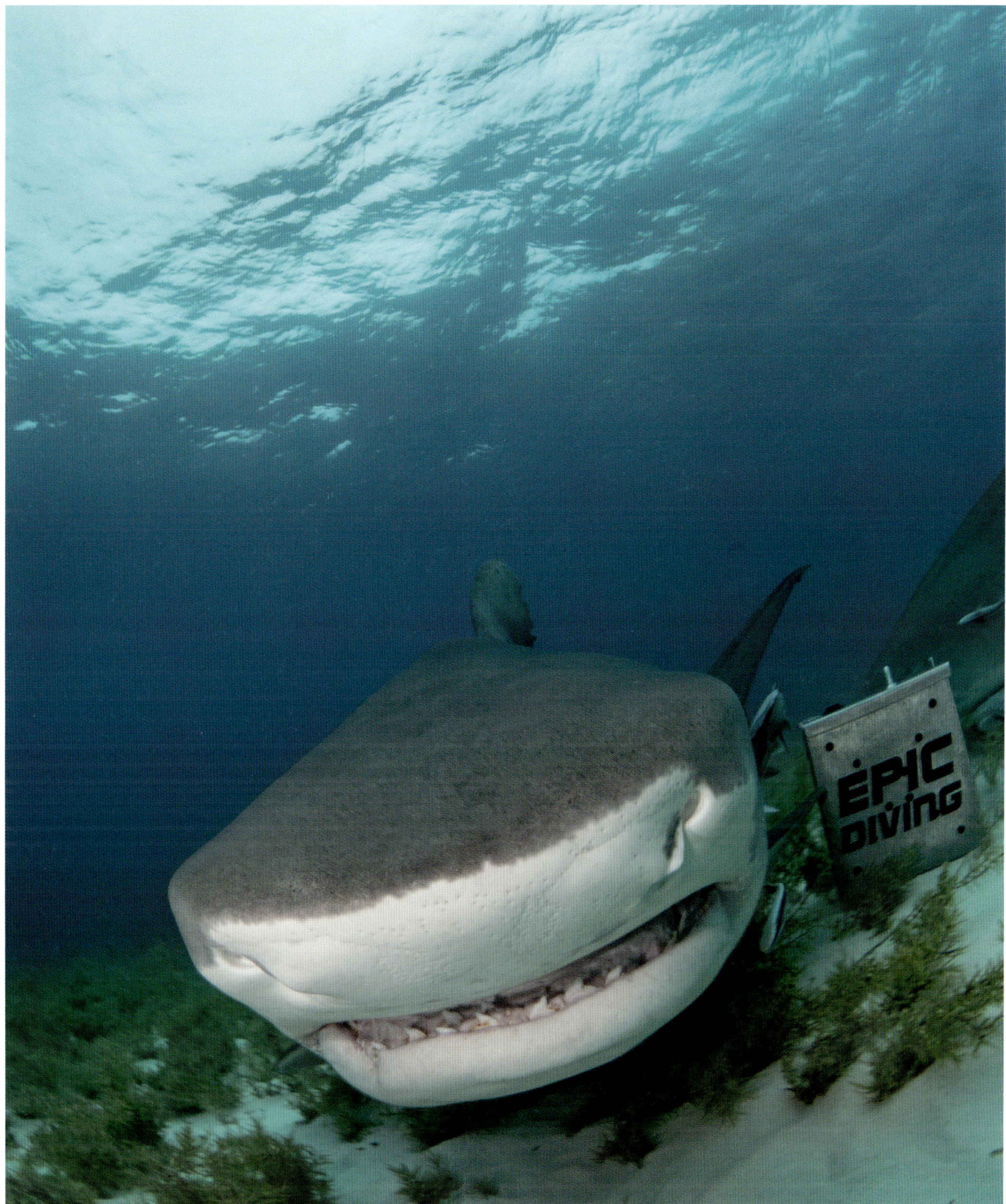
EPIC
DIVING

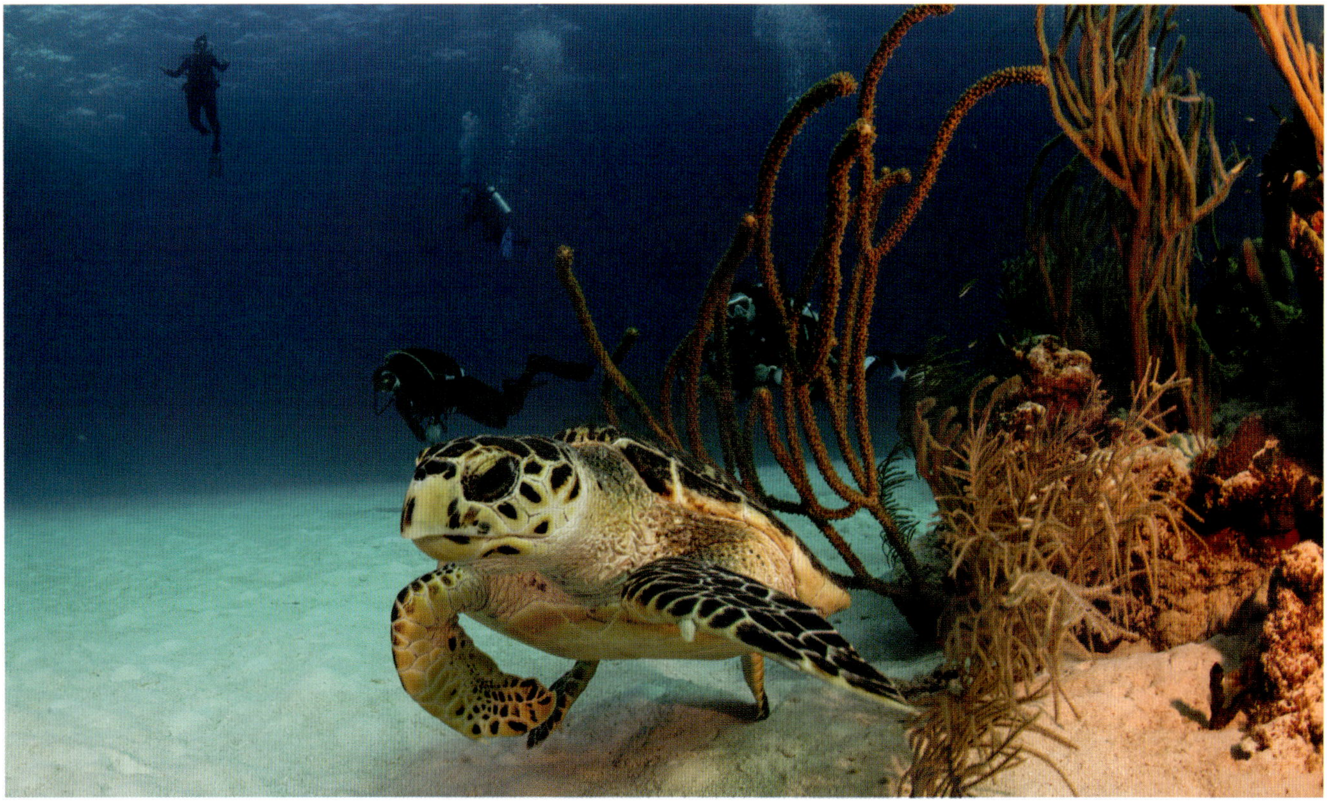

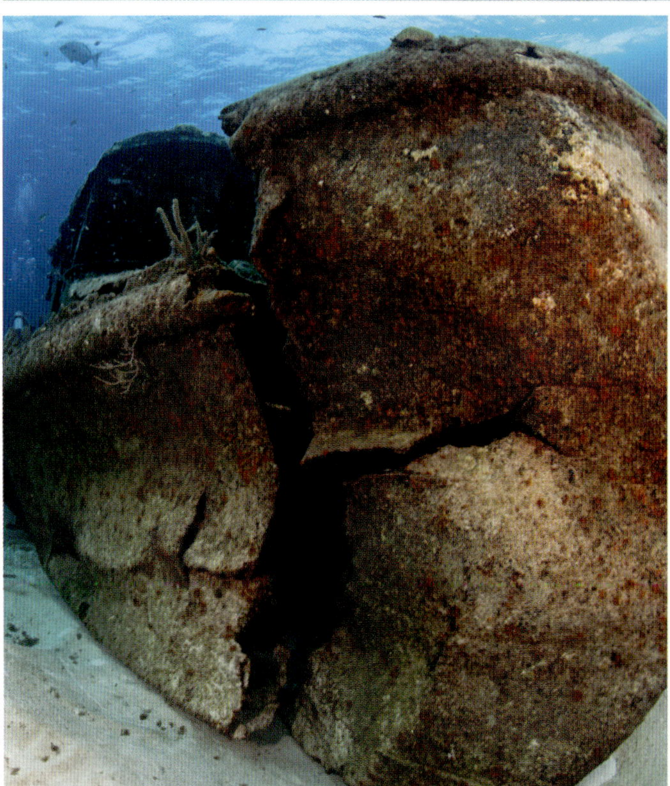

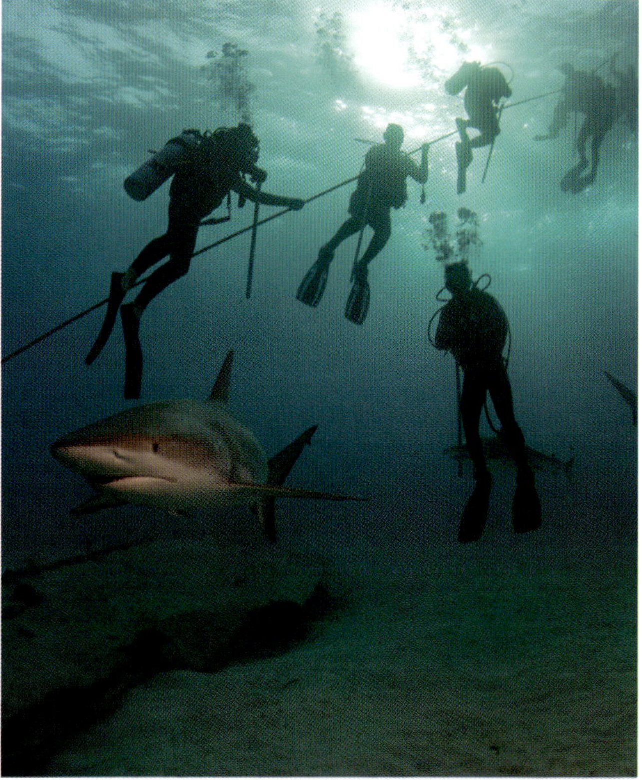

The dive was in fairly shallow water, between 8–15m (25–50ft), and the water temperature was a balmy 28°C (82°F). I knelt on the sand completely mesmerised by the sharks swimming in front of me. Soon we had a second Tiger Shark that wanted to get closer and closer to the bait box and the divers. I had heard that these sharks are very used to divers and not at all shy in coming in close, and this proved to be true, with both Tiger Sharks making close passes in front of us or over our heads.

All too soon our dive guide was signalling for us to head back to the surface. The Tiger Sharks seemed to know the routine and drifted off into the distance, but the Lemon Sharks followed us up the line and stayed at the back of the boat as we changed over tanks and got ready to do it all again. With a long ride home we knew that we would have limited time on our second dive with the sharks, but I was already delighted with the encounters we'd had on the first dive, so I was much more relaxed second time around, enjoying the return of the Tiger Sharks swimming over the sand and seagrass at this incredible dive site. Finally, after many years of trying, my dream of seeing Tiger Sharks had come true.

OPPOSITE TOP A turtle on a coral finger in Grand Bahama.

OPPOSITE BOTTOM LEFT Grand Bahama has some great wreck dives.

OPPOSITE BOTTOM RIGHT Divers make their way up the line at the end of a shark dive.

BELOW Young Tiger Sharks have the stunning 'tiger stripe' pattern on their sides.

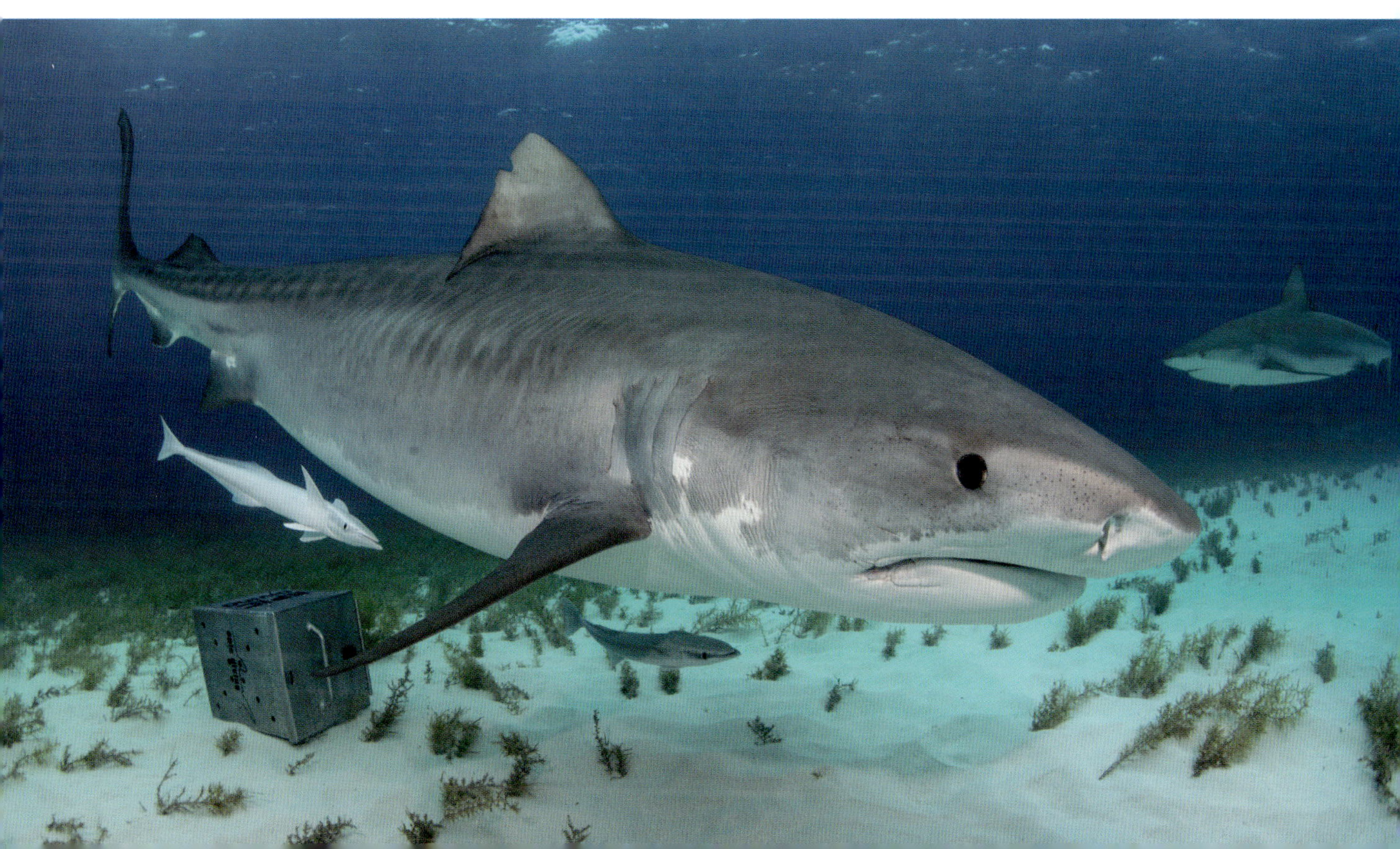

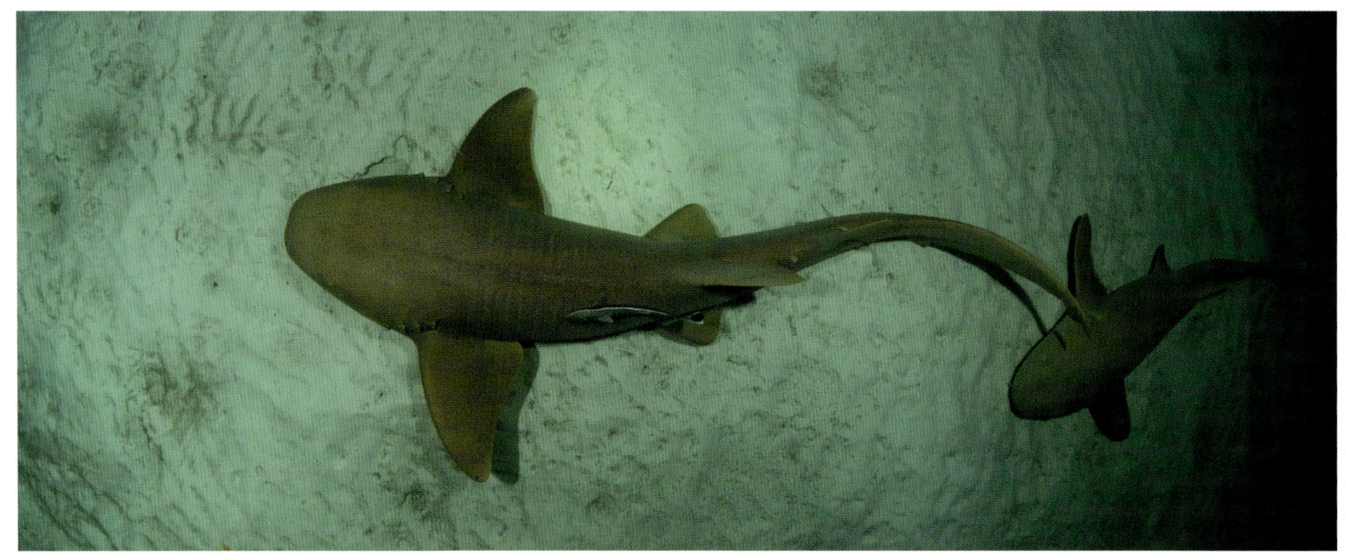

18 SHARK DIVING AT NIGHT

BAHAMAS

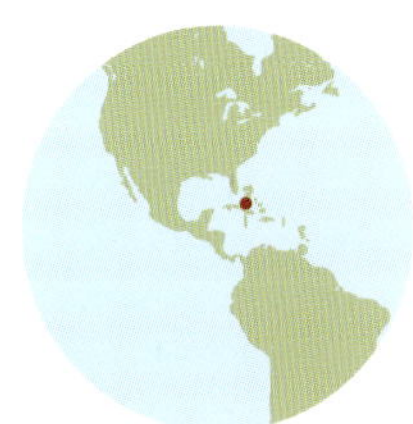

ABOVE Nurse Sharks hugged the seabed hoping to get in on the action.

OPPOSITE The sharks around us were only picked out when a torch light hit them.

NEW PROVIDENCE ISLAND is the most populated of the Bahamian islands and is also the location of the capital, Nassau. With its beautiful beaches, luxurious hotels and plenty to keep the tourists entertained it is also the focus of tourism in the islands. Sun worshippers join adrenalin junkies who love water sports. The Bahamas offer year-round warm weather and with regular inter-island flights it is the perfect Caribbean destination for multi-island tours. You can kayak through the mangroves, snorkel with stingrays, visit the swimming pigs of the Exumas, covet the huge shoreline houses of the rich and famous, or just enjoy the local culture and food. There is so much to do here, but there is nothing wrong with just lounging by the pool for while too.

GENERAL DIVING

The Bahamas are graced with warm clear turquoise-blue water. The name 'Bahamas' translates as shallow waters and the depth of most of the water around

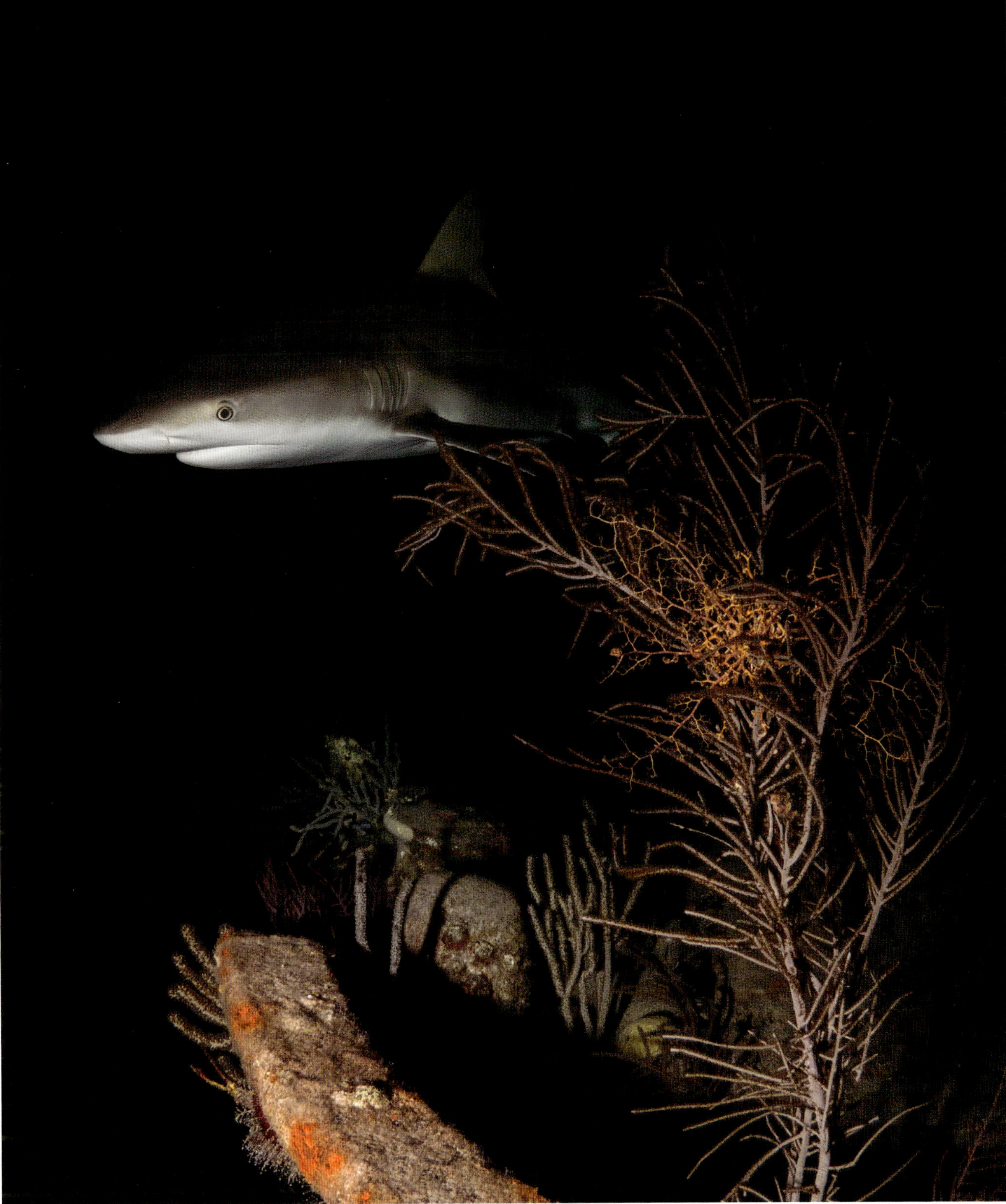

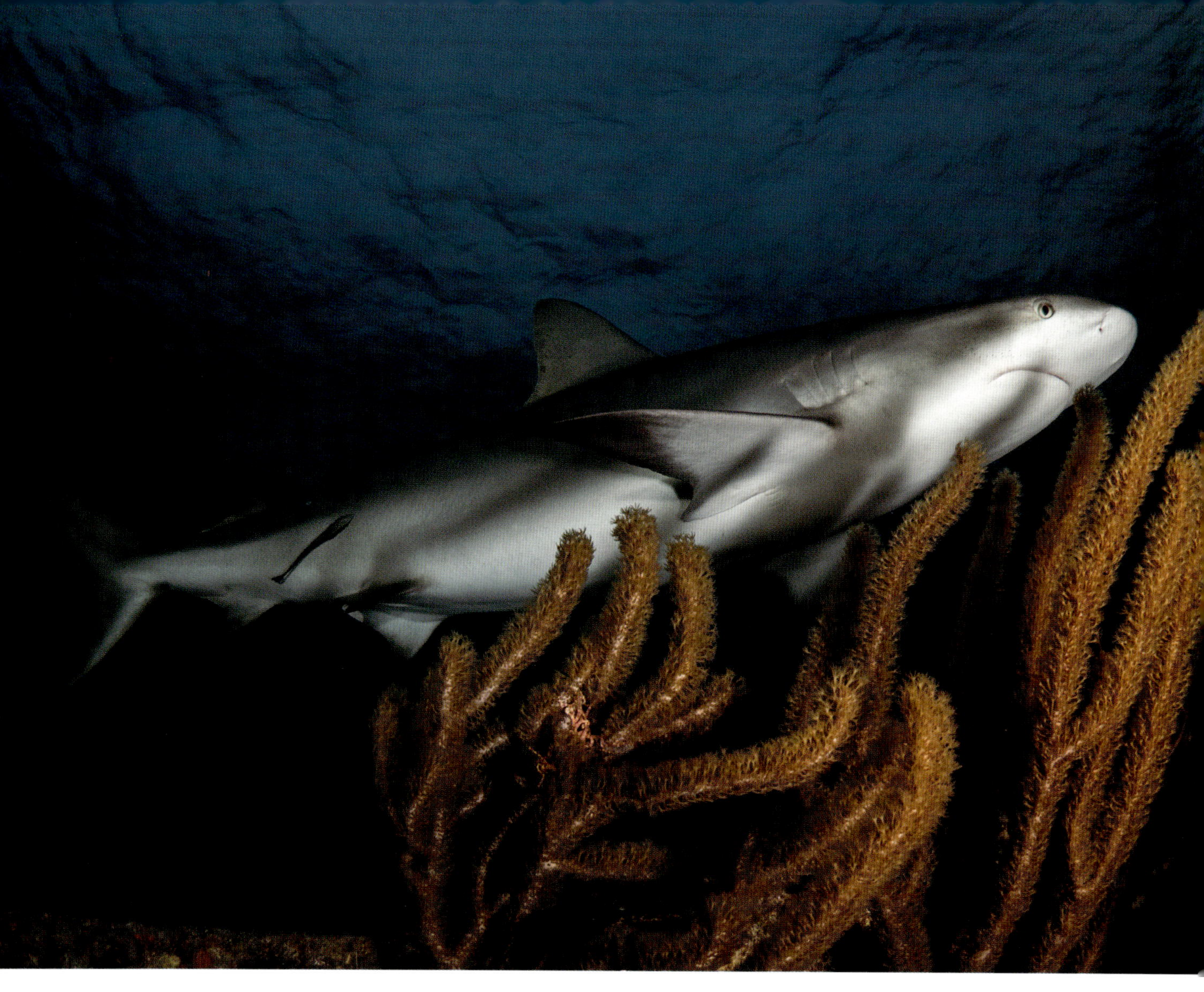

the islands is less than 12m (40ft). Sandy bays and palm trees top off this image of tropical perfection. Underwater this paradise continues with vibrant reefs, amazing shipwrecks and the highlight (sometimes the only reason) for many divers – sharks. Underwater scenes for numerous movies have been filmed here, including several of the James Bond films. The reefs teem with fish; turtles cruise past and eagle rays swoop over the sandy seabed in search of prey. Many of the wrecks here have been sunk deliberately as artificial reefs, to provide new homes for the marine life and encourage new corals to grow. The waters around Nassau are a playground for divers.

ABOVE A Caribbean Reef Shark swims over the coral-encrusted wreck.

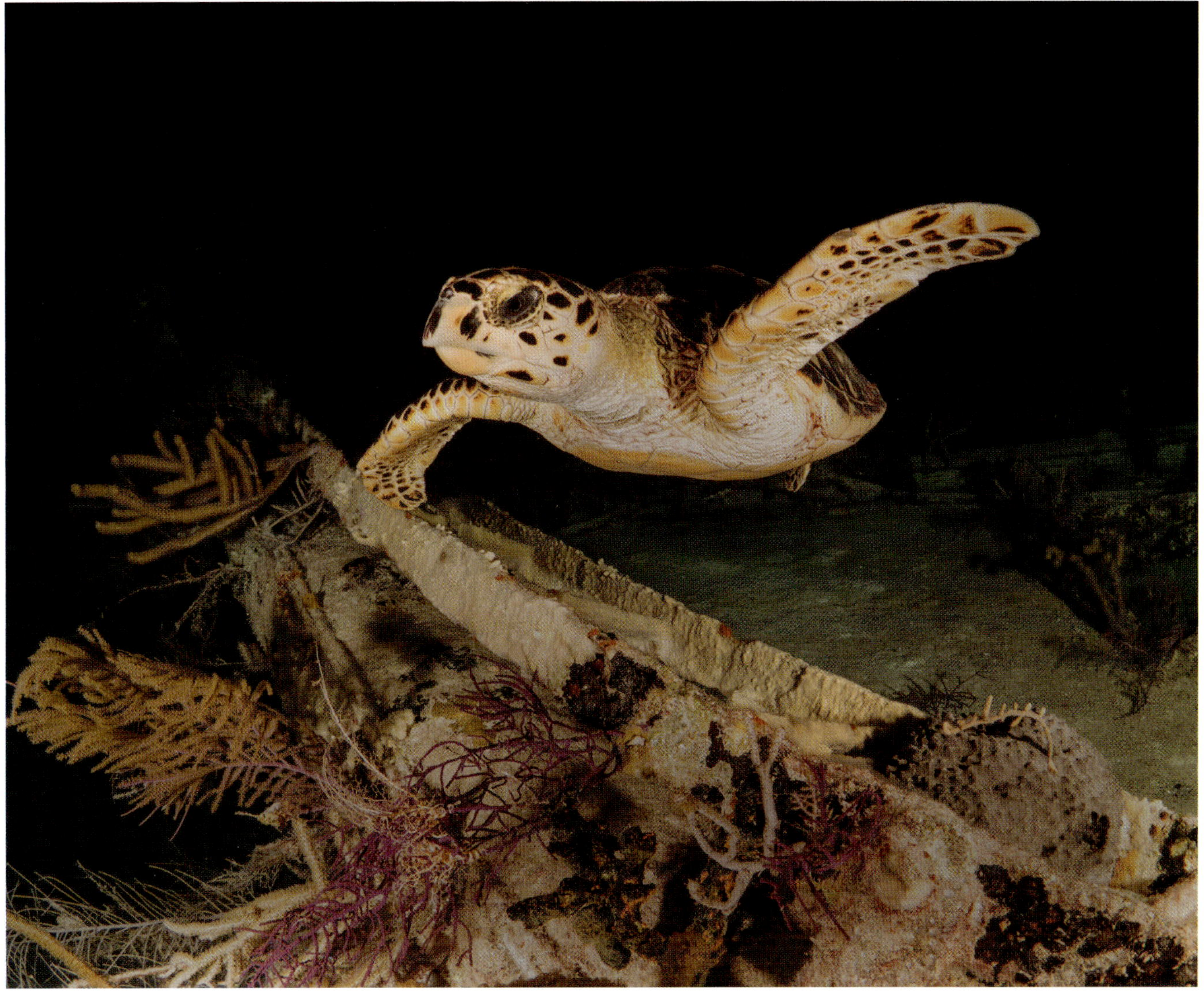

If you love sharks then this is an incredible place to dive, with early shark dives being pioneered here. Now you can dive on shallow wrecks as dozens of Caribbean Reef Sharks swim around you. Many of the shark dives will see a bait box – a metal box filled with fish scraps – taken to the seabed to attract the sharks. Sometimes on a specialist dive the guide will even feed the sharks small scraps. Several dive operators offer specialist shark-diving courses where you can learn about the sharks and also how to feed them safely while wearing a chainmail suit. All of these shark dives give the audience incredible close up encounters with one of the ocean's top predators, and hopefully recruit them to the growing band of shark lovers and advocates of protecting them worldwide.

ABOVE A turtle swims over the wreck at the start of the night dive.

THE WILD DIVE

Over the years we have done many shark dives, but on a recent trip to Nassau the owner of a dive shop asked us if we might be interested in trying a baited shark night dive. We jumped at the opportunity, always keen to try something new, especially when it comes to diving with these beautiful but misunderstood animals. So as the afternoon came to an end, we started to prepare our equipment and cameras. Our guide helped us load up onto the boat and then we sat around to listen to his safety briefing. The dive guides would be taking a bait box, suspended on a line attached to the boat, to a shallow sandy area where bright lights would give us a view of the sharks. While this was all being set up we were to descend down the mooring line to a shipwreck below and spend some time diving the wreck. We excitedly got into our gear and made a giant stride into the dark water.

The water was pitch black and only the light of our torches picked out the edges of the wreck below. Sweeping my light across the bow of the wreck I caught a turtle in my beam and headed over to take a closer look. As I readied my camera to take a shot, my torch beam picked out the shape of a reef shark just off the bow of the wreck. It was exhilarating to know that there were probably 20 or 30 Caribbean Reef Sharks very close by, and yet without lighting them, you would never know. After taking a shot of the turtle I swung my torch round in a wide arc and caught the bright eyes of several other sharks around me. As the other divers reached

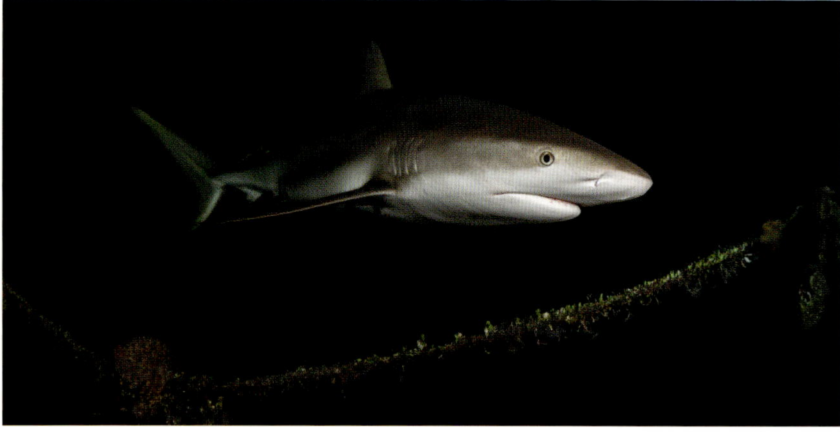

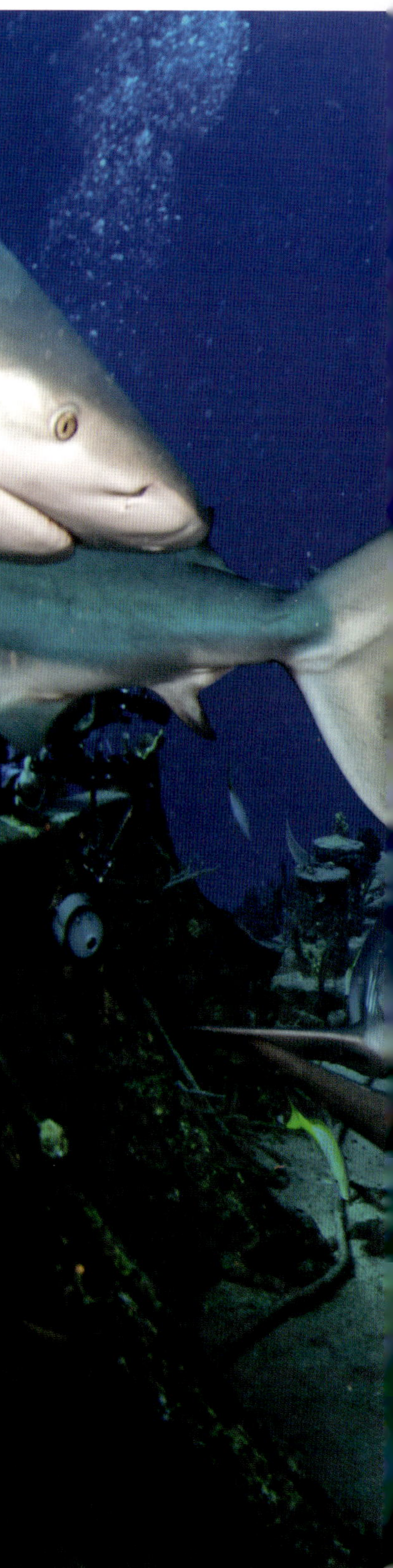

ABOVE A shark turns in front of me.

RIGHT The daytime shark feed on the same wreck as the night dive.

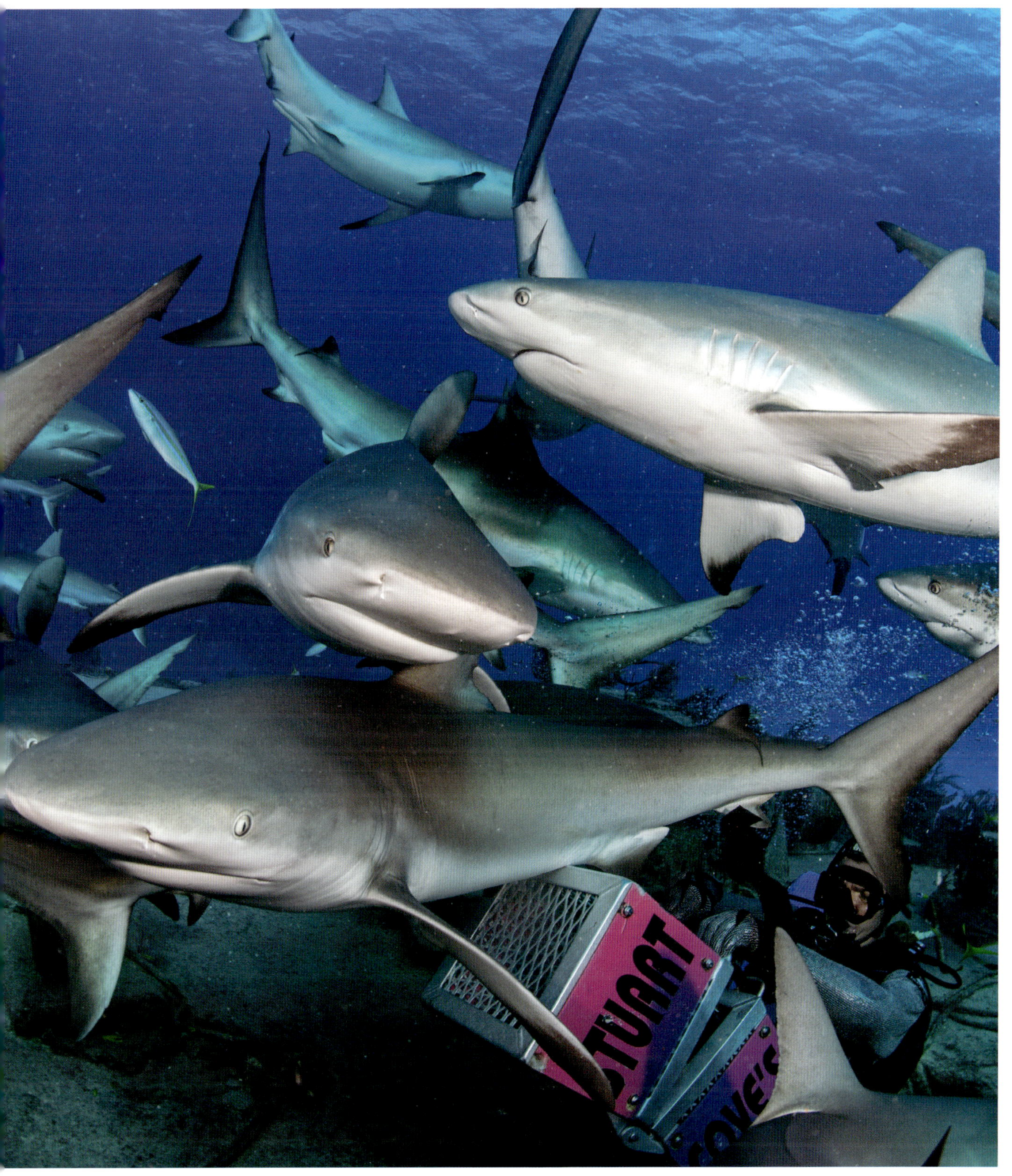
STUART

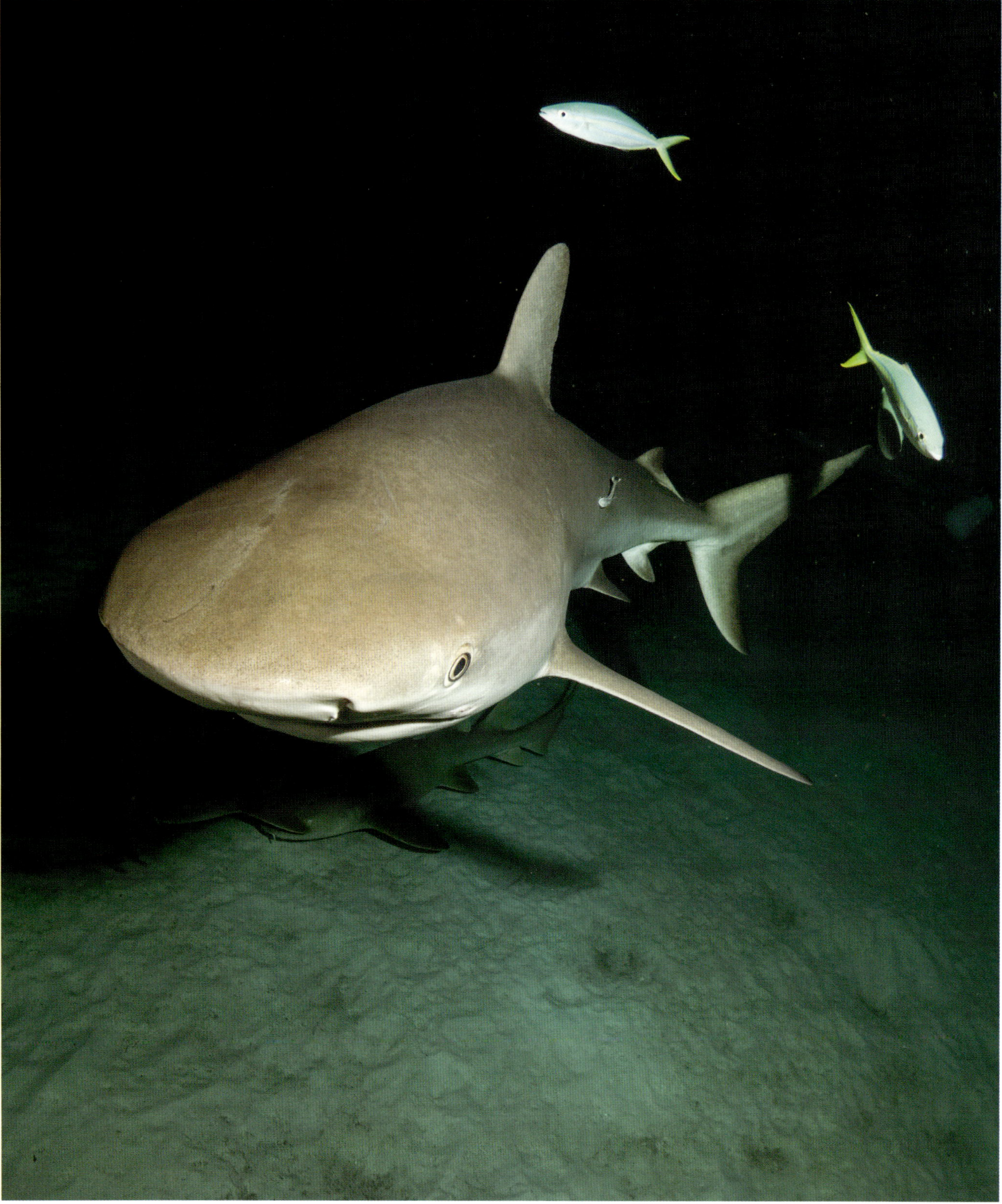

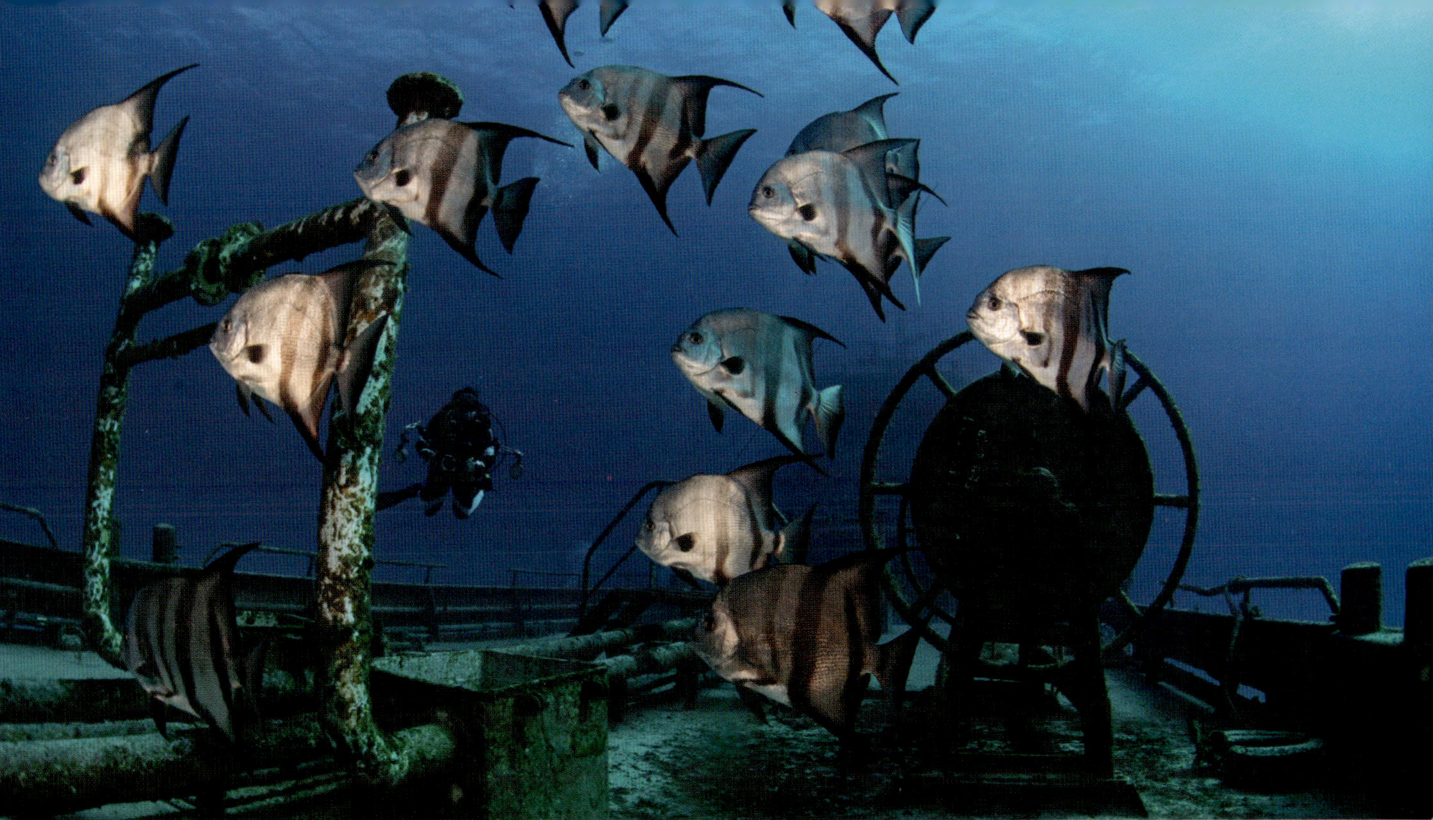

the wreck more light beams penetrated the darkness and lit up the sharks which were swimming around us. The guides beckoned us over to the sandy 'arena' as the bait box was lowered into position. It was an incredible spectacle and I wanted to get closer. The powerful light they had lowered into the water pointed vertically down and created a wide circle of light directly below. I signalled to the guide, asking if I might approach and he gave me the OK sign. I finned slowly into the light and closer to the action. Soon I had Caribbean Reef Sharks all around me, getting very close and then turning at the last minute avoiding banging into me. The light seemed to pick out their sleek lines perfectly against the black background of the night sea. It was amazing, or as my buddy put it, 'bonkers'.

Dives like this always come to an end way too quickly. It was like no other shark-diving experience I have done before. The darkness certainly added to the excitement, as well as giving everyone on the dive a unique shark experience. I reluctantly headed back up the line on the instruction of the dive guide, all the while watching the sharks circling below in the light. After an extra-long safety stop, I could not prevaricate any more and had to climb up the boat ladder and back into the real world. The others were already on the boat and the buzz and chatter of excited divers filled the warm night air. This would be a dive that we would all talk about for a long time to come.

OPPOSITE Sharks came in closer the nearer I got to the bait box.

ABOVE This area is also famous for its wrecks.

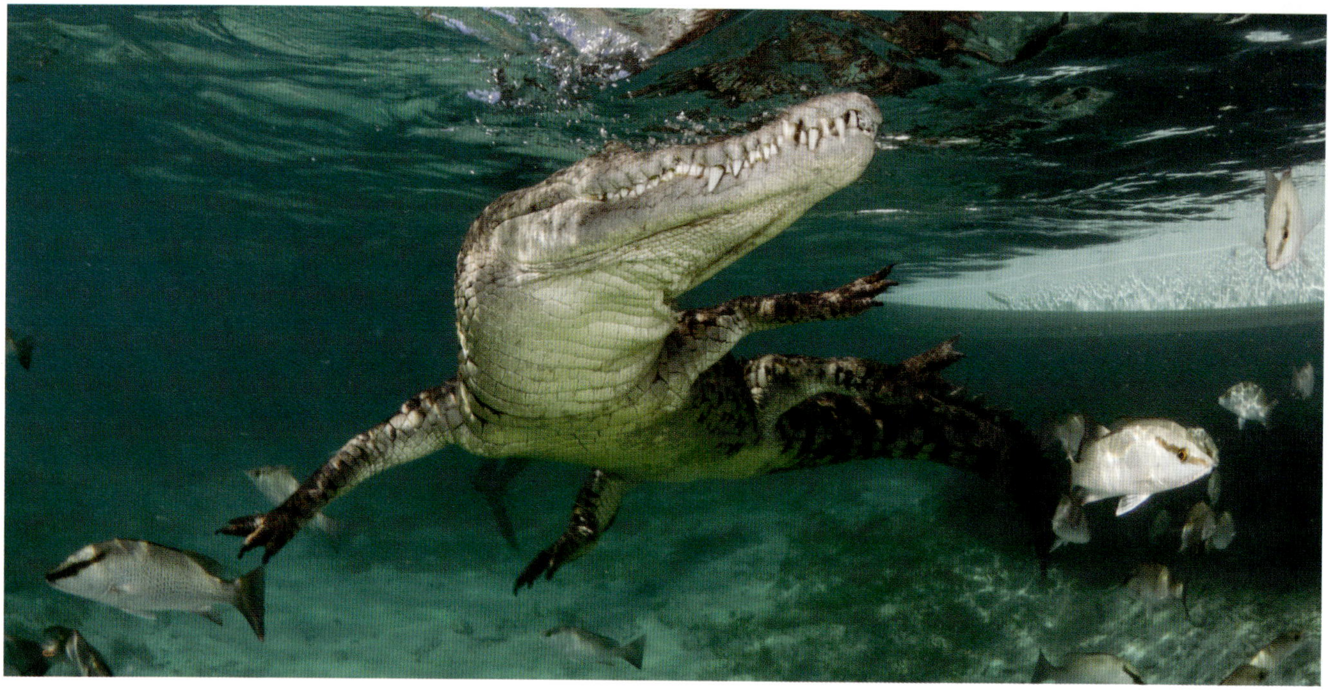

19 WHAT BIG TEETH YOU HAVE!

CUBA

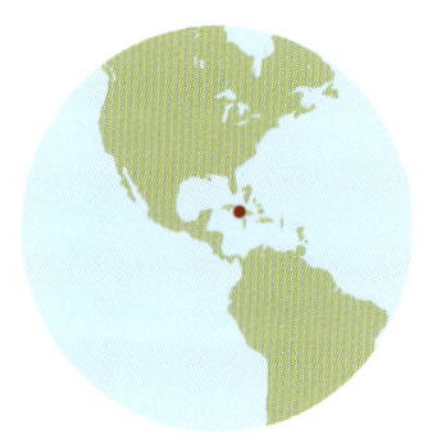

ABOVE It is an incredible experience to be close-up with a croc in shallow clear water.

CUBA IS A CARIBBEAN ISLAND with a difference. It has a unique history and this seeps through everything you see, so while you are there make sure that you visit Havana and take time to soak up the atmosphere. The island is situated in the northern Caribbean, just a little over 160km (100 miles) off the coast of Florida, at the point where the Caribbean meets the Gulf of Mexico and the Atlantic Ocean. The country has an exciting vibe and Havana is a city not to be missed, with its classic cars, famous cigars, crumbling yet beautiful buildings, music-filled streets and incredible nightlife and restaurants. However, move outside of the capital and there is plenty more to explore, with white sandy beaches lining the coastline, dramatic mountains further inland and sugar cane and tobacco plantations on the flatter lands. The landscape is ever-changing.

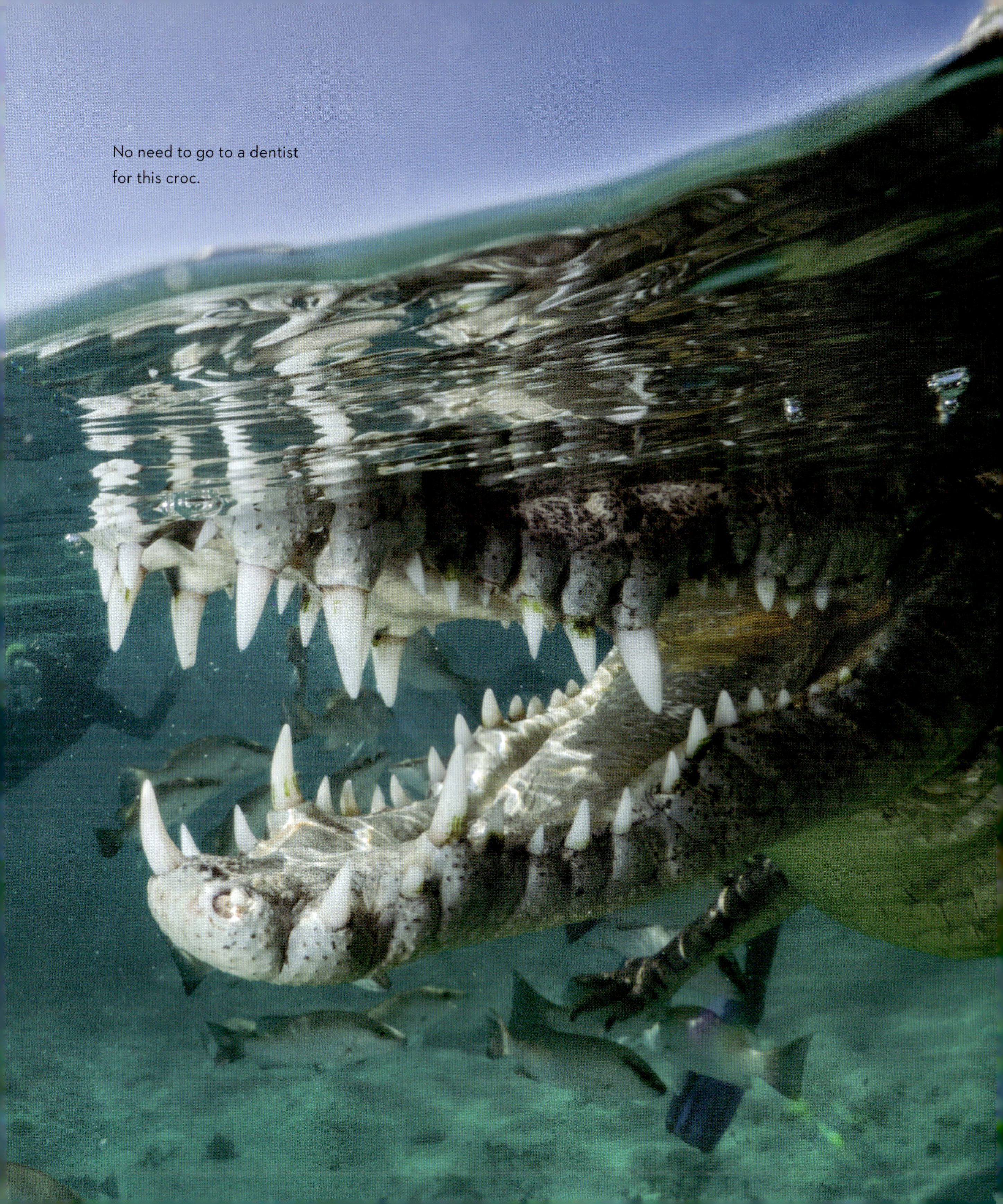

No need to go to a dentist
for this croc.

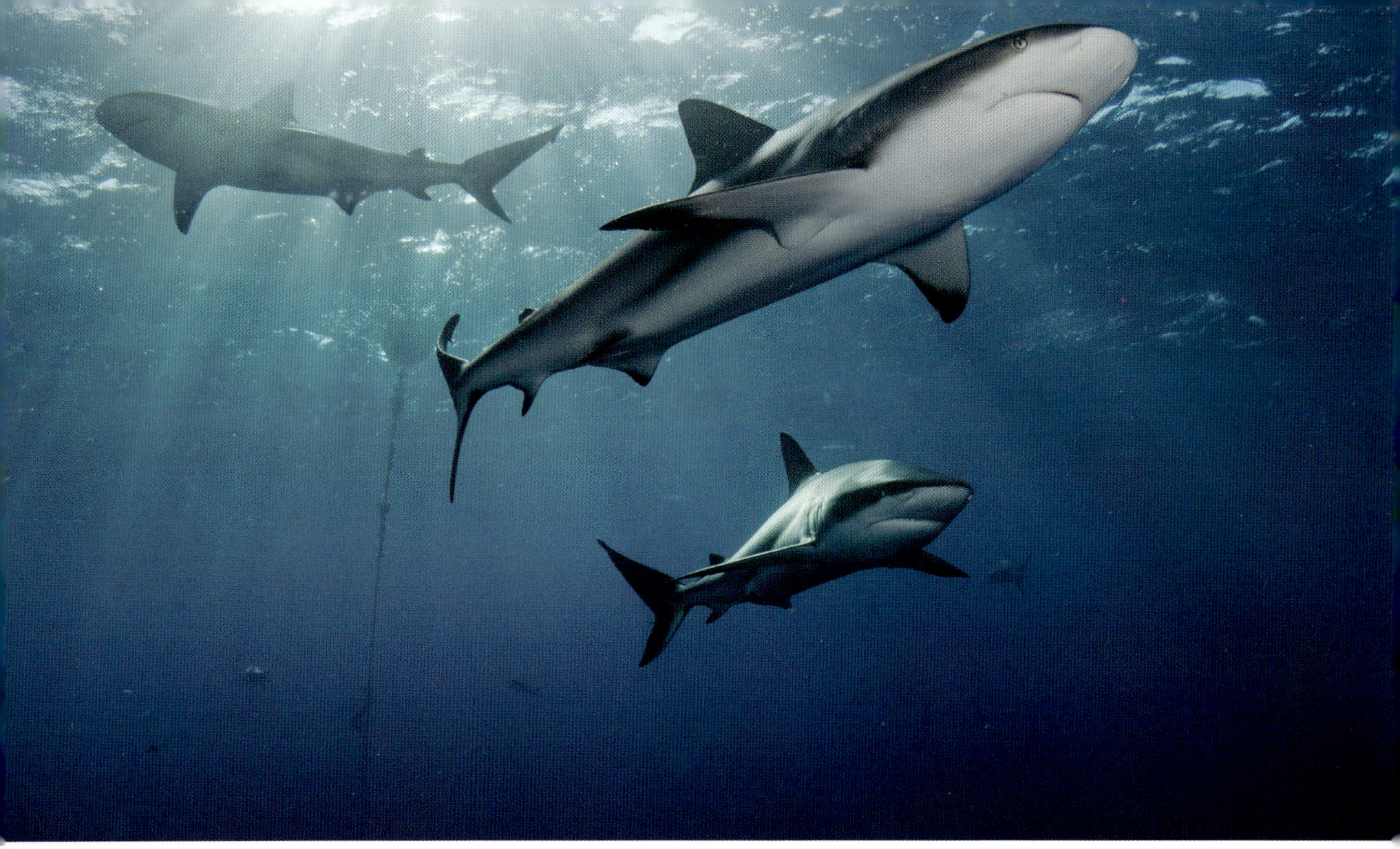

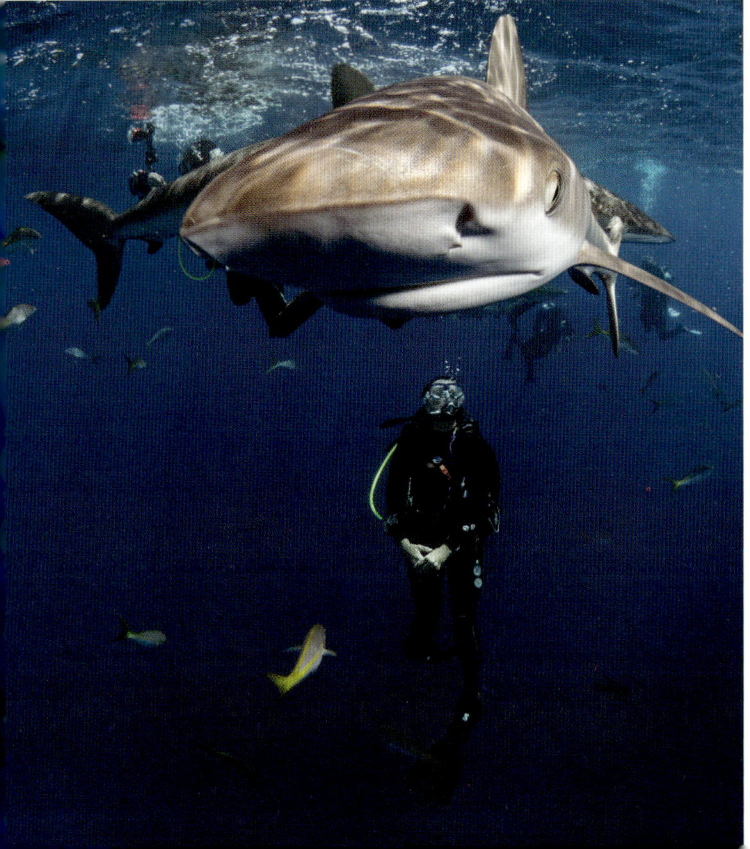

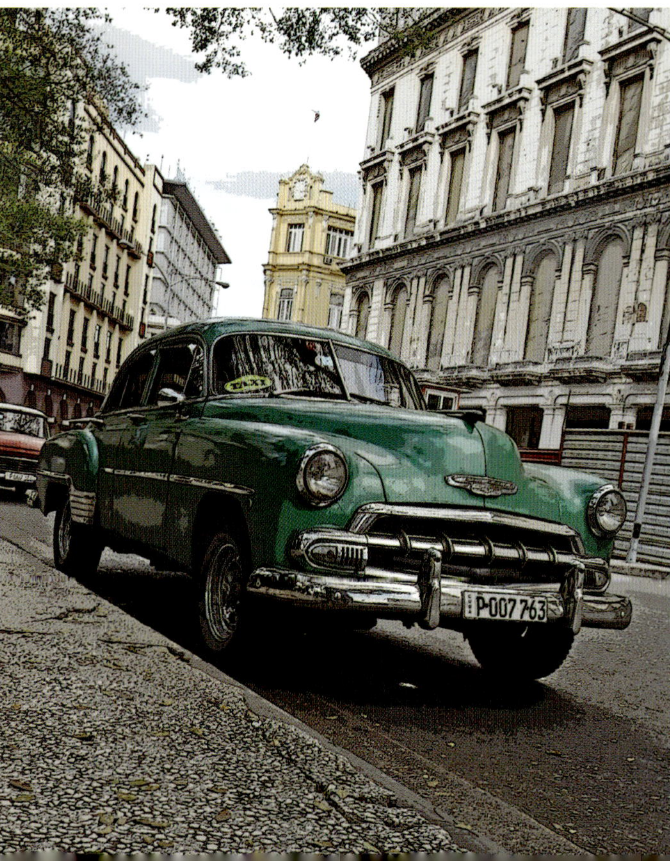
TAXI
CUBA
P-007 763

GENERAL DIVING

The diving around the mainland of Cuba offers classic Caribbean diving with vibrant reefs and marine life. However it is the more remote islands that offer a spectacular location for divers with a sense of adventure. Jardines de la Reina, which translates as Gardens of the Queen, is an archipelago in the southern part of Cuba situated 80km (50 miles) off the mainland and only 130km (80 miles) north of the Cayman Islands. It is made up of more than 200 small islands and is the largest protected marine park area in the region. Limited numbers of people are allowed to visit Jardines de la Reina and it is rumoured to have been Fidel Castro's favourite diving spot. It takes about four to five hours by boat to get there. There are no buildings, so for the duration of your stay you have to sleep on the boat or in the floating hotel that is permanently moored there.

OPPOSITE top Sharks in the Cuban sunshine.

OPPOSITE BOTTOM LEFT A shark comes in for a closer look.

OPPOSITE BOTTOM RIGHT A classic car in Havana.

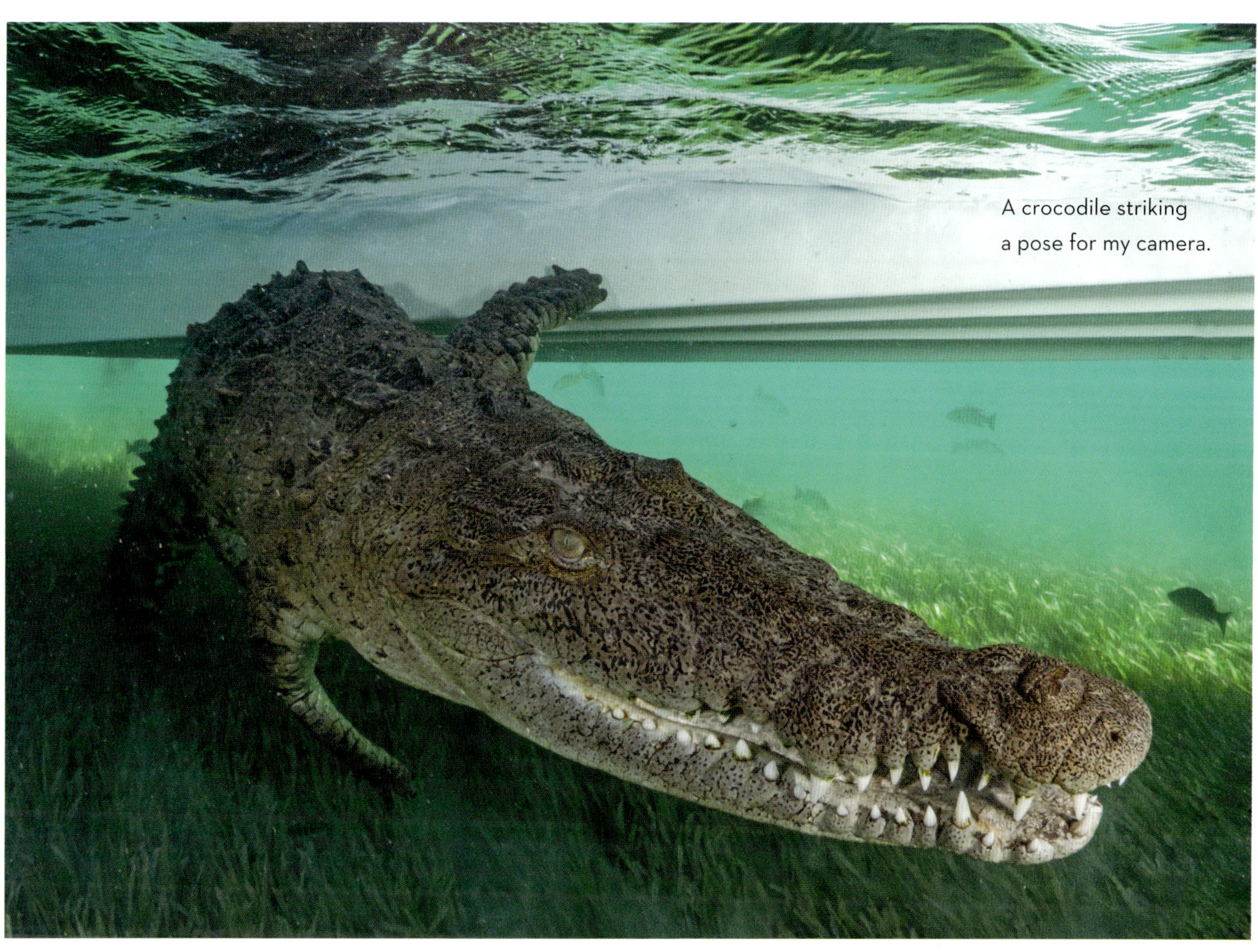

A crocodile striking a pose for my camera.

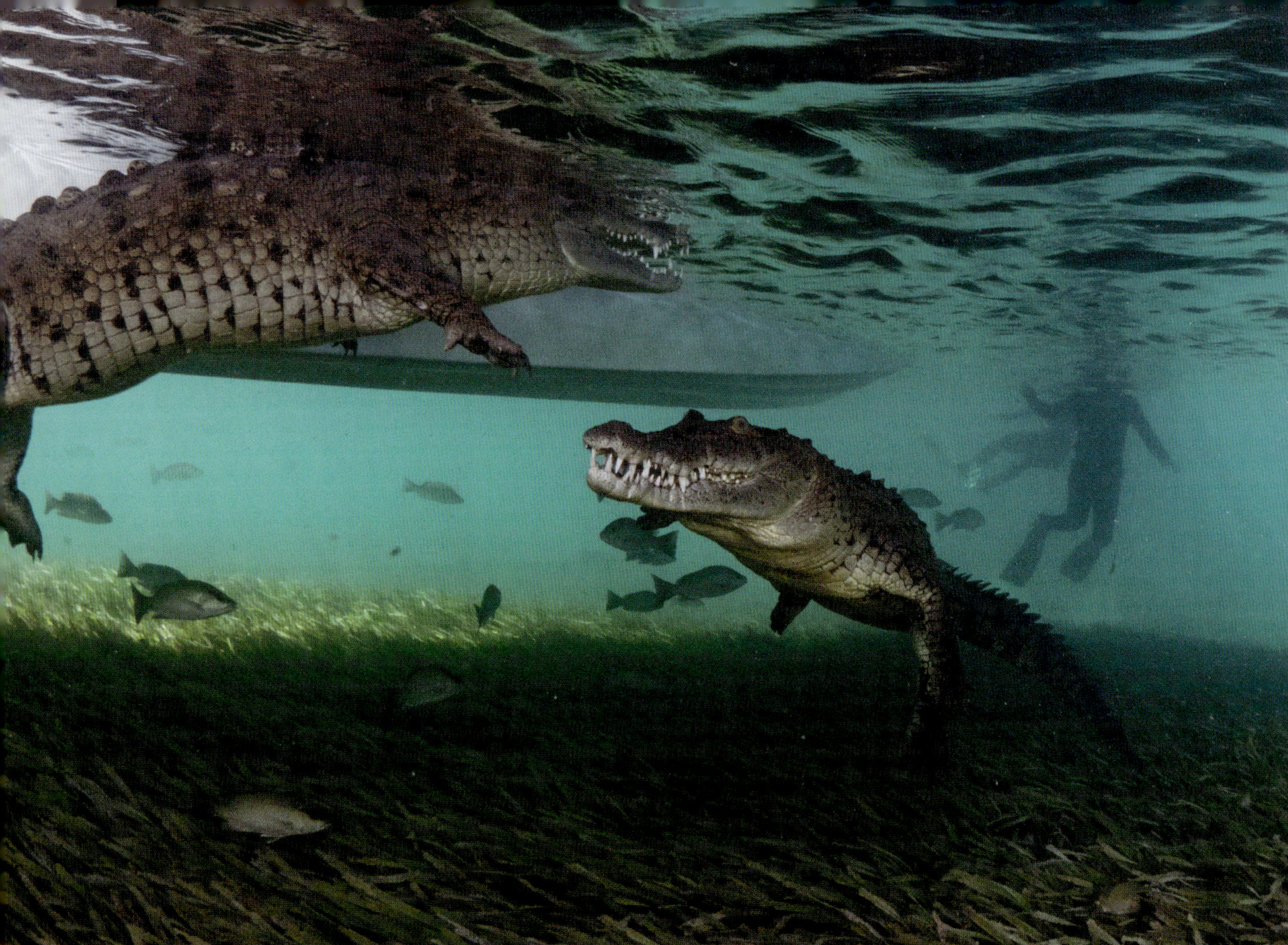

ABOVE Two crocs join in the fun under the boat.

RIGHT A close-up experience with an American Crocodile.

The reefs here are in pristine condition as commercial fishing is prohibited, and divers can delight at healthy coral and sponges teeming with life. However, it is the sharks that most divers come to see. Healthy reefs and restricted fishing mean that populations of sharks have rebounded and both Caribbean Reef Sharks and Silky Sharks are abundant. You might even catch sight of a Whale Shark at the right time of year. Caribbean Reef Sharks and Nurse Sharks will join you on every dive and Silky Sharks will greet you on your return to the boat. If baited, you can get large numbers of both in shallow water that has seemingly endless visibility.

THE WILD DIVE

Another attraction that brings divers from all over the world to Jardines de la Reina is the population of American Crocodiles. They tend to be found in the shallows

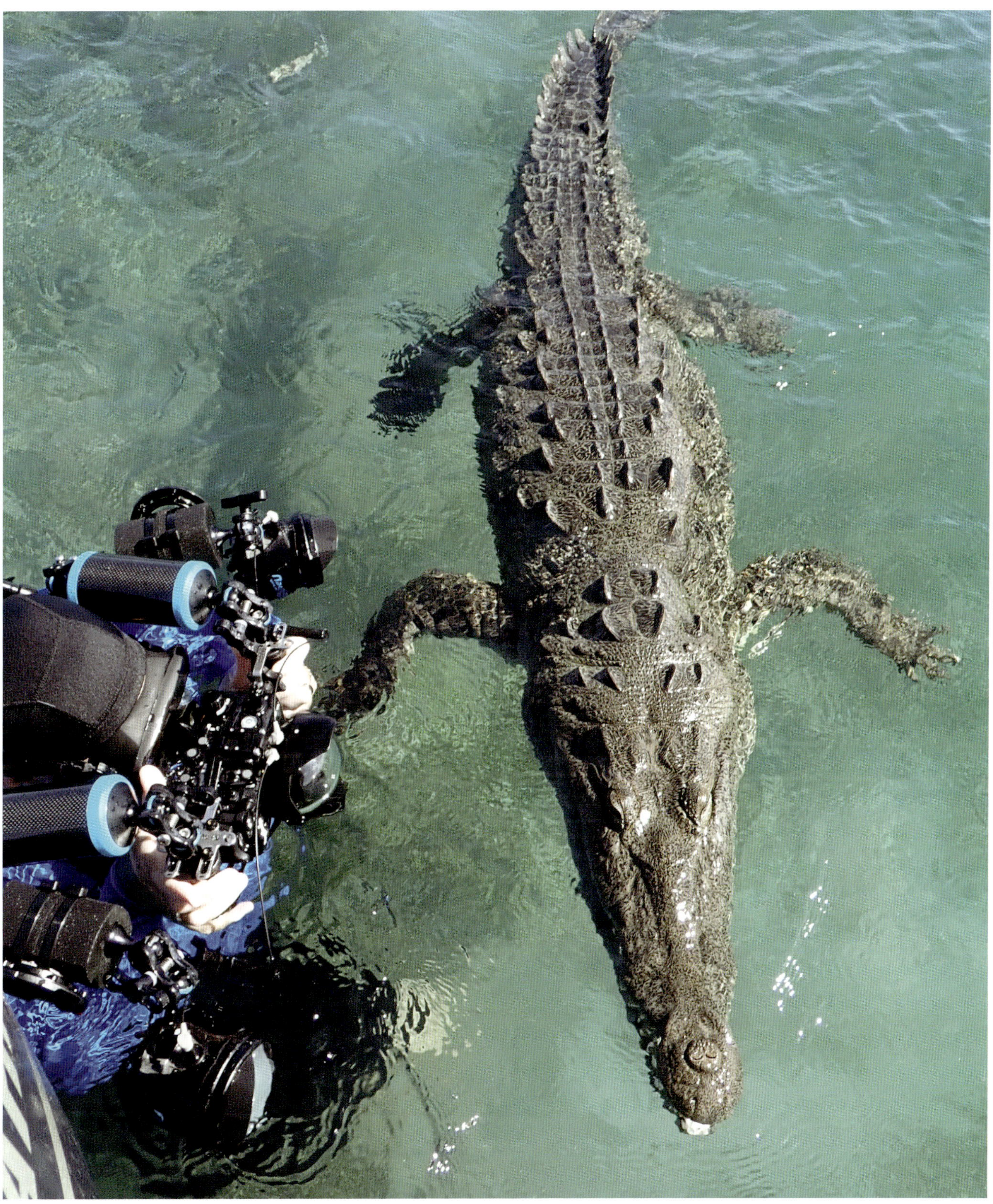

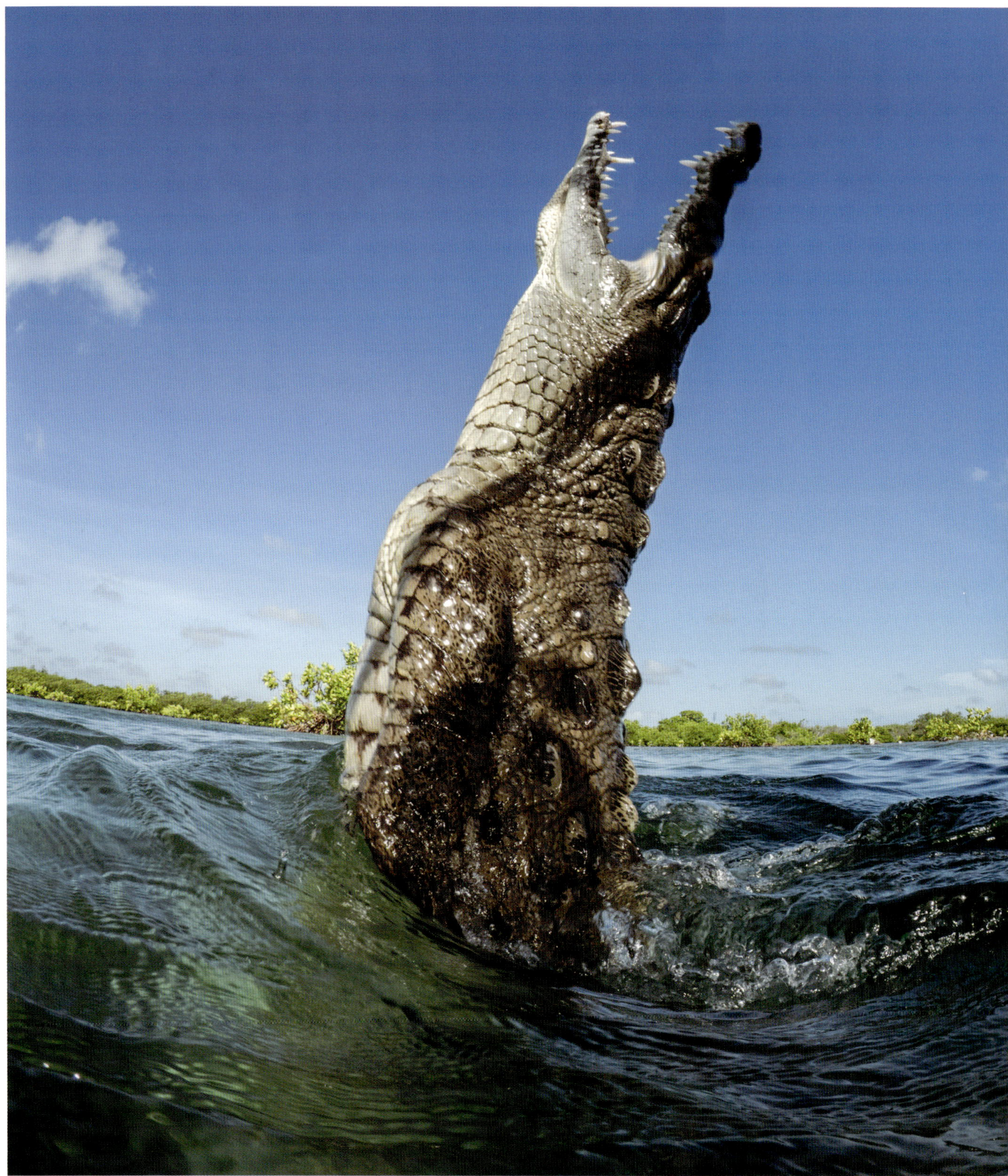

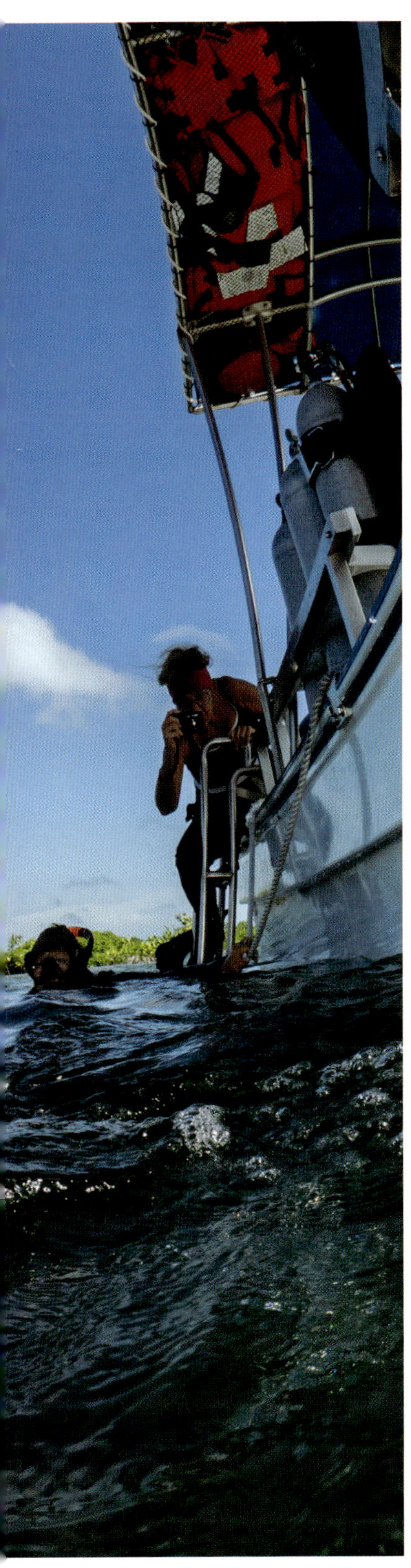

near to mangroves, sunning themselves while resting on the seagrass beds, their nose and eyes just breaching the surface. It is also here that they hunt a large but very cute rodent called the Hutia which lives in the mangroves. If you are lucky, and your guide is confident that the crocodile is a known individual and seems in the mood, you can get into the water to snorkel with these prehistoric creatures. I was hopeful that I would be lucky and get the chance to experience this for myself.

On our way back from a dive on the reef the captain slowed, bringing us closer to the shallows, and pointed. At first I found it hard to see, but then as my eyes adjusted there it was – barely visible above the water was a 4m (13ft) crocodile. I grabbed my camera and mask and snorkel and slipped over the side of our small boat. Edging closer and closer, soon I was face to face with this magnificent creature. The biggest challenge was to avoid disturbing too much sand as this would create a cloud that would obscure our view of the crocodile. The croc lay in the water, unperturbed by the presence of me and the one other person brave enough to have jumped in. I glanced back at the boat to see the others hanging over the side taking photos of the croc and then saw the guide with a piece of chicken tied to a string. He swung it over the boat and the croc instantly sprung into life, leaping into the air to try to catch the snack. On its second attempt it used me as a lever to get more height, propelling its back legs onto my thigh to gain an advantage, and this worked as it snatched the chicken piece and gulped it down in one go. I was left slightly dazed, having had such a close encounter with this magnificent beast. Unfazed, the crocodile settled back into sunbathing by the side of the boat so I carried on taking more images, amazed at the big white teeth in front of me.

My buddy and I spent a good while in the water with this crocodile and some of the rest of the group plucked up some courage and joined us. It was only later that the guide leaned towards me in a conspiratorial fashion and said: "We have never seen this particular croc before, so it was very brave of you to get in with it!" I realised that in the excitement I had not waited for the guide to give us the go ahead to even get in the water and we had just been up close and personal with a croc that had likely never seen a diver before.

LEFT A croc uses me for leverage to get to into the air.

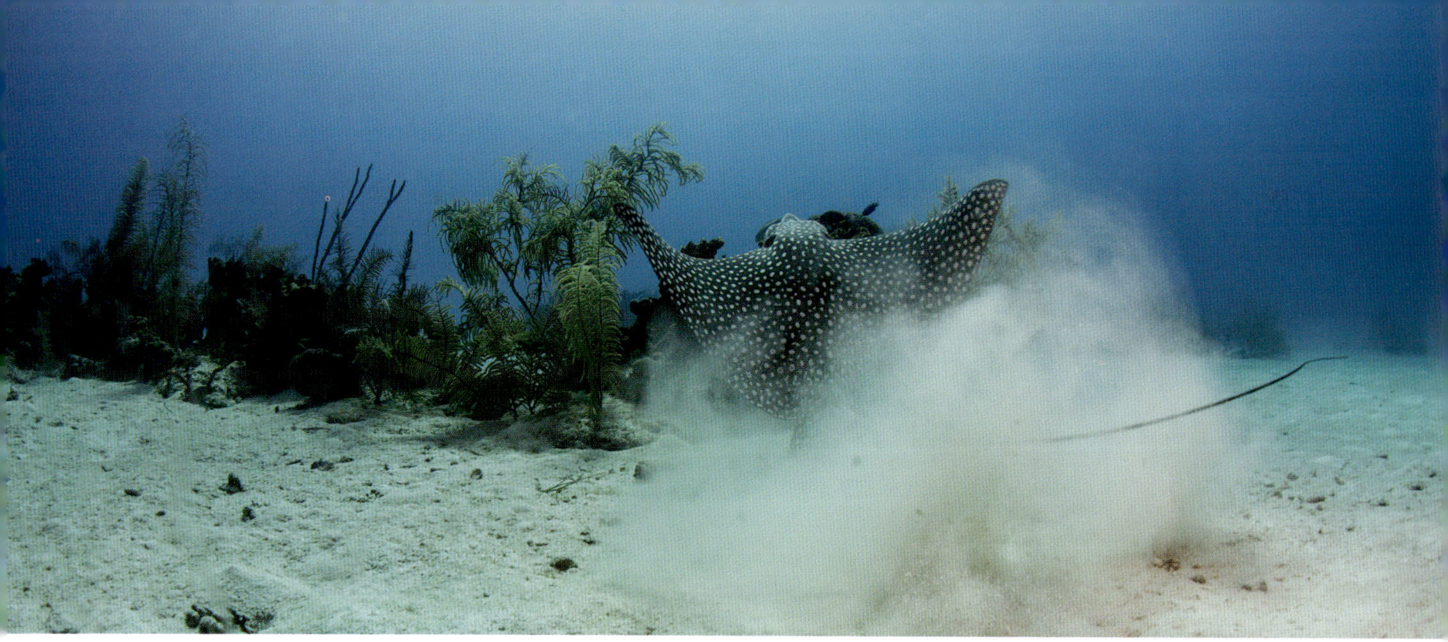

20 SOARING EAGLES

CAYMAN ISLANDS

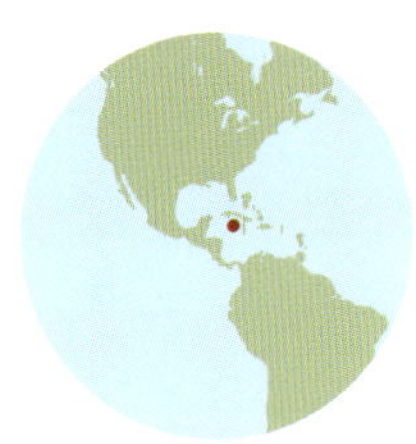

ABOVE The eagle ray swoops over the reef.

OPPOSITE A diver explores inside the wreck of the *Kittiwake*.

GRAND CAYMAN IS THE BIGGEST AND MOST POPULATED island of the Cayman Islands. The three islands that make up the Cayman Islands are located in the Caribbean, to the south of Cuba between Jamaica and the Yucatan Peninsula. Huge numbers of tourists visit these islands each year, but most come on the cruise ships and only see a small glimpse of what they have to offer. With only a few hours available to take in a snapshot of the local nature, history and culture, it seems a shame that they are missing out on proper exploration of an island with an infamous history. Grand Cayman is a popular destination for divers and tourists alike, with beautiful white sandy beaches and clear blue water, where you can enjoy sunshine all year round.

GENERAL DIVING

There are a host of dive operators on Grand Cayman catering for all types of diver from those just learning to dive to the most experienced technical divers from all over the world. One of the sights that the island is famous for is its stingrays that rest in shallow sandy areas. These creatures, closely related to sharks, will approach

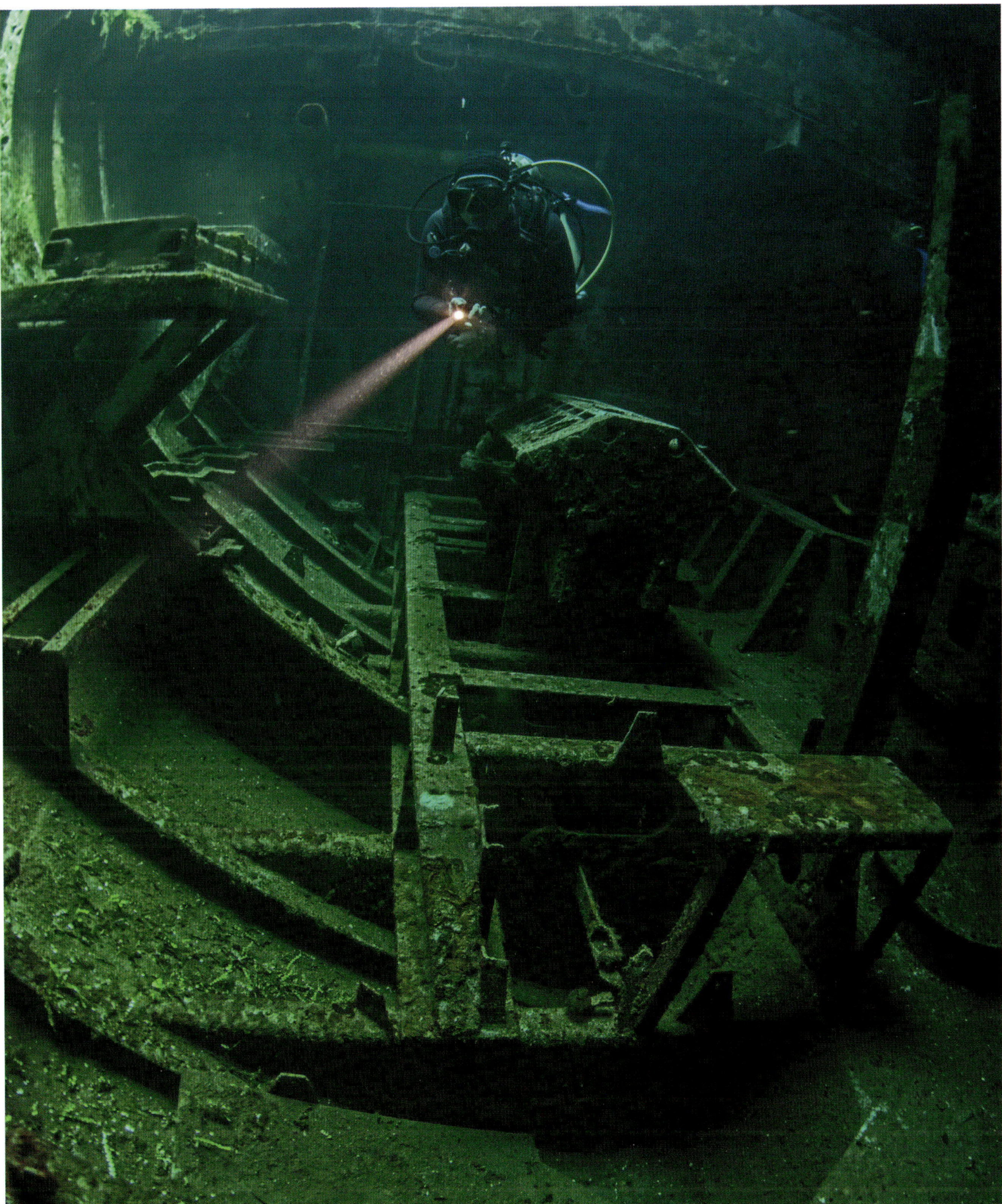

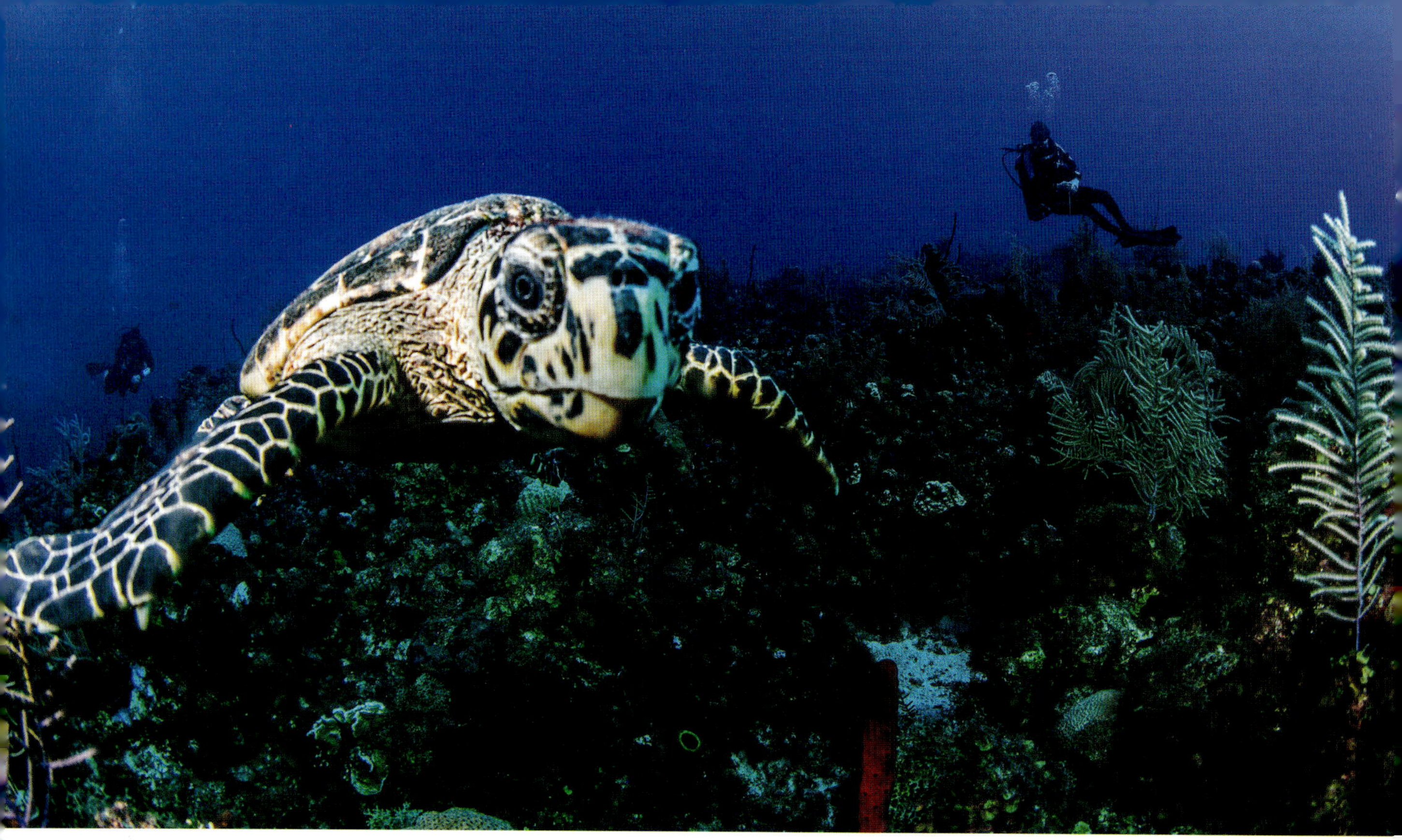

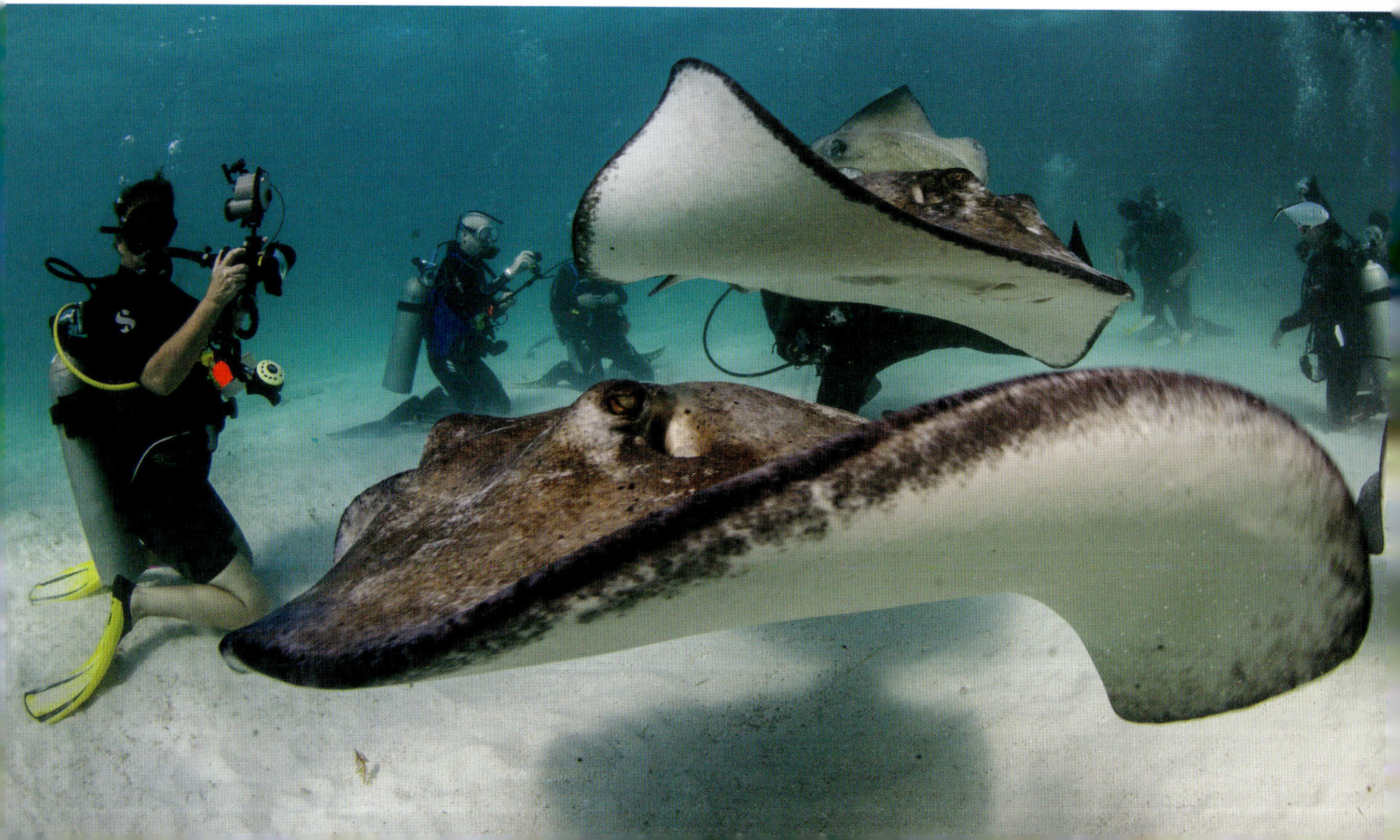

snorkelers and divers as they are used to being hand fed by some of the local dive operators. For a more natural experience, snorkel out to any shallow sand bank and you are likely to be rewarded as these rays rest on top of the sand or hunt for their food that hides beneath.

Perhaps the most famous underwater attraction that lures divers from all over the world is the *Kittiwake* wreck. The USS *Kittiwake* was a United States Navy submarine rescue boat that was in service from 1946 to 1994. Following a lot of negotiation with the US government it was prepared and deliberately sunk as an artificial reef off Seven Mile Beach in 2011. The wreck used to sit upright in the water in between 5–20m (16–66ft) until in 2017 Tropical Storm Nate knocked her over onto her side. Then and now she is a popular dive, and is home to a host of marine life and with plenty of easily accessible penetration areas for the more adventurous and well-qualified diver. There is even a recompression chamber inside the wreck that would have been used for the rescued submarine crews.

The reefs off Grand Cayman are healthy and full of fish life. The reef diving can vary from shallow patchy reef with sandy areas between the coral bommies, to exhilarating wall dives where the corals cling to a near-vertical wall that heads down to the bottom of the Caribbean Sea. Turtles and nurse sharks are common sights on the reef and they seem particularly unafraid of visiting divers, approaching them and circling them as they explore. Like most of the Caribbean, the ecological balance of the reefs and their wildlife are under threat from lionfish. These adaptable creatures are not native to the Caribbean and have been artificially introduced, probably by people with fish tanks who released them when they became too big or aggressive for the tank and its other fish. The Cayman Islands have been at the forefront of culling the species by taking them out of the ocean where they are then offered on the menus of many of the local restaurants.

OPPOSITE TOP
A friendly turtle checks out the camera.

OPPOSITE BOTTOM
The Stingray City dive enables you to see these rays up close.

THE WILD DIVE

One of my all-time favourite marine animals has to be the Spotted Eagle Ray. It is majestic, has a beautiful face and appears so graceful as it 'flies' through the water. The wonderful spotted pattern on its back makes it stand out from many other rays and it has a goofy expression that always makes me smile. These beautiful rays can sometimes be hard to find as they swim across the reefs, and so it is always a real treat to spot one. They can occasionally be seen hunting in pairs, with the lead ray swimming low across the sand, disturbing it and revealing any prey that may have been below the surface. The second ray then snatches the food before they swap roles.

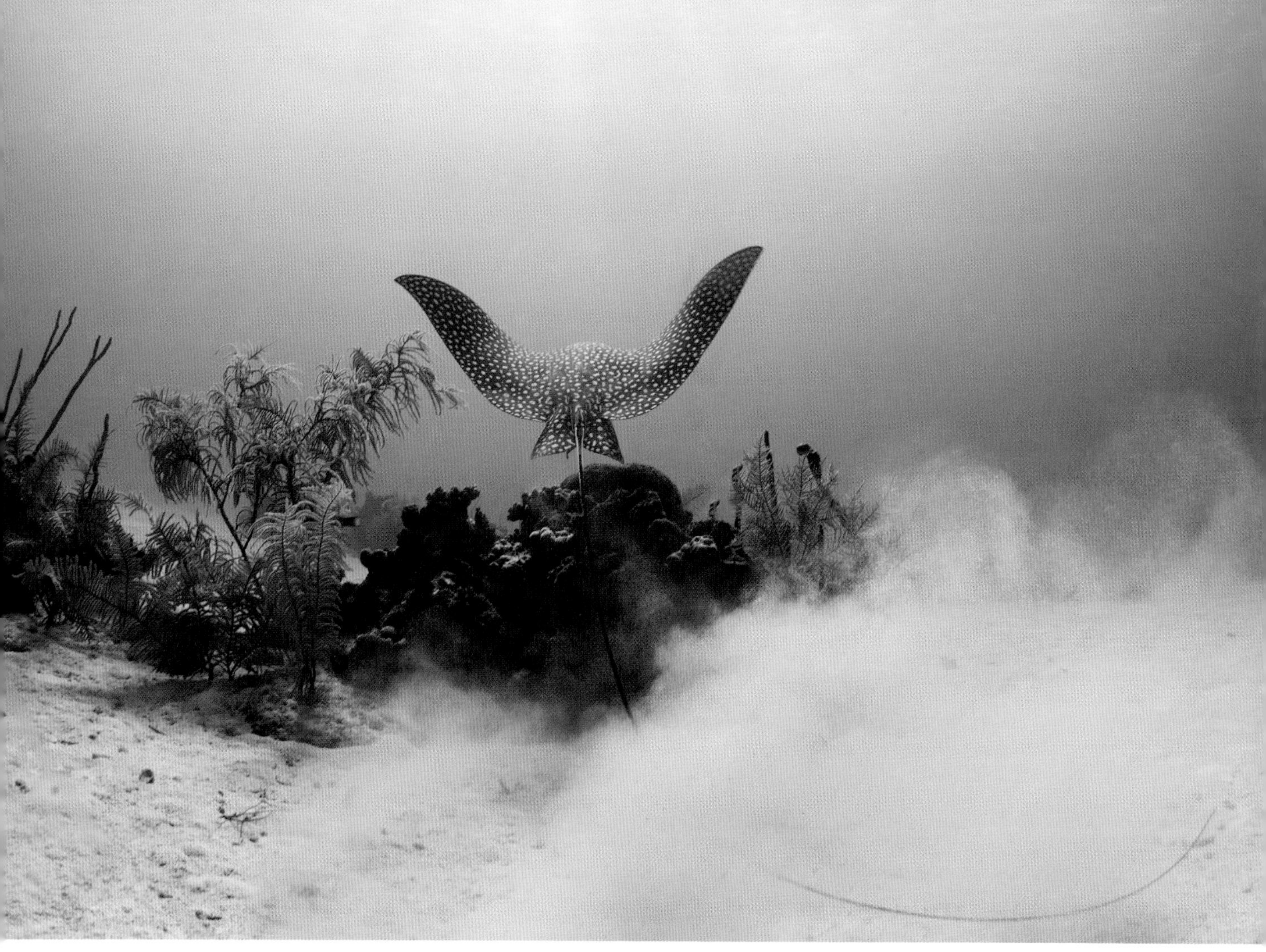

ABOVE An eagle ray, with a very long tail, is disturbed and swims away from the group.

OPPOSITE A lobster and a green moray eel compete for the same hole.

I was diving a shallow patchy reef in Grand Cayman and had just finished photographing a moray eel trying to evict some lobsters from a hole so that it could take up residence there itself, so it had already been a good dive. Then I saw an area of disturbed sand and decided to investigate. Something had kicked up the sand into a cloud that hung over a few small clumps of soft corals. I approached very slowly and was delighted to see that it was a lone hunting eagle ray which had disturbed the sand to try to find food below. I wanted to get a head-on photograph of the eagle ray and so very carefully and slowly took a wide arc around it. Unfortunately, my strange movements had attracted the attentions of another dive group, who then saw what I was trying to photograph and rushed towards it at full speed, sticking out their action camera before them. Before I was able to get into position the eagle ray had been

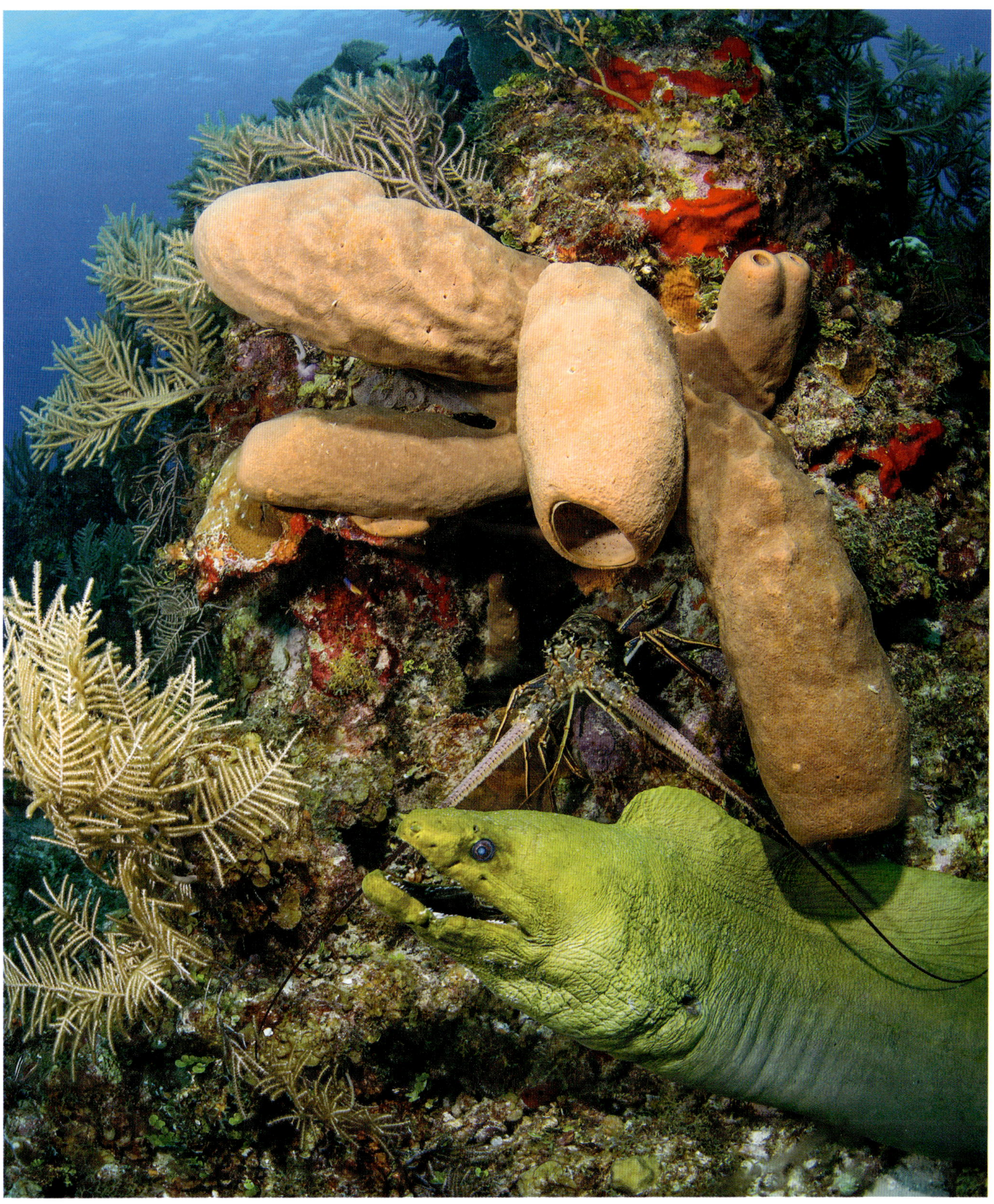

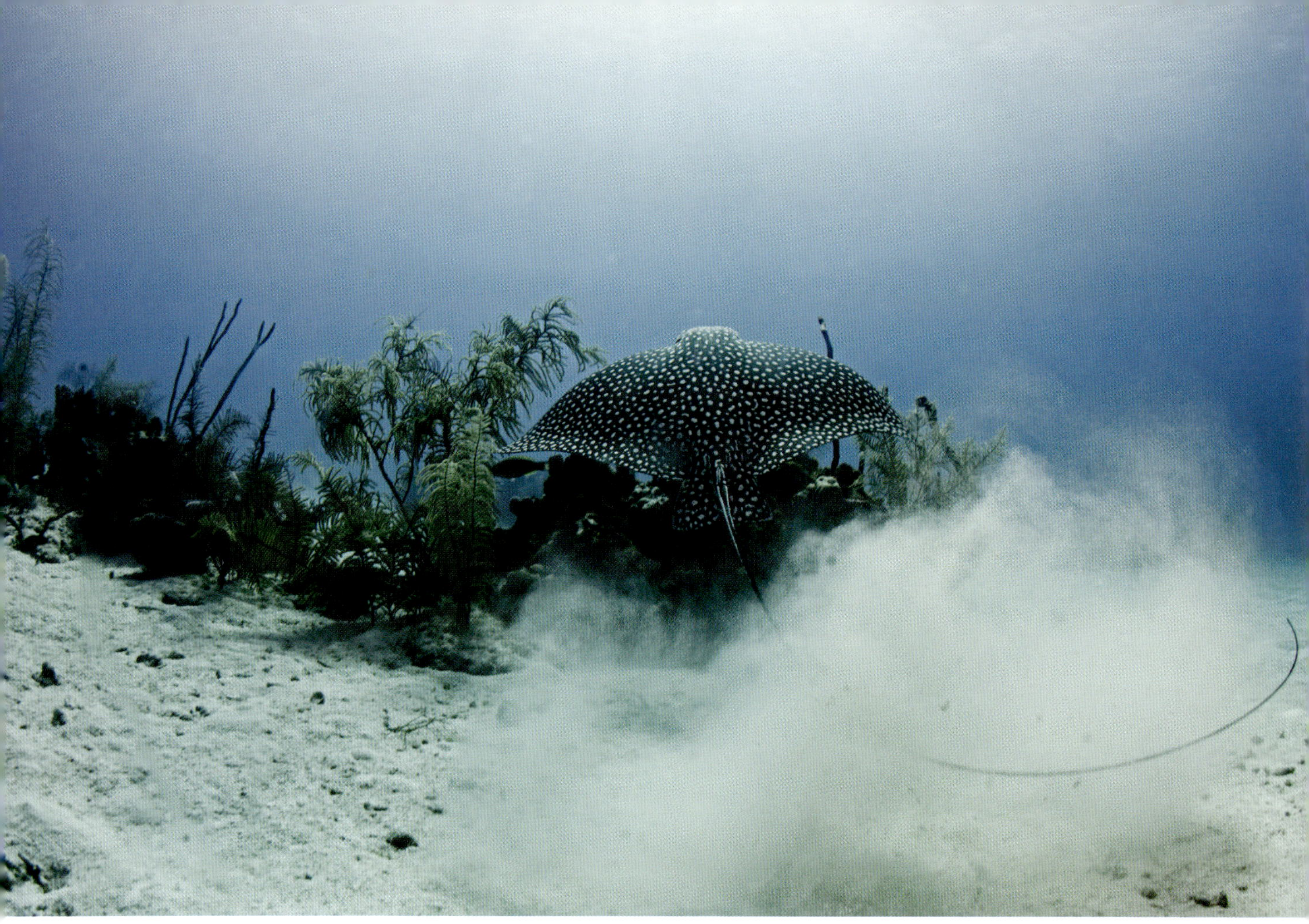

spooked by the other group and took off away over the reef. I cursed under my breath and fired a few shots of the ray departing. At least I could prove to my buddy that I had seen it!

Usually I would tell my underwater photography students that they should never take a photo of a fish or any other marine life from behind. But every now and then these rules are made to be broken and that evening, as I reviewed my images whilst sitting outside with a nice cold beer and a lionfish sandwich, I came across the images I had taken of the Spotted Eagle Ray swimming away. I was completely taken aback as the angle actually showed off the gorgeous pattern on its back, the kicked-up sand gave the viewer a hint of motion and you could see the incredibly long tail very well indeed. The best of these images has since won a major competition and also made the front cover of a magazine. Non-divers often ask me what type of bird it is – "Eagle" is always my reply.

ABOVE The eagle ray disturbs the sand on take-off.

OPPOSITE The *Kittiwake* when she stood upright.

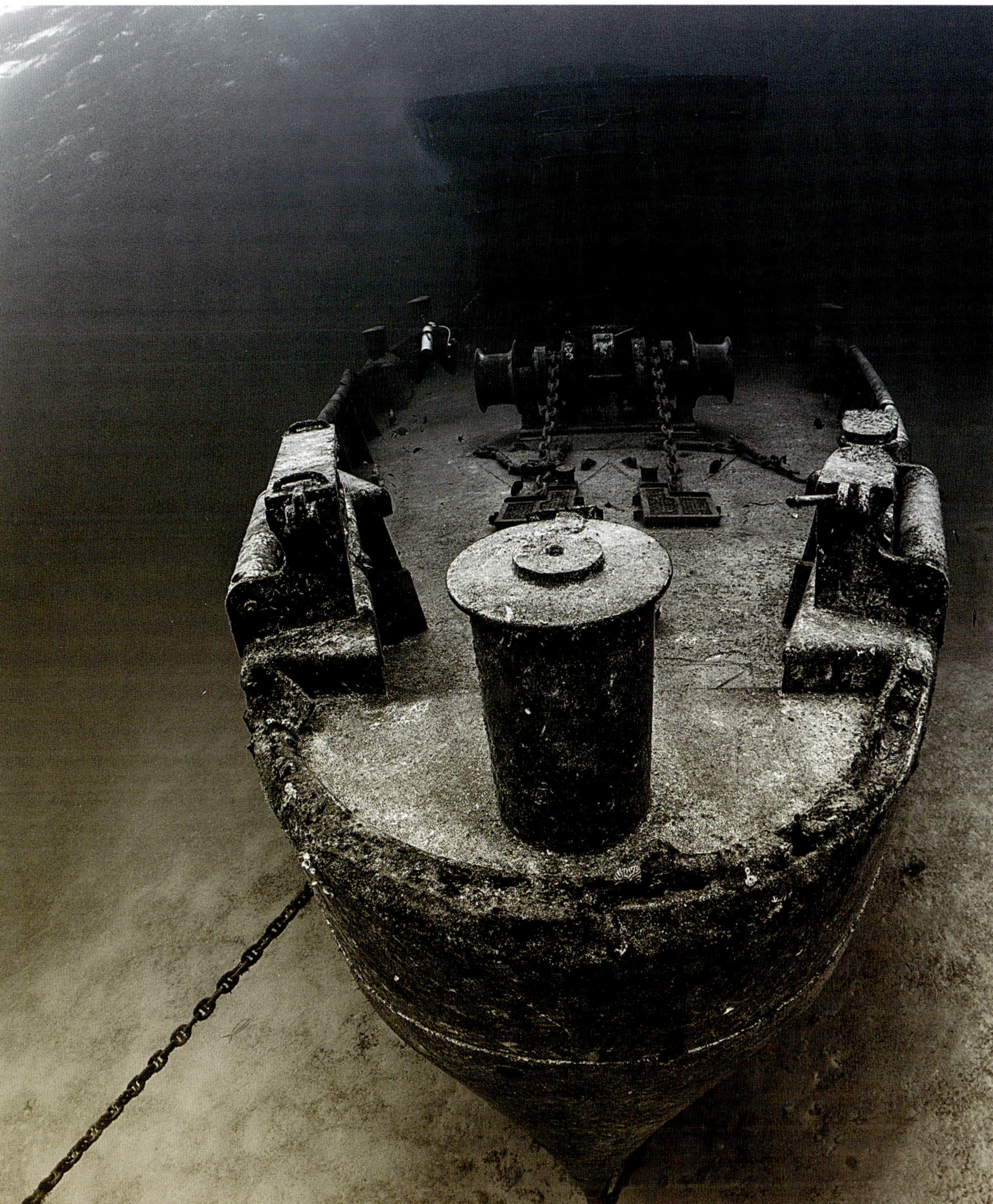

21

WILD SUBMARINE RIDE

BARBADOS

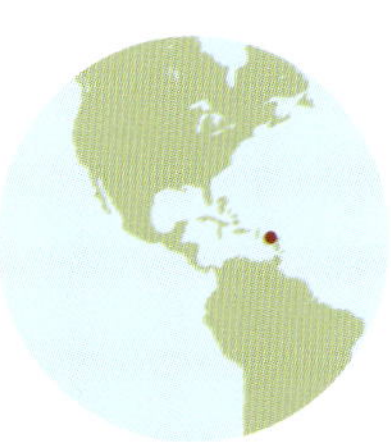

OPPOSITE There are plenty of shallow wrecks to explore in Carlisle Bay.

BELOW A diver approaches the submarine hovering over the reef.

THE BEAUTIFUL ISLAND OF BARBADOS sits out on its own on the very eastern edge of the Caribbean island chain. It is a part of the Lesser Antilles of the West Indies, and its unique isolated location means that it sits outside the main hurricane belt that traverses the Atlantic Ocean from the continent of Africa. It is regarded as the spiritual home of cricket in the West Indies, has the oldest rum distillery in the world, and offers its visitors beautiful sandy beaches, warm clear turquoise waters, lush forests, local culture and a vivid nightlife. Barbados is the epitome of a Caribbean island.

GENERAL DIVING

Barbados is the ideal place for the recreational diver as it offers a host of different types of diving regardless of certification level. Carlisle Bay is a small natural harbour in the south-west of the island, which is sheltered from the prevailing wind coming across from the Atlantic. It is a veritable playground for divers and

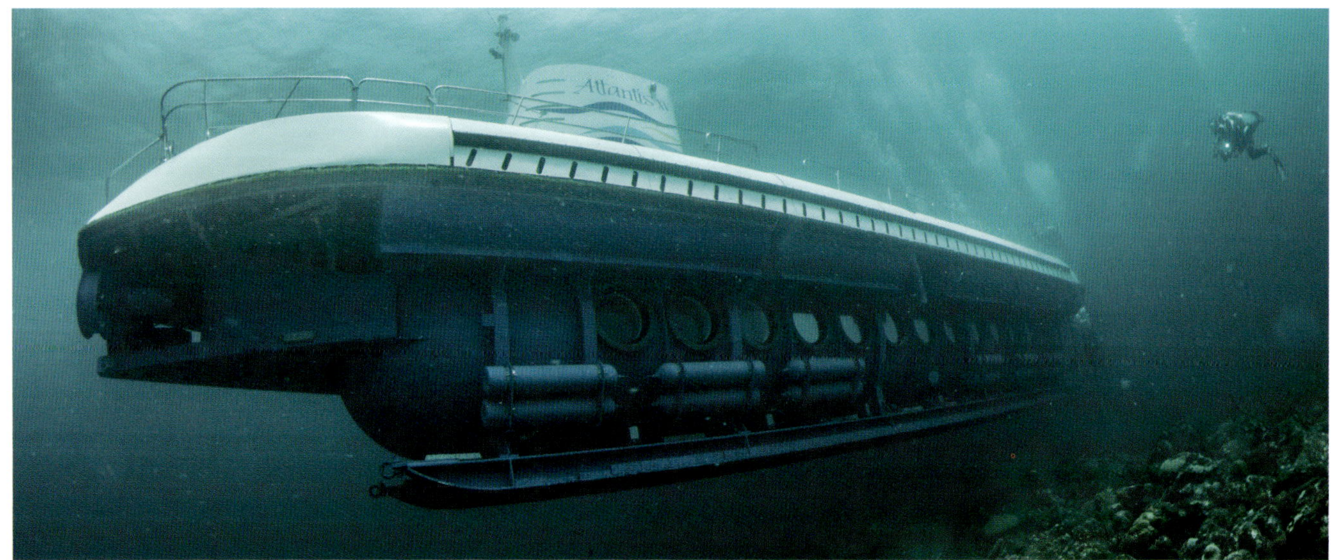

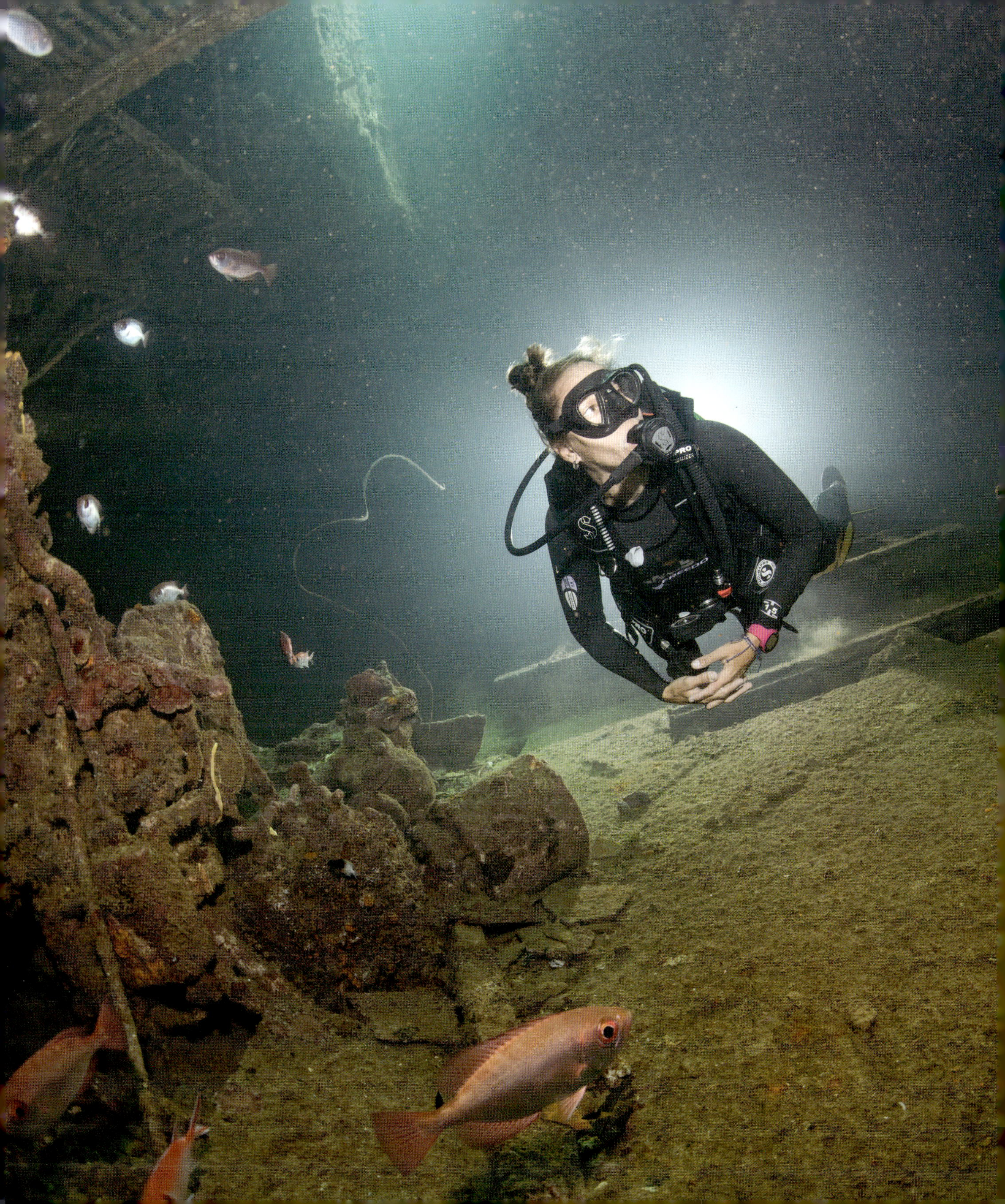

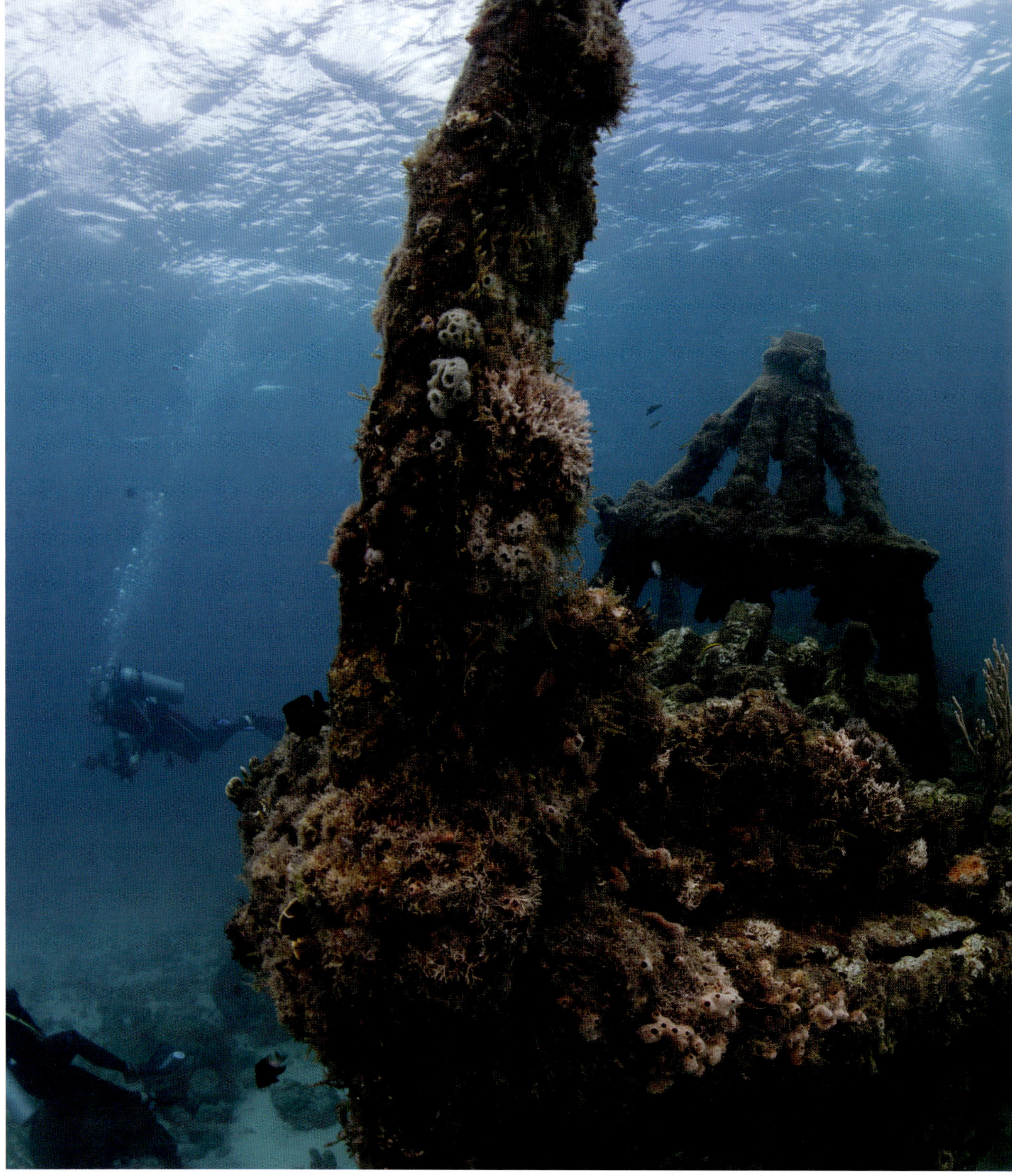

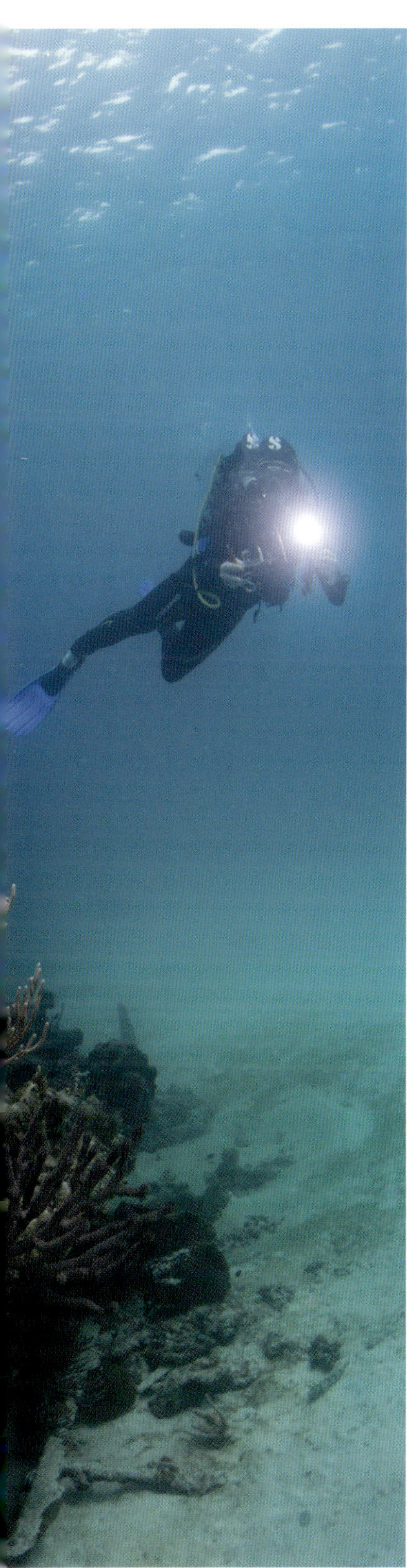

snorkelers alike, with a series of wrecks which have been sunk deliberately close together to form a marine park playground area with plenty of interest. Here you can explore wrecks, see schooling fish in and around these sunken boats, watch turtles feed on corals and even find seahorses clinging on to soft corals, swaying in time with the gentle movement of the sea.

For divers who like larger, deeper wrecks to dive on, the *Stavronikita* is an impressive site and one that attracts rust-lusters of all levels. It is a bit further out to sea than the wrecks of Carlisle Bay and is covered in bright sponges and corals as a result of 50 years of submersion. There are several piers and jetties which afford shelter to various marine creatures and the dives offer a relaxed chance to find some smaller critters hiding among the growths on these structures. Barbados has more than 90km (50 miles) of coral reefs. These range from shallow bommies to wall dives, covered in both hard and soft corals. Huge barrel sponges dominate the shallow reef tops and the island has the second largest breeding population of Hawksbill Turtles in the Caribbean.

THE WILD DIVE

As part of the annual local diving festival, Dive Fest, we were attempting to try out as many different diving experiences as we could manage, and each day had offered us incredible marine life encounters as well as the chance to join in with the conservation initiatives that are, for the most part, intended to inspire and encourage the local schoolchildren to take ownership of their environment. One morning we asked our guide and local dive shop owner Andre what was in store for us and he simply replied: "You'll see." He immediately got back onto his mobile phone, talking animatedly to whoever was on the other end. We got on with getting our diving gear and our cameras together onto the boat and as we pulled out of the bay we questioned again what we might be doing, so that we could prepare our cameras appropriately. This time he answered: "We are going to be diving on a submarine." He must have seen our faces drop as we imagined getting into a submarine to be taken around the reefs, rather than diving, and he then clarified:

LEFT A diver on a wreck in Carlisle Bay.

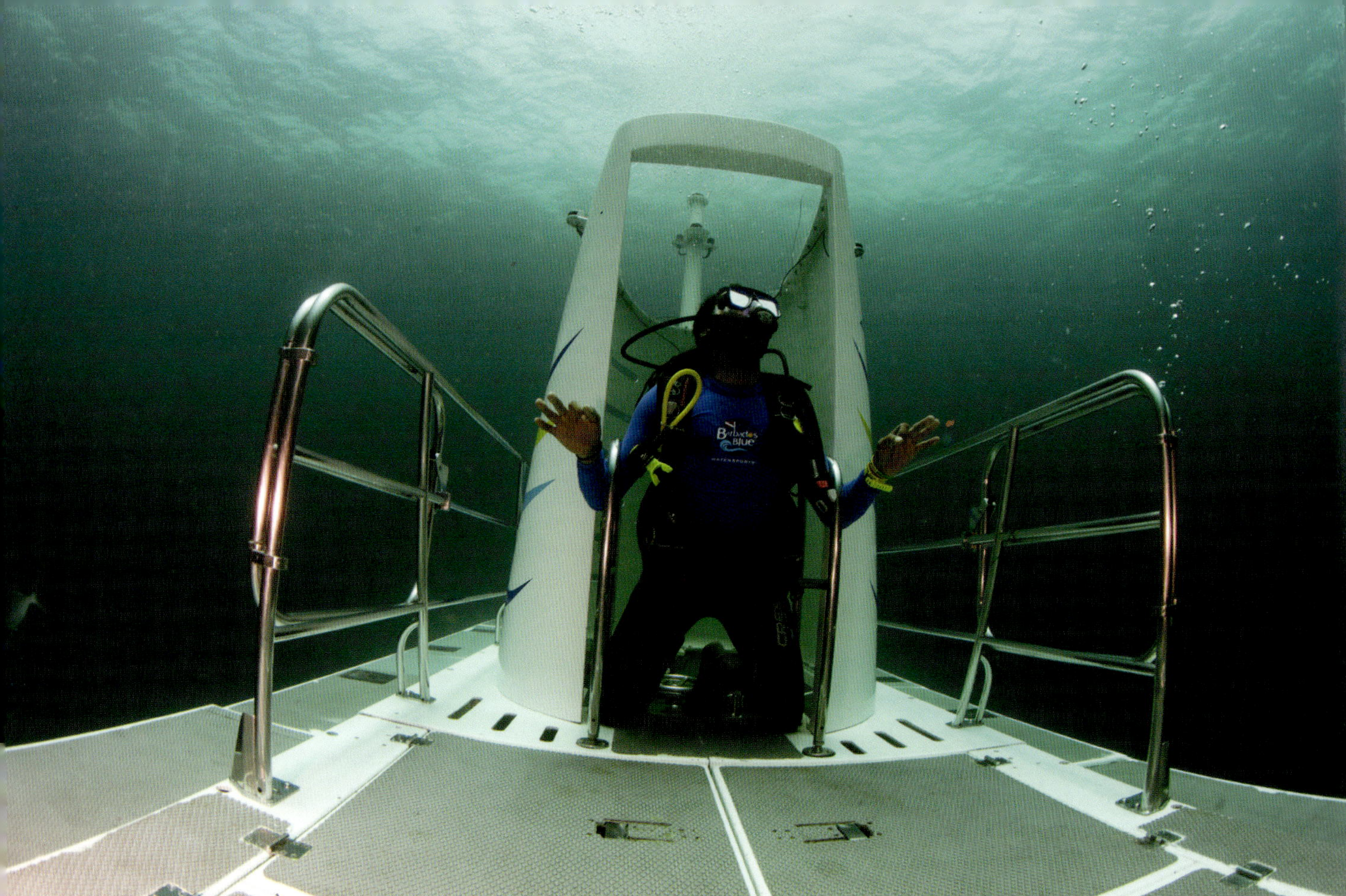

ABOVE Andre enjoys the ride once the journey is a little calmer.

ABOVE RIGHT Better hold on tight!

OPPOSITE A diver and the submarine occupants enjoy the reefscape.

"We will be diving outside the submarine, looking in and high-fiving to the kids on board." The submarine company had got a group of school kids on board as a part of DiveFest to allow them to see the underwater world that surrounds their island for the very first time. Our mood was much better and we listened carefully to the safety instructions on where to avoid when diving close to the submarine. We watched the submarine dive beneath the water and waited for a radio signal to tell us we were safe to approach and then dived down to meet them. We swam round to the front of the vessel to give the captain a wave and let him know we were there, and then proceeded to wave at the children on the inside. They were very excited to see us and the whole dive was a joyous experience for us, as well as for the kids. Back at the surface on our dive-boat, Andre was back on his phone excitedly talking at double speed. The submarine surfaced nearby and started the process of unloading one set of kids and loading another from the adjacent tender. Andre told us we were going to dive the sub again – but this time we were actually going to ride it to the bottom of the Caribbean.

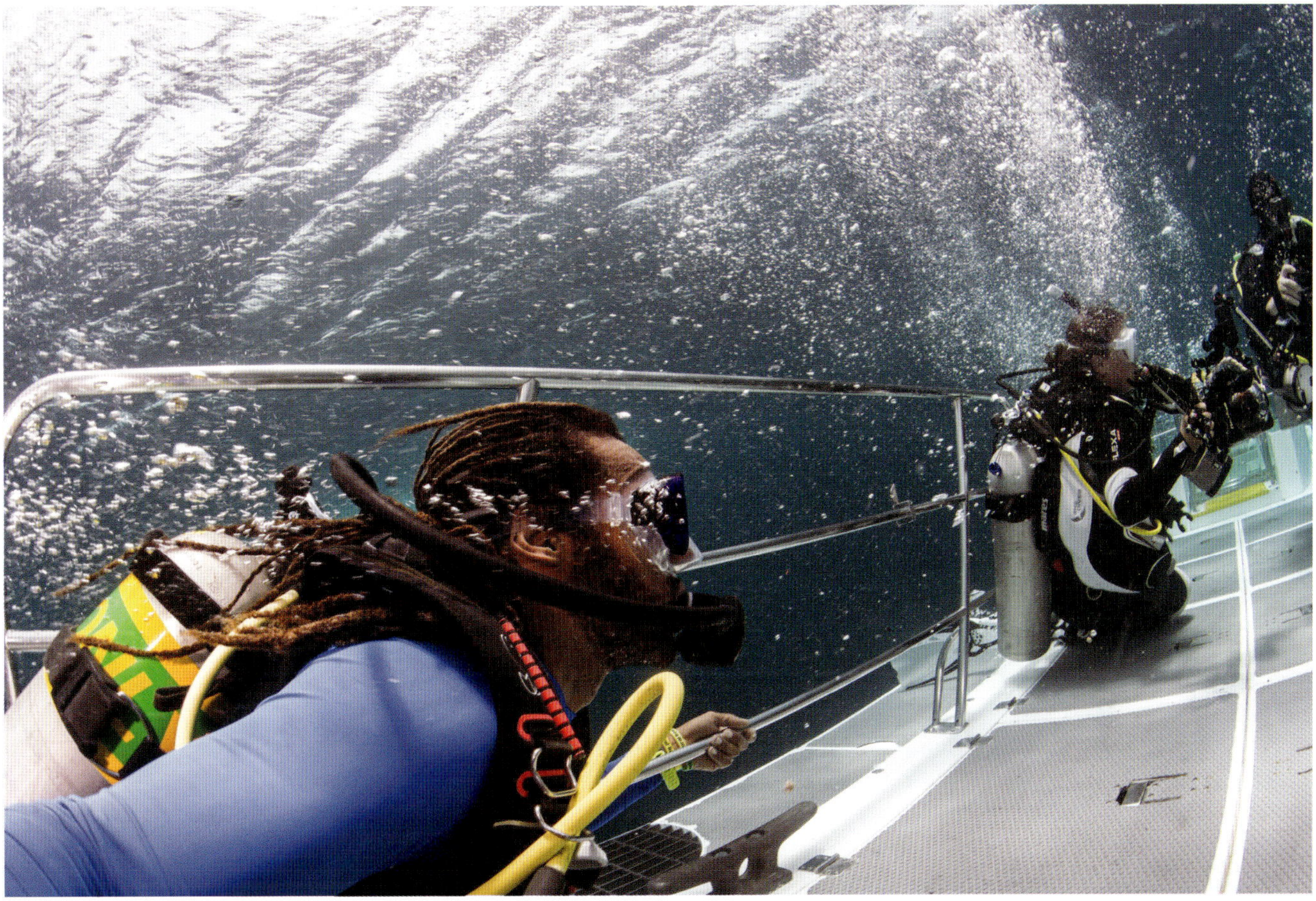

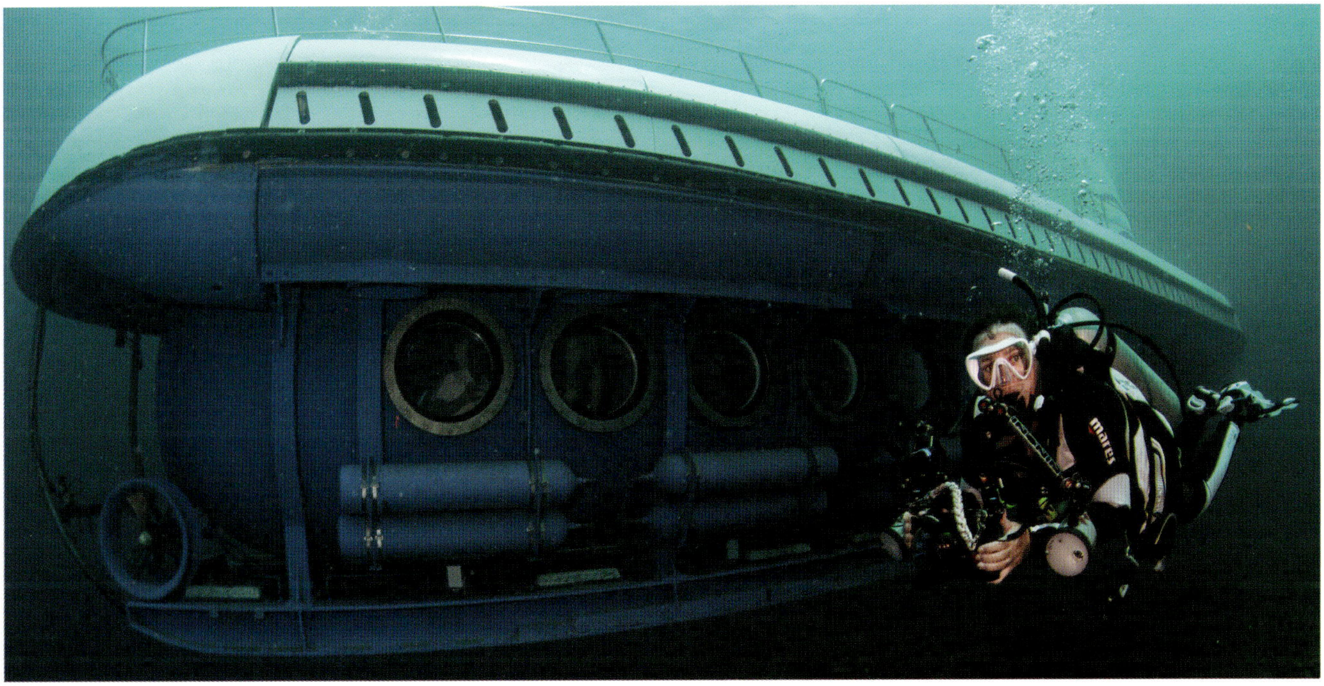

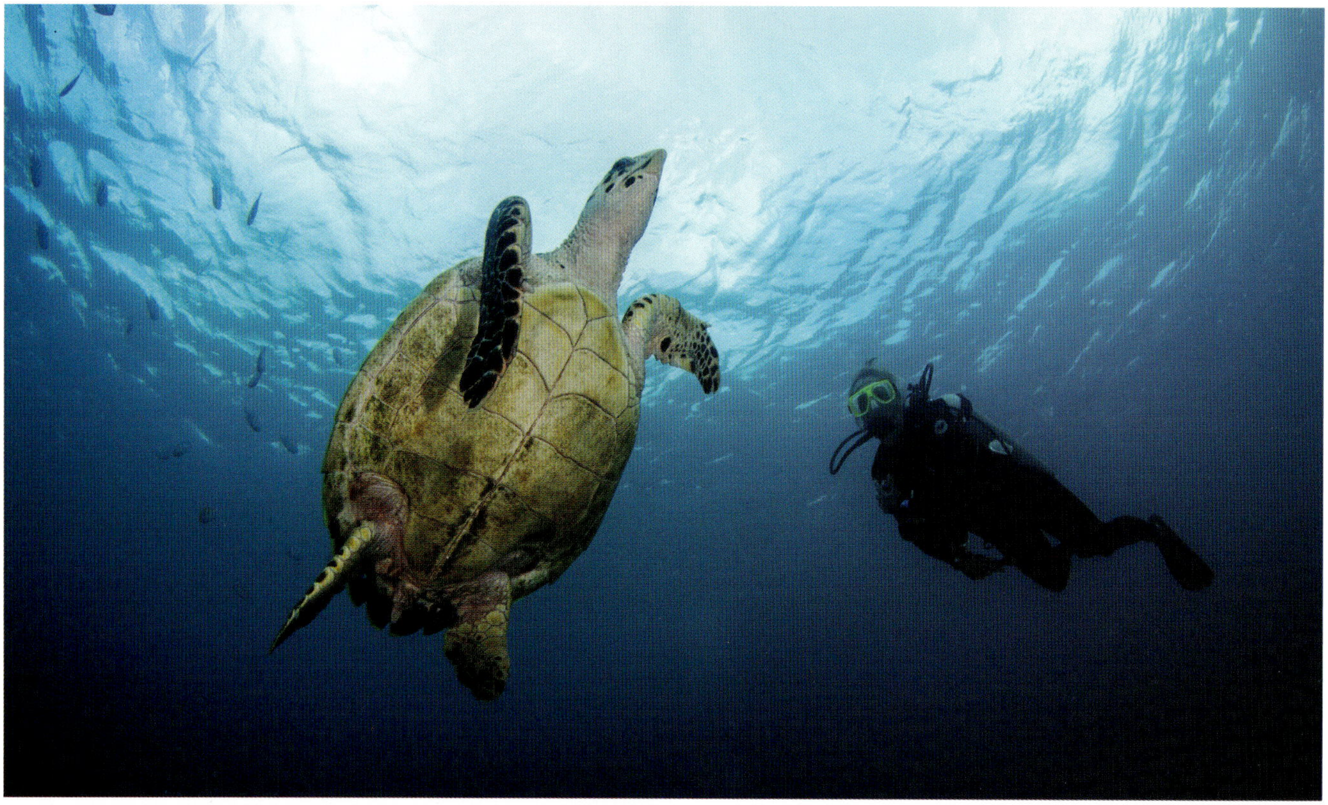

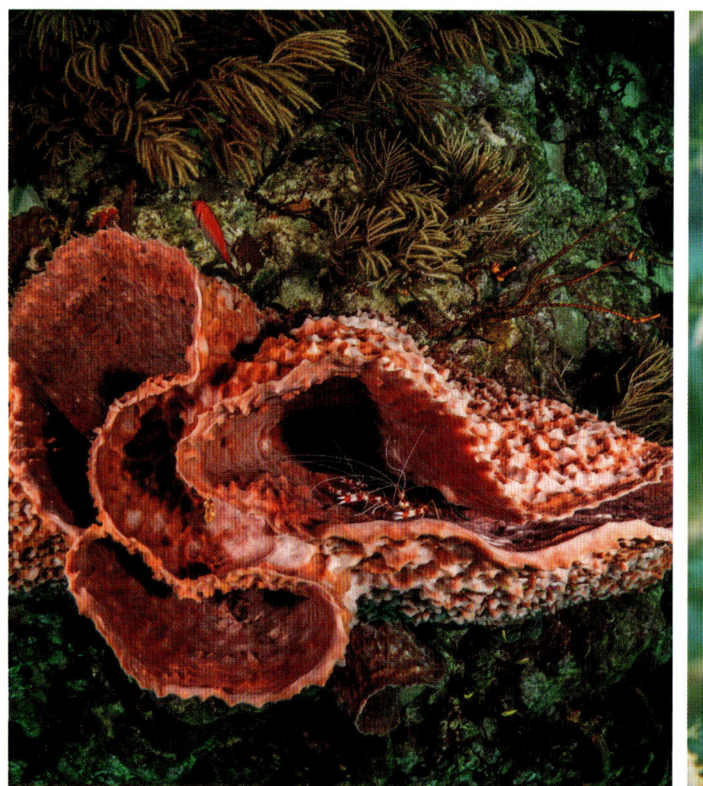

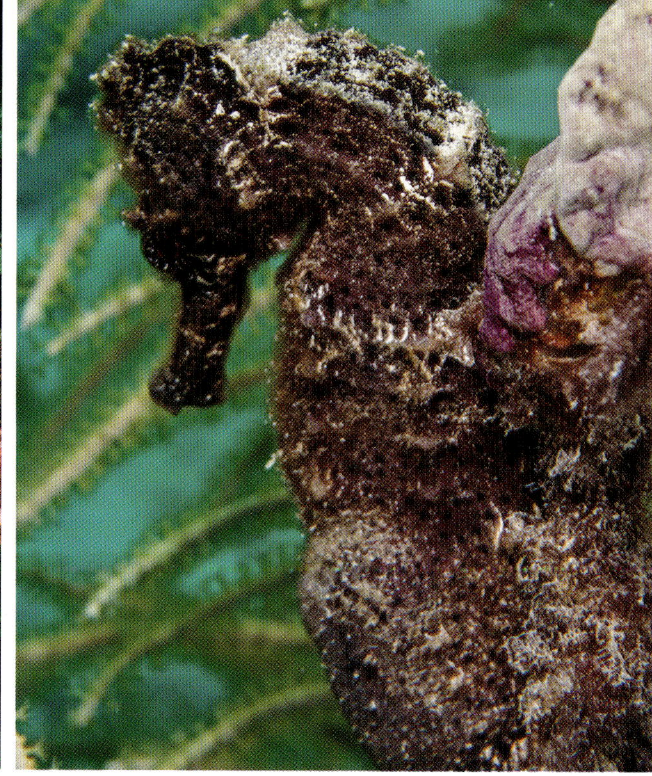

"Wait... What?" we asked. The plan was to get on top of the submarine while it was still at the surface, sitting on the deck in our diving gear. Then we would hold on and ride it as it descended to a depth of about 18m (60ft). We were suddenly very excited at the prospect of this, and why not? Who wouldn't be? We got into our gear and then climbed from our boat onto the top of the sub. Again we listened to a safety briefing and then the support boat gave the all clear. The submarine started to make a loud noise as it began to expel air from its ballast tanks. This sent frothing water all around us and slowly the sub started to descend. We had to hold on tightly for the first few moments as the water and bubbles streamed past us in an experience that has to be unique. We started to doubt our sanity, but then the water cleared and the sub seemed to slow down and give us time to enjoy the ride.

Once the submarine had stopped and hovered just a few feet above the seabed we slipped off the side of the vessel and made our way to the port holes where the school kids were. As with the previous group, they made fist bumps to us as we approached. Hopefully after this experience these youngsters will appreciate what they have on their doorstep, and perhaps some will become involved in diving, swimming and conservation all year round.

OPPOSITE TOP Turtles are abundant in the waters around Barbados.

OPPOSITE BOTTOM LEFT Huge barrel sponges on the reef are home to bright crustaceans.

OPPOSITE BOTTOM RIGHT A shy seahorse hides amongst the soft coral.

BELOW Andre greets the kids onboard the submarine.

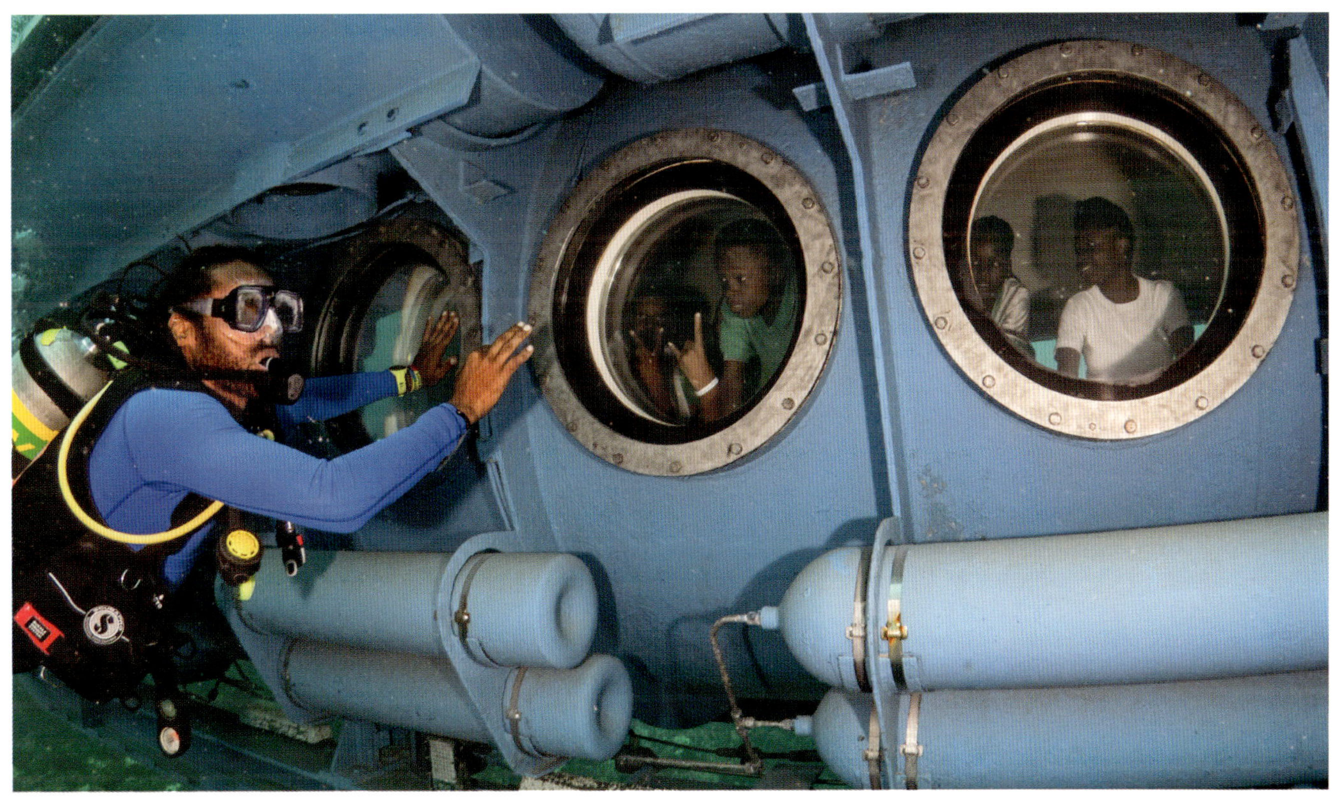

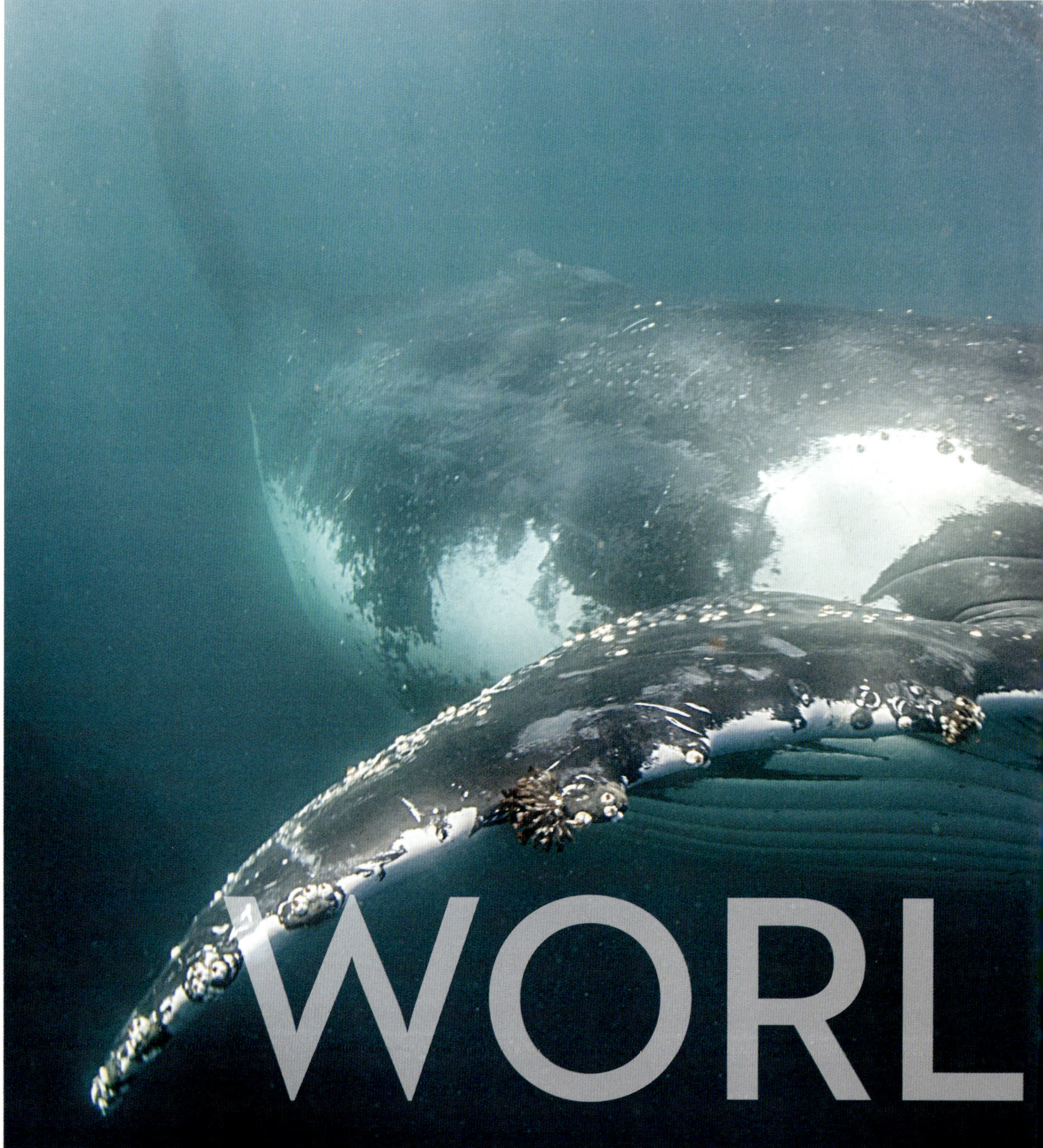
WORL

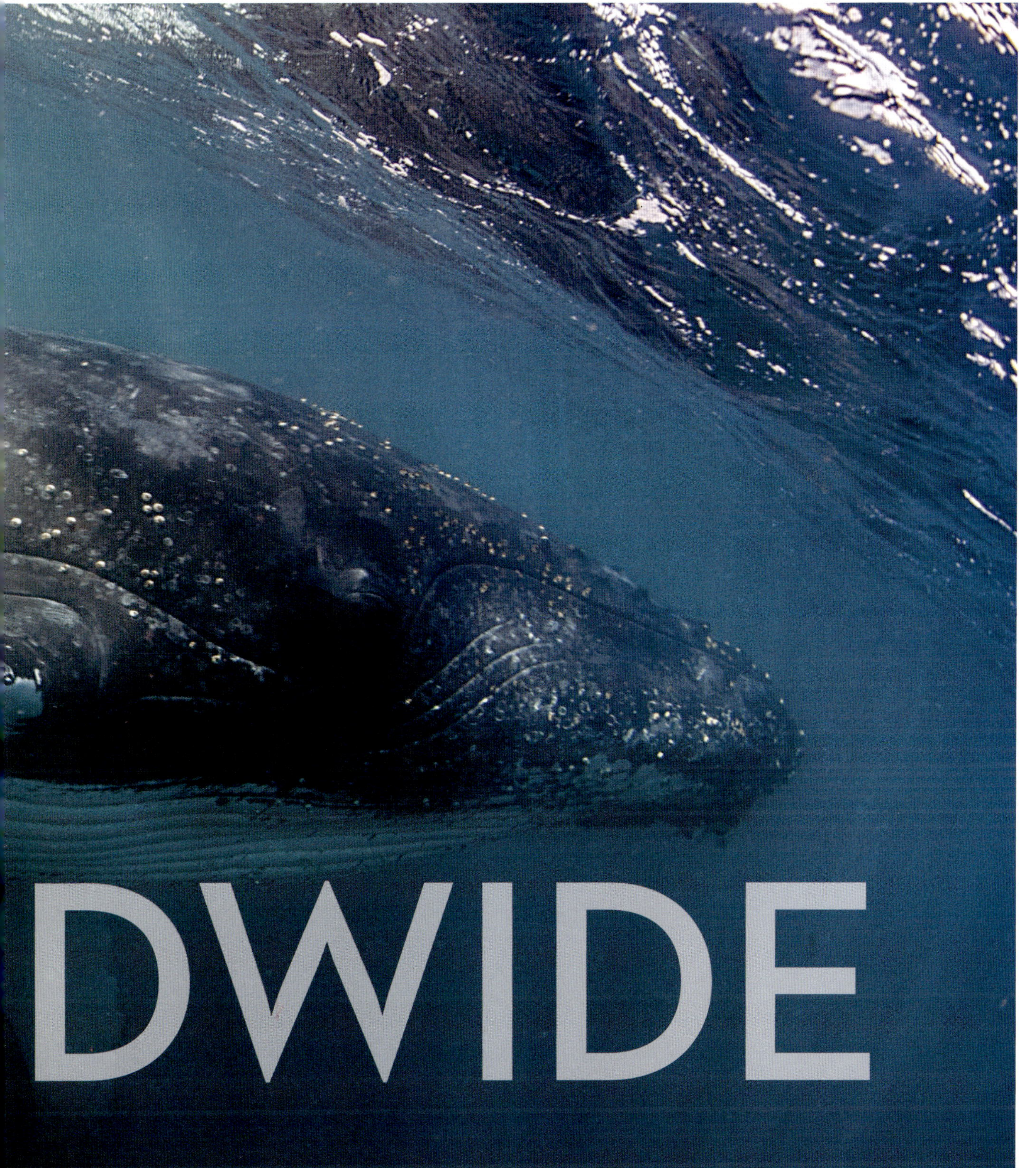
DWIDE

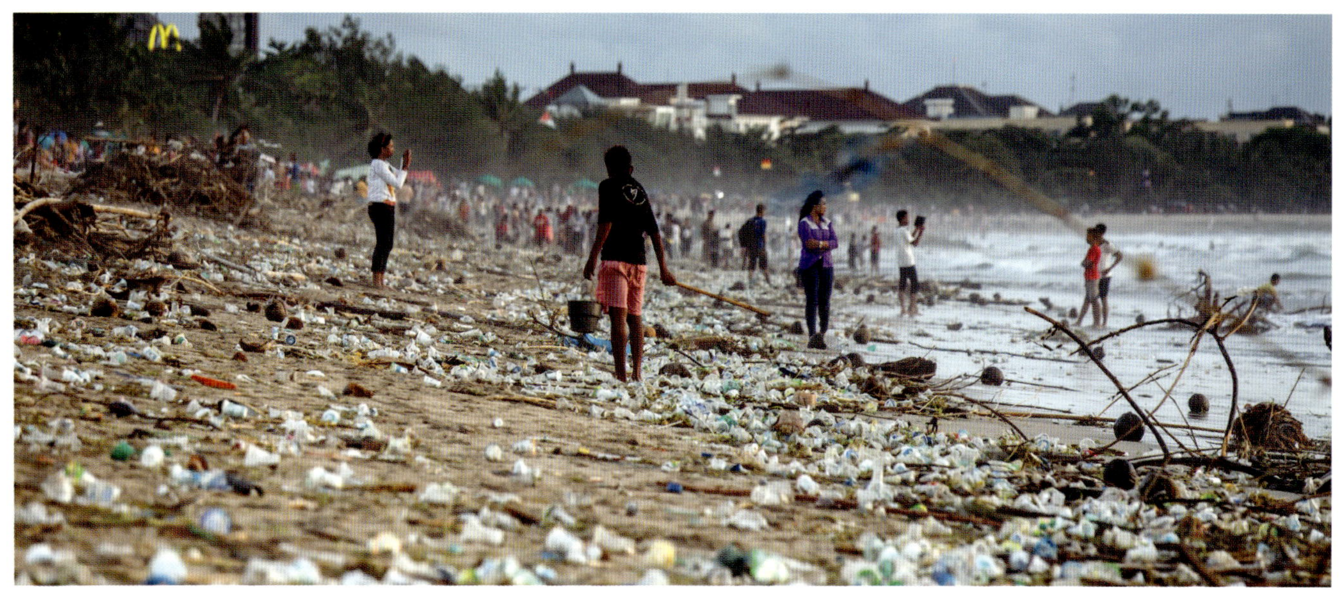

22 OCEANS FULL OF PLASTIC

MARINE ENVIRONMENTS GLOBALLY

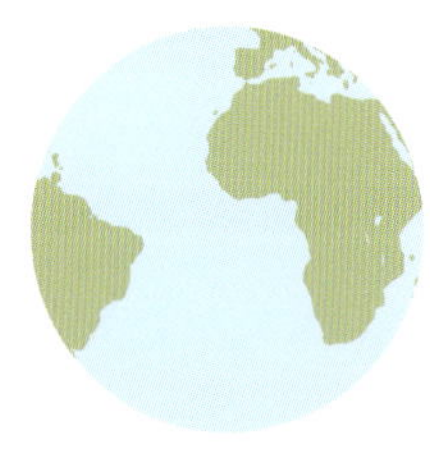

ABOVE Huge numbers of plastic bottles wash up on a beach.

OPPOSITE A turtle mistakes a plastic bag for a jellyfish and starts to eat it.

WE HOPE THAT YOU HAVE ENJOYED READING about some of our favourite and more unusual dives from all around the world. We have been privileged to experience some of nature's most incredible underwater sights, diving and snorkelling with beautiful marine species. However, if future generations are going to be as lucky as us and get to see these reefs and creatures, we have an obligation and a need to protect them.

One thing that has become very obvious to us in recent times is the need to reduce the amount of plastic getting into our oceans, and for this to happen we all need to change our consumer habits. We must move away from our addiction to single-use plastics wherever we can. Some things are very easy for us to do, and if done by everyone can have a huge impact. These include simple ideas such as giving up plastic straws, drinks sold in plastic bottles, plastic bags, and throw-away plastic plates, cups and cutlery. Stop releasing helium balloons, as there are so many less damaging ways to celebrate an achievement. Buy food that is not wrapped in

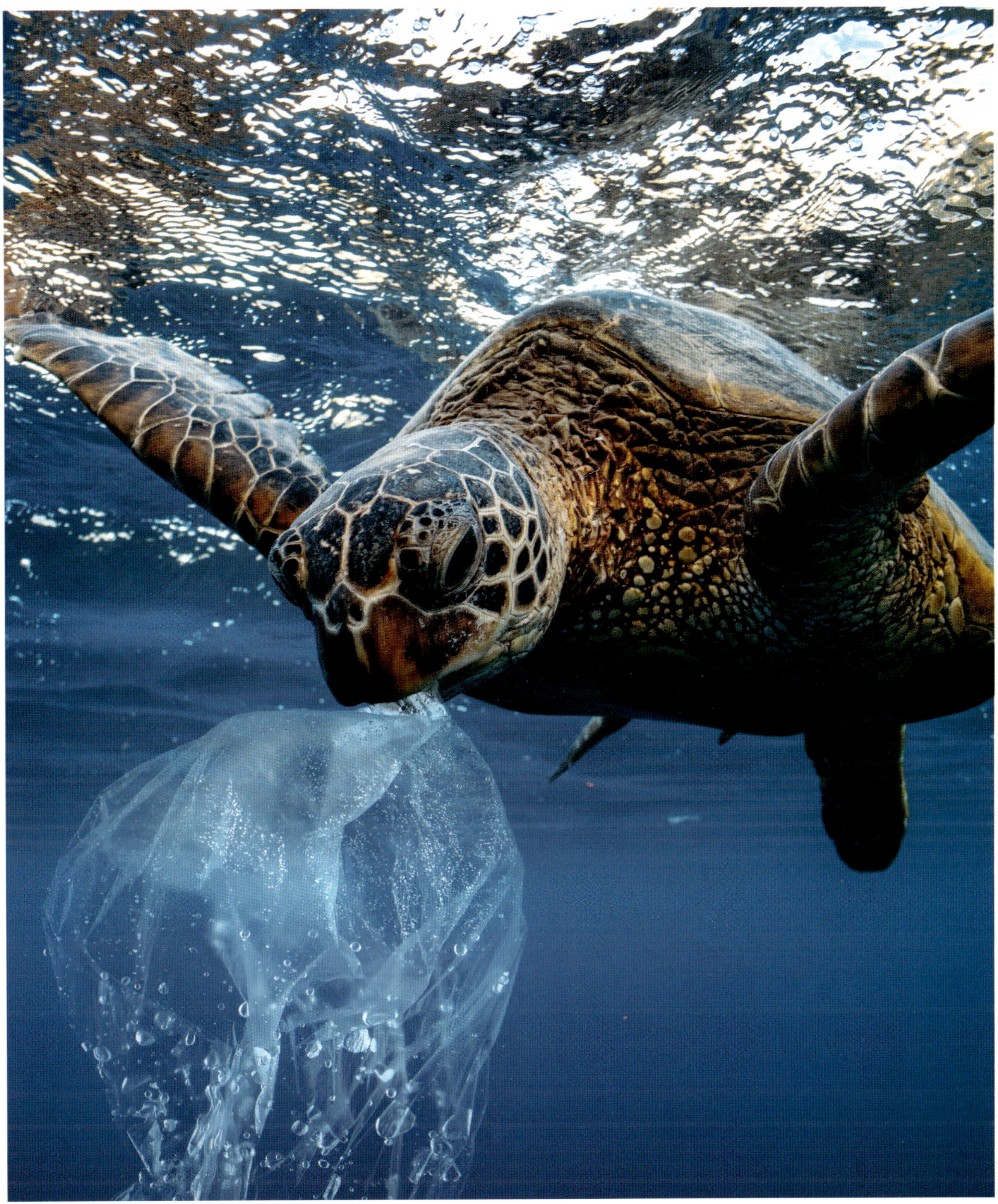

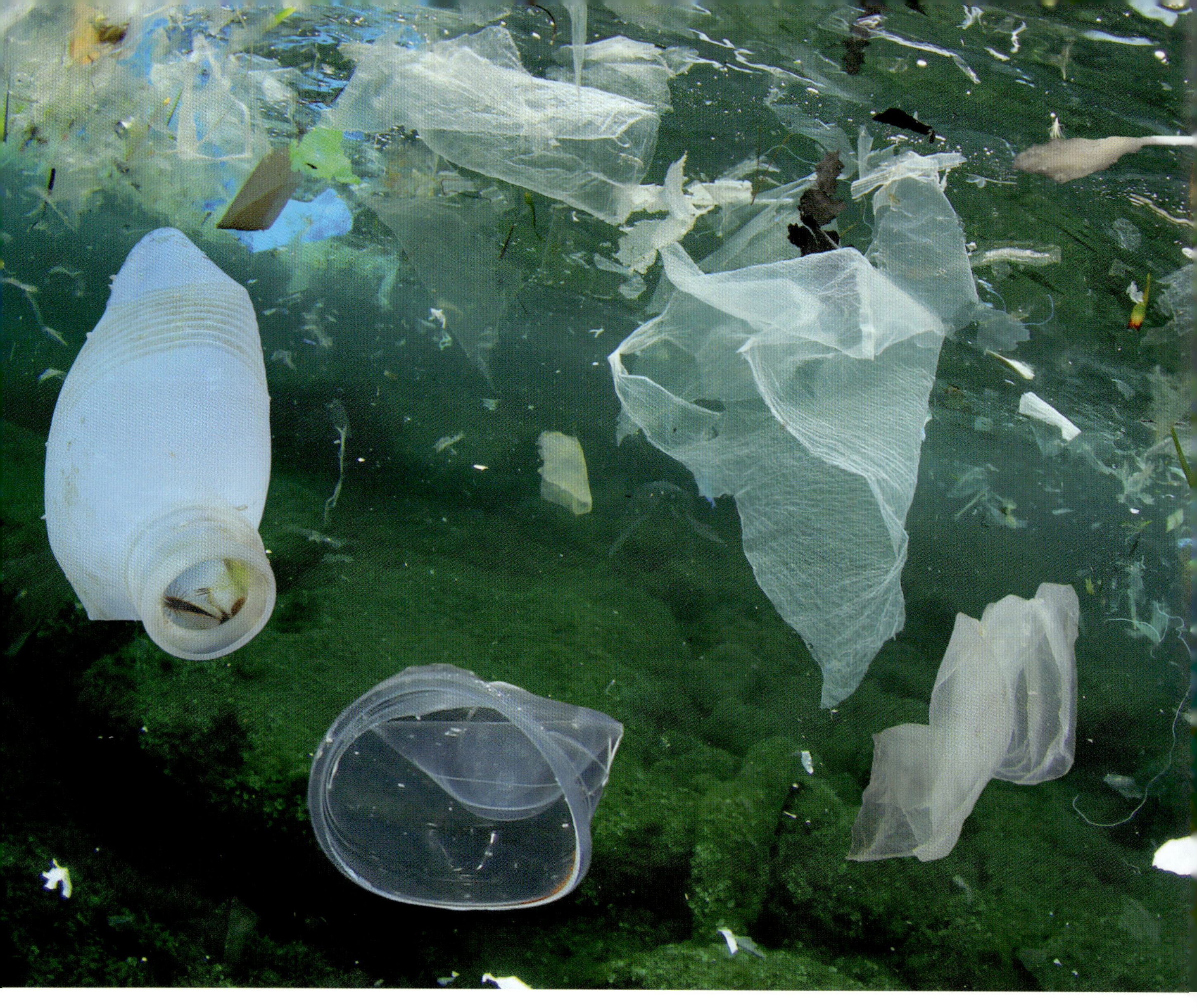

plastic, and tell the managers of your local stores that this is unsustainable. Pick up any plastic pollution that you see while at the beach or on a dive, and it helps if you carry a mesh bag to place the plastic garbage in. Pressure also needs to be increased on the polluting companies and governments to get to work and find solutions for the huge amount of plastic waste we all consume.

ABOVE It is said that there will be more plastic than fish in the sea by 2050.

Plastics can persist in our seas and oceans for more than 2,000 years, slowly breaking into smaller and smaller pieces. At the time of writing more than 8 million tonnes of plastic end up in the oceans each year. The UN estimates that that there are 51 trillion microplastic particles in the ocean – 500 times more than the number of

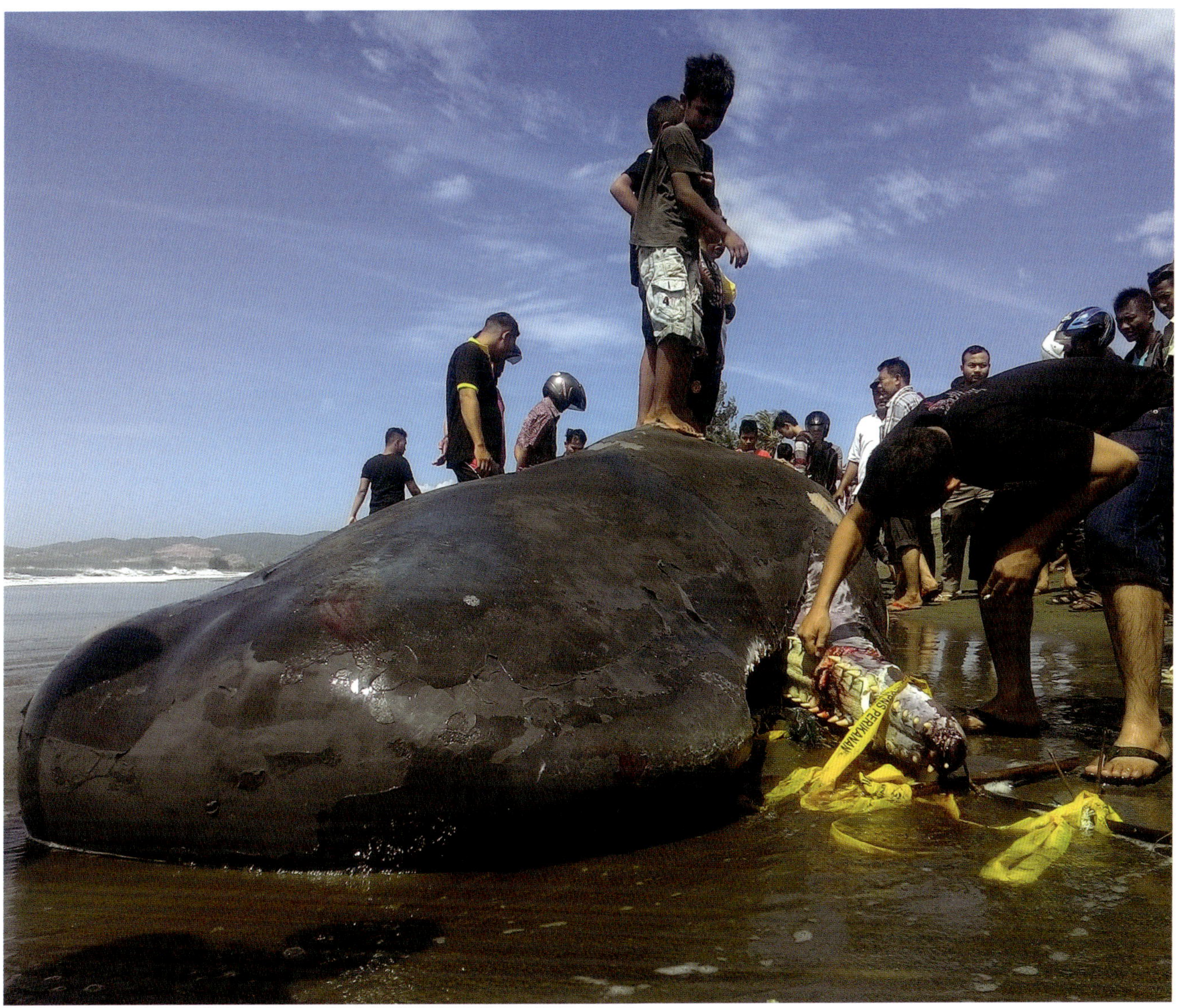

stars in the galaxy. Plastic debris outweighs plankton by a ratio of 36 to 1. Oceans are predicted to contain more plastic than fish by 2050. More than 5 trillion plastic pieces weighing over 250,000 tonnes are afloat at sea. In July 2017 a plastic waste patch bigger than the area of Mexico was discovered floating in the Pacific Ocean. Plastic products leach toxins that are now found in most people. Exposure to these toxins is linked to infertility, cancers and many other health problems. Discarded fishing nets continue to trap marine life and kill a wide range of species indiscriminately.

All of the animals we encounter underwater are being affected by plastic pollution. Turtles mistake drifting plastic bags and balloons for jellyfish and eat

ABOVE More and more dead whales are found with vast amounts of plastics in their stomachs.

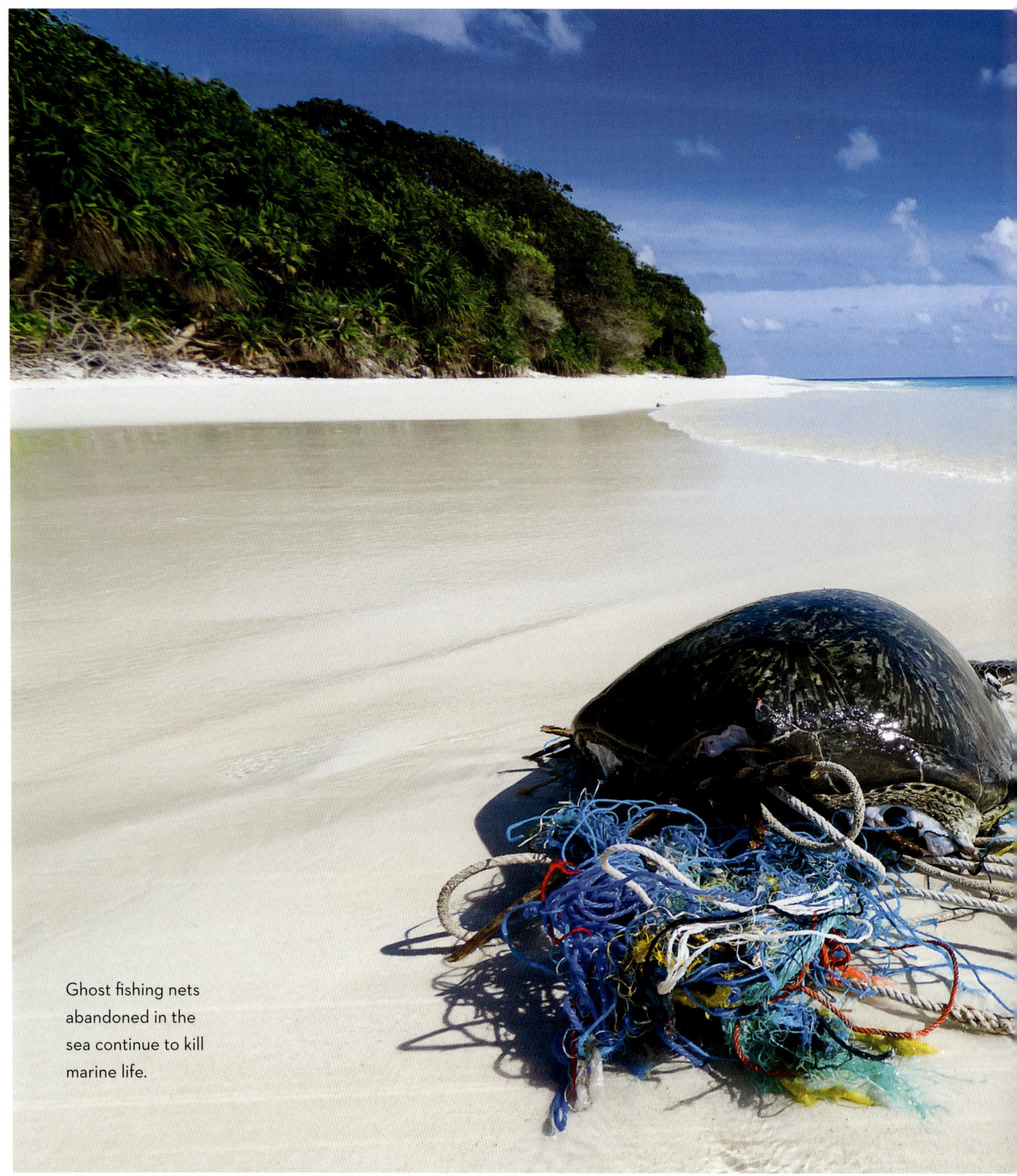

Ghost fishing nets abandoned in the sea continue to kill marine life.

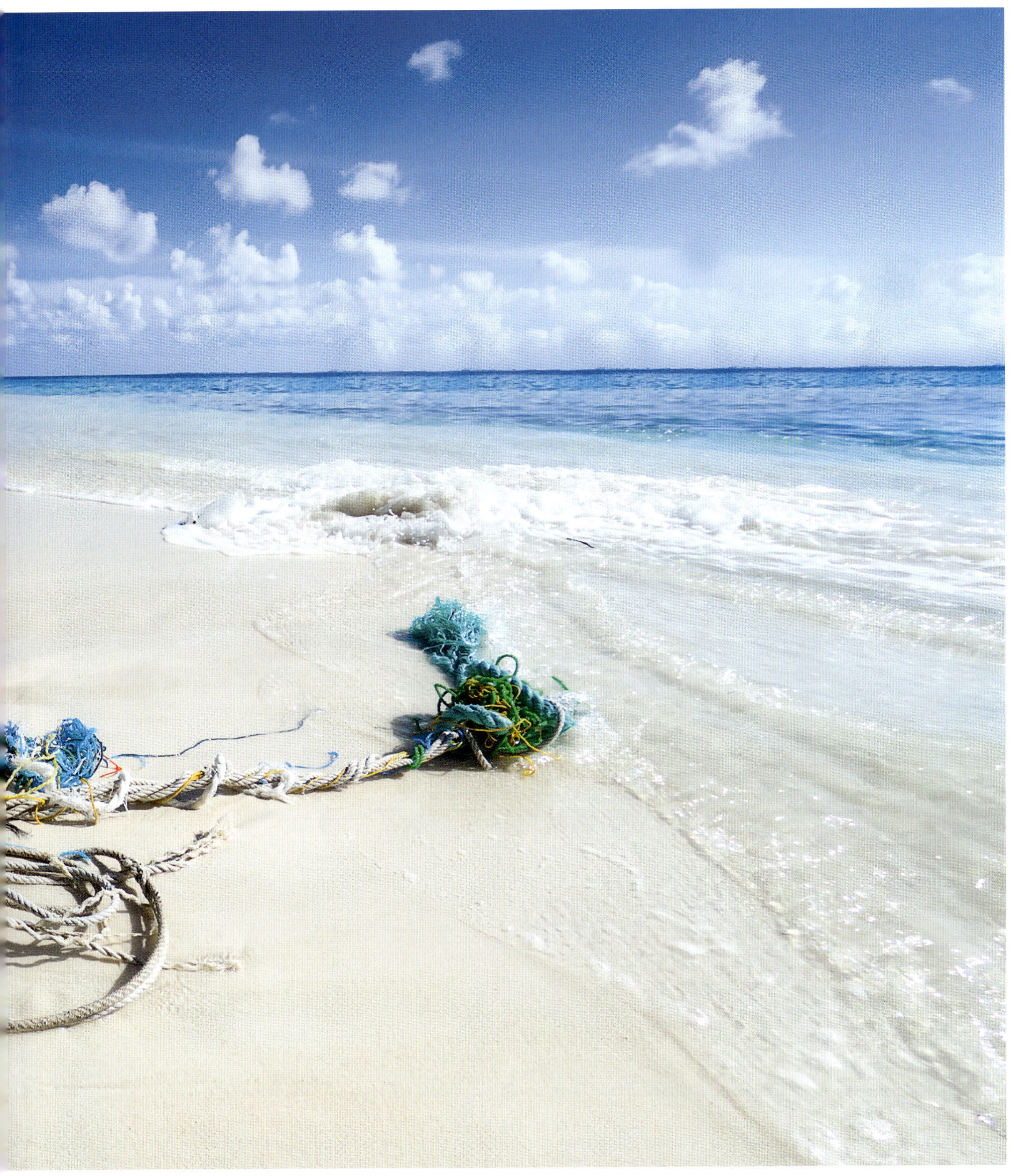

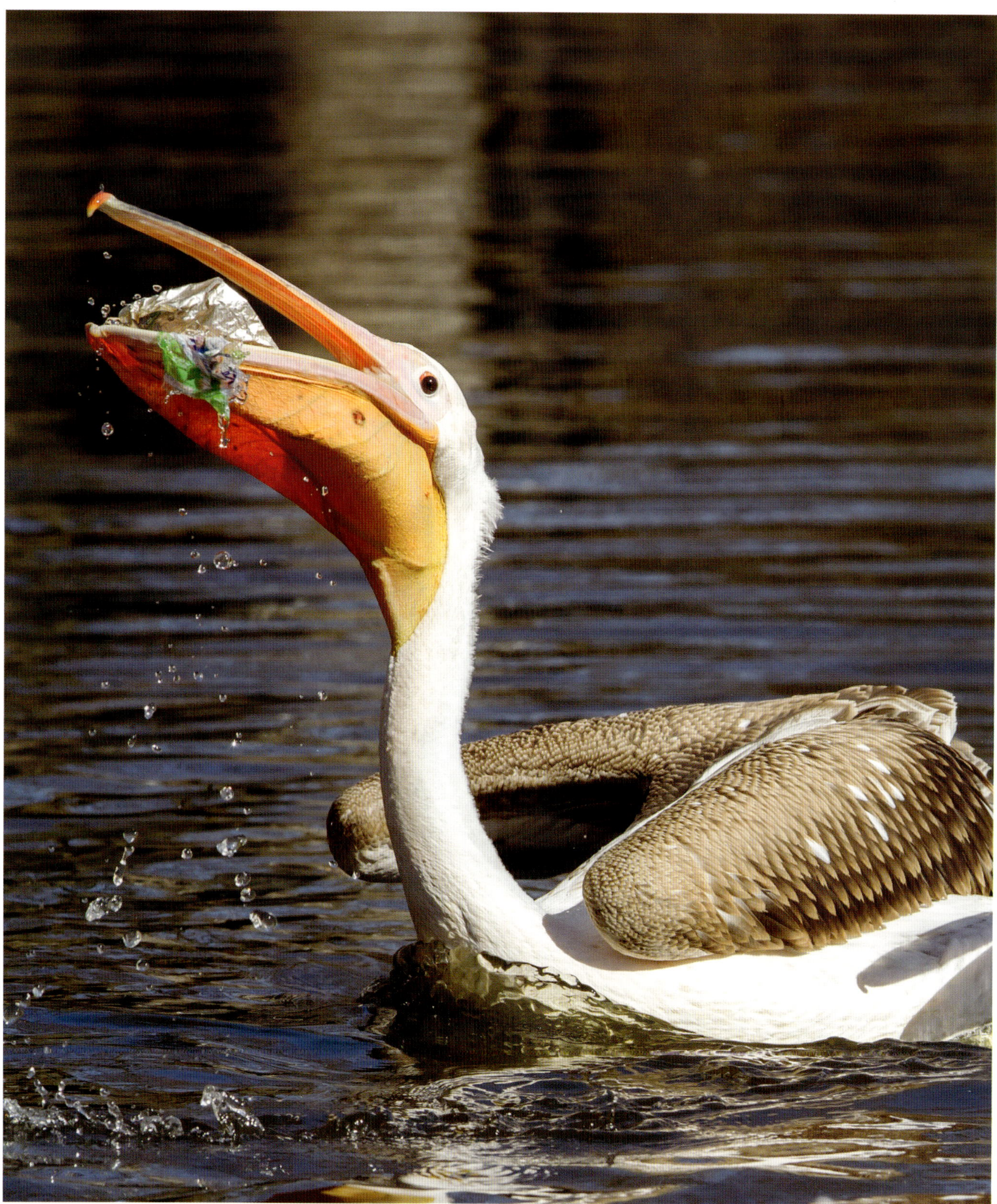

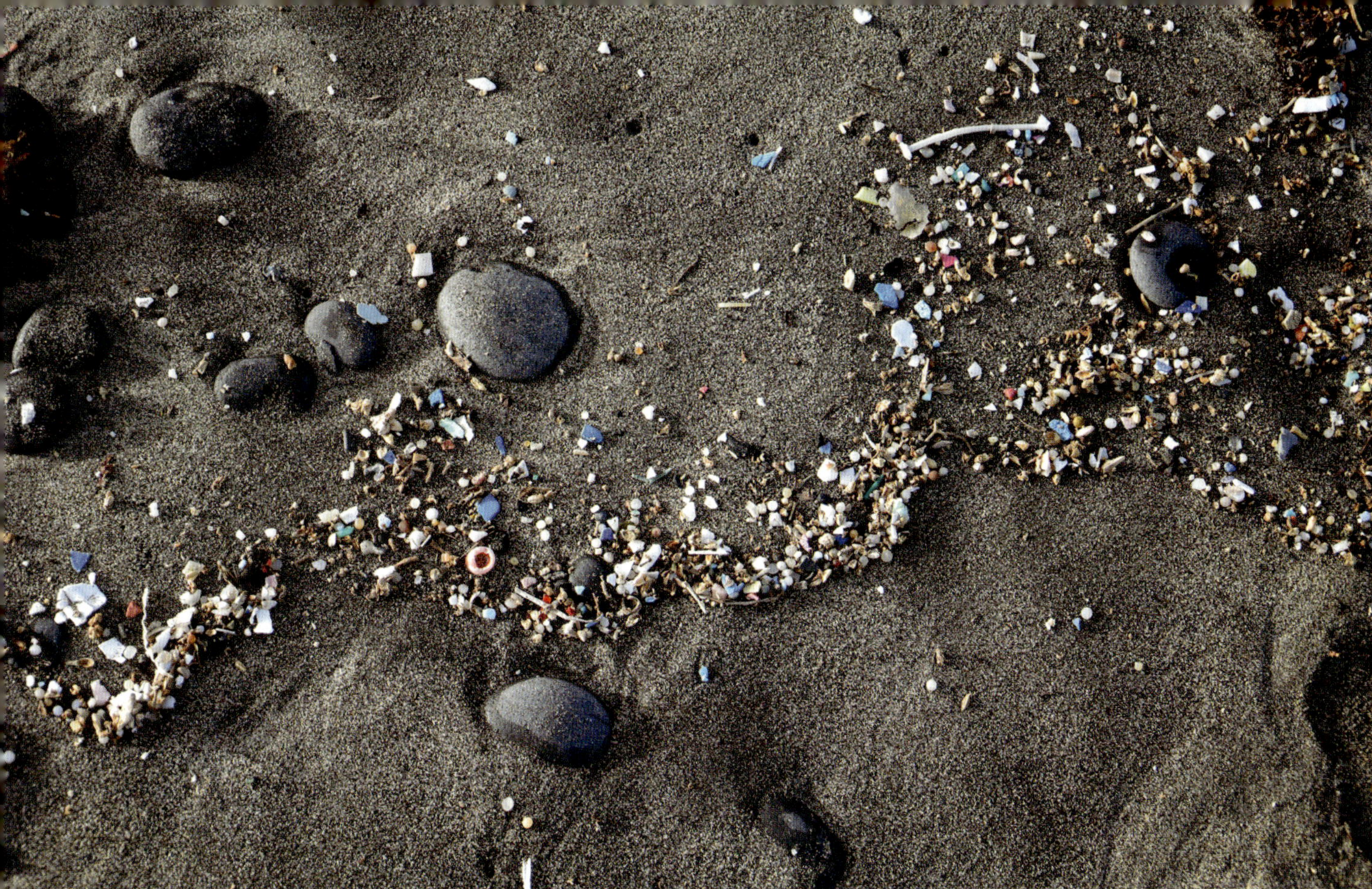

them, and this then blocks up their digestive system and eventually kills them. The filter-feeding sharks and rays featured in many chapters of the book ingest plastics as they scoop up small fish, crustaceans and plankton as they swim. Whales and dolphins are being found washed up on beaches with stomachs full of plastic. Birds are feeding their chicks plastic waste and fish are eating microplastics. Reefs are dying due to plastic smothering them. Discarded ghost fishing nets continue to capture and kill marine life such as turtles, seals, sea lions and fish. There is no avoiding the fact that this is a huge issue which needs to be resolved as quickly as possible.

There are lots of great organisations, charities and individuals who are striving to try to clean up our oceans and we applaud their work. Others work to educate the young people by going into our schools to spread the word and talk about the problem of plastic pollution. It is these passionate people who will save the oceans and seas from the plastic nightmare that has built up over the past few decades. We must see an end to the reliance on single-use plastics in all areas of our lives if we are to save our oceans and the planet for future generations.

OPPOSITE Birds also mistake plastic pollution for food and even feed it to their young.

ABOVE Plastic pollution breaks down into smaller and smaller pieces, but never goes away.

ACKNOWLEDGEMENTS

There are always so many people to thank on a project such as this one. Tourist boards, dive operators, hotels and equipment providers are all essential to us being able to get out there and do the dives.

Firstly, we would like to thank our sponsors: Mares, Nauticam and Olympus, without whom this would not be possible.

Here is a list of people that gave us a helping hand along the way:

- Australian Dwarf Minke Whales: Mike Ball Dive Expeditions (mikeball.com)
- Azores: Azores Tourist Board (www.visitazores.com)
- Barbados: Barbados Tourism Marketing Inc (visitbarbados.org/about-us); Barbados Blue (divebarbadosblue.com); Atlantis Submarines Barbados (www.barbados.atlantissubmarines.com)
- California: Visit Long Beach (visitlongbeach.com); Sundiver International (http://sundiverinternational.com/)
- Kefalonia: Scuba Hellas (scubahellas.com)
- Cuba and Grand Cayman: The Scuba Place (thescubaplace.co.uk)
- Fiji: Tourism Fiji (www.fiji.travel/)
- Guadalupe and Socorro: Nautilus Liveaboards (nautilusliveaboards.com)
- Philippines: Philippines Department of Tourism (itsmorefuninthephilippines.co.uk); Magic Resorts (https://magicresorts.online/); Atmosphere Resorts and Spa (atmosphereresorts.com)
- Bahamas: Bahamas Tourism (bahamas.com); Stuart Cove's Dive Bahamas (https://stuartcove.com/); Neal Watson's Bimini Scuba Center biminiscubacenter.com); Epic Diving (epicdiving.com); Reef Oasis Dive Club (reefoasisdiveclub.com/diving-bahamas/)
- Egypt: Egyptian Tourism Authority (www.egypt.travel); Camel Dive Club and Hotel (cameldive.com)
- St Helena: St Helena Tourism (sthelenatourism.com); Dive St Helena (divesainthelena.com), Ellen Cuylaerts (www.ellencuylaerts.com)
- Plastic Seas and Oceans: SeaStraw (seastraw.co.uk)
- Nick Barrett for his patience and proof reading

INDEX

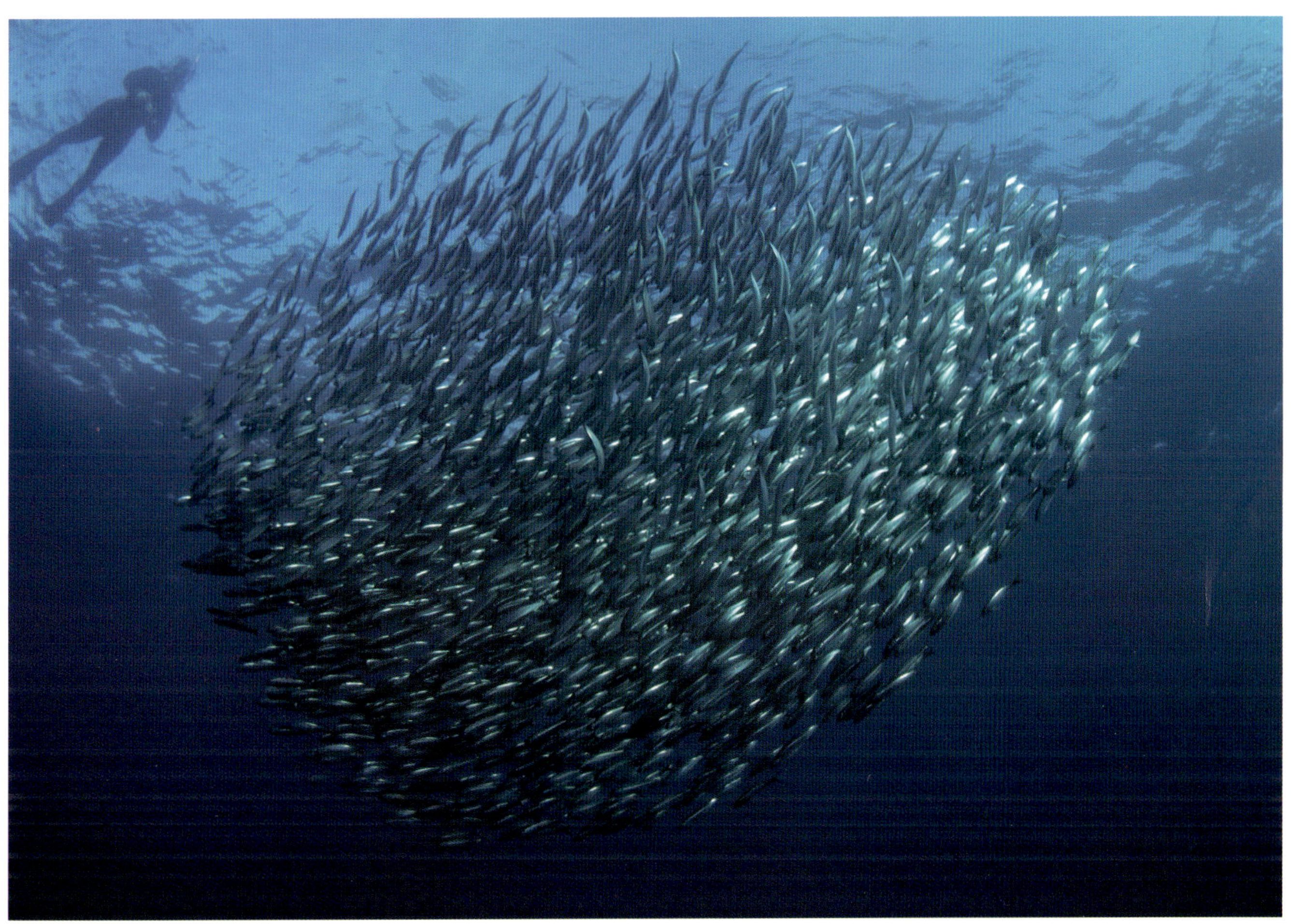

PHOTOGRAPHIC CREDITS

All images by Nick Robertson-Brown and Caroline Robertson-Brown except for the following:

- Ellen Cuylaerts: Whale Shark images on pages 13 and p17.
- Mike Ball Dive Expeditions: Dwarf Minke Whale images on pages 104–111.
- Shutterstock.com (individual photographer names in brackets): plastics in oceans images on pages 212 (Maxim Blinkov), 213 (Willyam Bradberry), 214 (Rich Carey), 215 (Ary Malemdiwa), 216–217 (Mohamed Abdulraheem), 218 (MyImages – Micha) and 219 (Susanne Fritzsche).

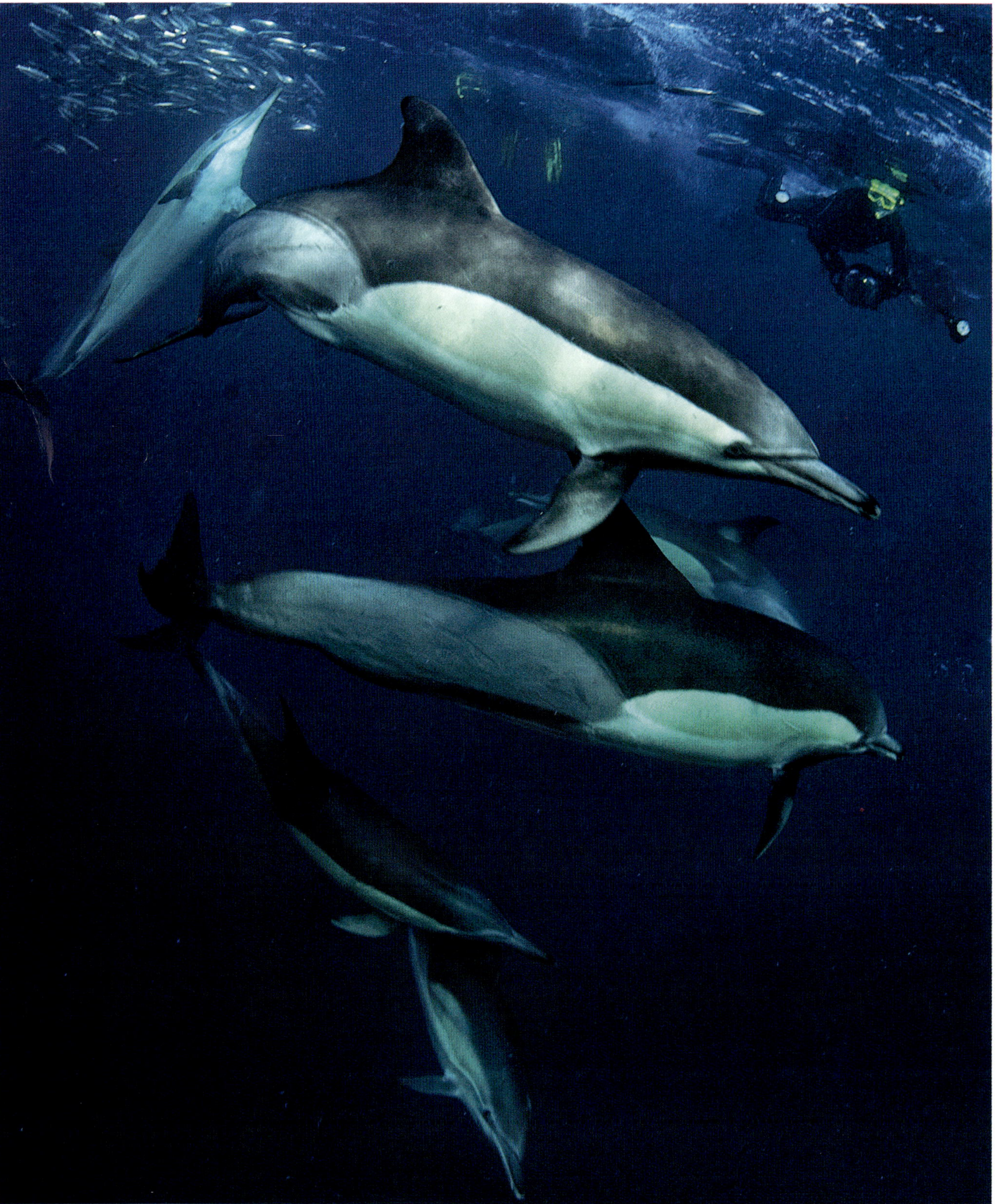

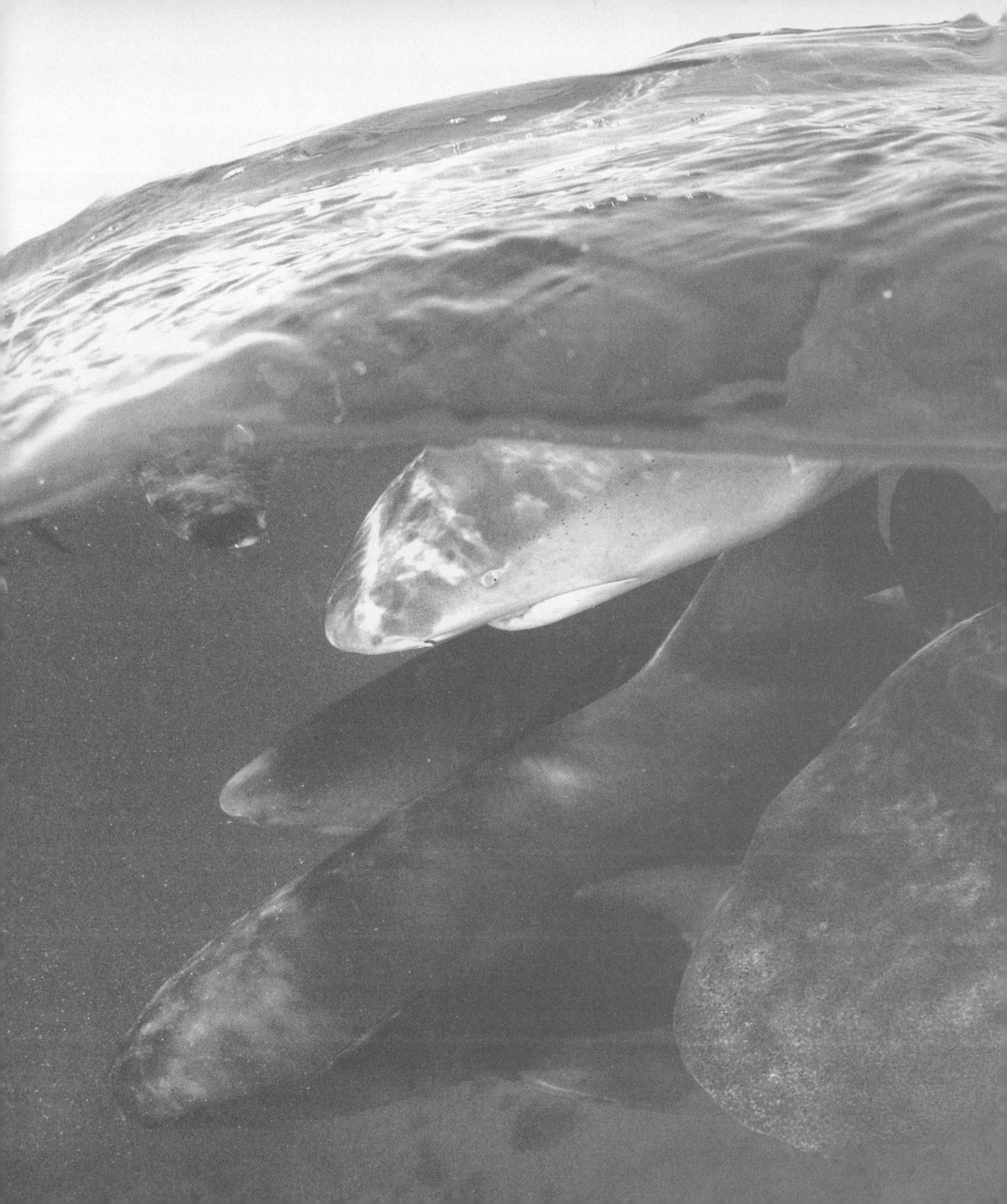